Lecture Notes in Mathematics

A collection of informal reports and seminars
Edited by A. Dold, Heidelberg and B. Eckmann, Zürich

234

Aldo Andreotti
University of Pisa, Pisa/Italy

Wilhelm Stoll
University of Notre Dame, Notre Dame, IN/USA

Analytic and Algebraic Dependence of Meromorphic Functions

Springer-Verlag
Berlin · Heidelberg · New York 1971

AMS Subject Classifications (1970): 32–02, 32 A 10, 32 A 20, 32 J 10, 32 F 99

ISBN 3-540-05670-X Springer-Verlag Berlin · Heidelberg · New York
ISBN 0-387-05670-X Springer-Verlag New York · Heidelberg · Berlin

Offsetdruck: Julius Beltz, Hemsbach.

Contents

Preface

Let X and Y be irreducible complex spaces. Denote by $\mathfrak{K}(X)$ and $\mathfrak{K}(Y)$ the fields of meromorphic functions on X and Y respectively. Let $\varphi: X \to Y$ be a holomorphic map of rank n, where n is the dimension of Y. Then φ induces an injective homomorphism $\varphi^*: \mathfrak{K}(Y) \to \mathfrak{K}(X)$ onto a subfield $\varphi^* \mathfrak{K}(Y)$ of $\mathfrak{K}(X)$. The purpose of these Lecture Notes is to investigate the algebraic dependence or independence of elements of $\mathfrak{K}(X)$ over $\varphi^* \mathfrak{K}(Y)$; for instance: When is $\mathfrak{K}(X)$ an algebraic function field over $\varphi^* \mathfrak{K}(Y)$?

Historically, the case of a constant map has been extensively studied. The roots of the problem reach back to Weierstrass. After several attempts by distinguished mathematicians, Thimm [22] was the first to solve the problem rigorously. Later, Remmert [11] applied his proper mapping theorem to give a short and elegant solution. Siegel [15] utilized a generalization of the Schwarz Lemma for a new type of proof which could be extended to pseudoconcave spaces in [2]. An excellent historical report is given by Thimm [25] in the Weierstrass Festband.

The general case of a holomorphic map is attacked along both lines. Remmert's method naturally leads to the new concepts of a "quasiproper map" and a "quasiproper map of codimension k", which correlate the analytic geometry of the map with the algebraic structure of $\mathfrak{K}(X)$ over $\varphi^* \mathfrak{K}(Y)$. Siegel's method is applied to pseudoconcave maps, which yields algebraic dependence over a neighborhood of some cleverly chosen point of Y only. In order to globalize this dependence, the concept of a full map is introduced.

After this research was completed, Professor Karl Stein visited Notre Dame for two months. In a number of discussions with the second author, it became apparent, that the results obtained by the Remmert method can also be derived by Stein's method of complex basis for meromorphic maps. The details have still to be worked out. However, the globalization theorem and the pseudoconcave case remain inaccessible to Stein's approach.

The properties of the new concepts had to be studied in detail. Also it became necessary to clarify some "well known" concepts and properties of holomorphic maps to be used effectively here. So, Remmert's concept of rank is adjusted to non-normal spaces in §1. The well known concept of "chart" and "product representation" are formalized in §2. Since it is difficult to find a readable account of the concept and properties of a meromorphic function on a complex space in the textbook literature, the basic facts on meromorphic functions are collected and partially proved in §3. Some other material contained in the real or folklore literature is proved here (§5-§7) for clarification, for coherence and for the convenience of the reader.

The sophisticated expert may sneer at the inclusion of this and other introductory material and at the proofs of some facts "well known" to him and perhaps to him alone. As a plumber has to assemble and adjust his tools for a new job, so the mathematician has to ready his tools. Often enough, this task is neglected or omitted. Such a paper is difficult to read and may be even inaccessible to those less acquainted with the general subject area.

Although this monograph is essentially a research paper, it is self contained and should be accessible to a reader who is familiar with the basic concepts of the field as provided in an introductory course or a standard textbook, for instance, Narasimhan [9]. References will be given for material used above this level.

This work was began while both authors were visitors at Stanford University. It was continued and completed at the authors' home institutions. During these investigations the second author was partially supported by the National Science Foundation under Grant NSF GP 20139. The authors wish to thank these institutions for their help and support to make this work possible.

Aldo Andreotti
Pisa

Wilhelm Stoll
Notre Dame

Spring 1971

German Letters

A	B	C	D	E	F	G	H	I	J	K	L	M	N	O	P	Q

R	S	T	U	V	W	X	Y	Z

a	b	c	d	e	f	g	h	i	j	k	l	m	n	o	p	q

r	s	t	u	v	w	x	y	z

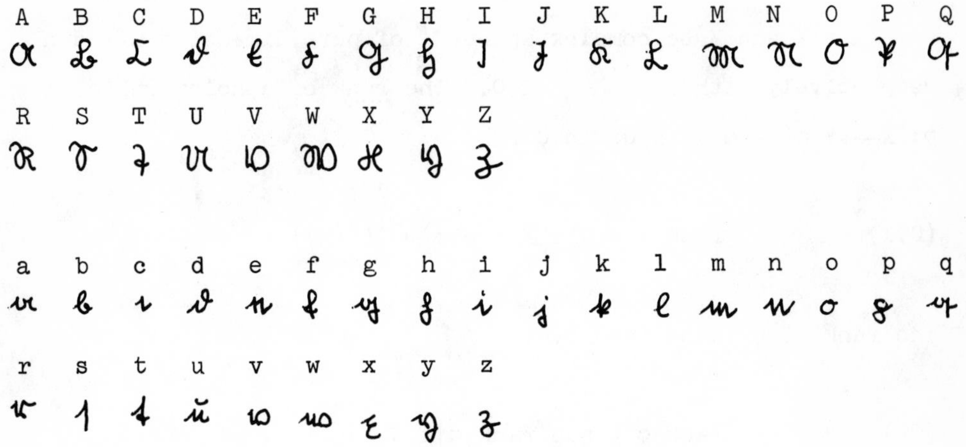

§0. Introduction

Let X and Y be complex spaces[1)] of pure dimension m and n respectively with $q = m - n \geqq 0$. The rank of a holomorphic map $\varphi: X \longrightarrow Y$ at $x \in X$ is defined by[2)]

$$\text{(0.1)} \qquad \operatorname{rank}_x \varphi = \dim_x X - \dim_x \varphi^{-1}(\varphi(x))$$

The rank of φ is defined by

$$\text{(02)} \qquad \operatorname{rank} \varphi = \sup\{\operatorname{rank}_x \varphi \mid x \in X\}$$

The holomorphic map has strict rank φ if and only if $\varphi|B$ has rank p for every branch[3)] B of X.

Let $\mathfrak{K}(X)$ be the ring of meromorphic functions on X. Let $f_1,\ldots,f_k$ be meromorphic functions on X. Let A be the largest open subset of X such that $f_1,\ldots,f_k$ are holomorphic on A. Consider the complex projective space $\mathbb{P}^k$ as the projective closure of the vector space $\mathbb{C}^k$. Then the closure Γ of

$$\{(x,f_1(x),\ldots,f_k(x)) \mid x \in A\}$$

in $X \times \mathbb{P}^k$ is a pure m-dimensional analytic subset of $X \times \mathbb{P}^k$. Define $\psi: \Gamma \to Y \times \mathbb{P}^k$ by $\psi(x,z) = (\varphi(x),z)$. The functions $f_1,\ldots,f_k$ are said to be φ-dependent if and only if $\operatorname{rank} \psi < n + k$. The functions $f_1,\ldots,f_k$ are said to be φ-independent, if and only if ψ has strict rank $n + k$.

Suppose that X is irreducible. Let $f_1,\ldots,f_k$ be φ-independent meromorphic functions on X. Then $k \leqq q$. Define $F = \{f_1,\ldots,f_k\}$.

Then

$$\mathfrak{K}_\varphi(X,F) = \{f \in \mathfrak{K}(X) \mid f_1,\ldots,f_k,f \quad \varphi\text{-dependent}\}$$

is a subfield of $\mathfrak{K}(X)$ and contains

$$\mathfrak{K}_\varphi(X) = \mathfrak{K}_\varphi(X,\emptyset) = \{f \in \mathfrak{K}(X) \mid f \quad \varphi\text{-dependent}\}$$

as a subfield. If X is irreducible, and if φ has rank n, then φ-independent meromorphic functions $f_1,\ldots,f_k$ exist such that $\mathfrak{K}(X) = \mathfrak{K}_\varphi(X,F)$ where $k \leqq q$.

Again, let X and Y be complex spaces of pure dimension m and n respectively with $m - n = q \geqq 0$. Let $\varphi: X \to Y$ be a holomorphic map of strict rank n. Then each meromorphic function f on Y lifts to a meromorphic function φ^*f on X. An injective homomorphism

$$\varphi^*: \mathfrak{K}(Y) \to \mathfrak{K}(x)$$

is defined. Obviously, $\varphi^*: \mathfrak{K}(Y) \to \varphi^*\mathfrak{K}(Y)$ is an isomorphism. Suppose that X and Y are irreducible. Then $\varphi^*\mathfrak{K}(Y)$ is a subfield of the field $\mathfrak{K}(X)$. The following problems shall be investigated.

<u>Problem 1.</u> When is $\mathfrak{K}_\varphi(X)$ finite algebraic over $\varphi^*\mathfrak{K}(Y)$?

<u>Problem 2.</u> When is $\mathfrak{K}_\varphi(X,F)$ an algebraic function field over $\varphi^*\mathfrak{K}(Y)$?

<u>Problem 3.</u> When is $\mathfrak{K}(X)$ an algebraic function field over $\varphi^*\mathfrak{K}(Y)$?

<u>Problem 4.</u> If $f_1,\ldots,f_p$ are φ-dependent in $\mathfrak{K}(X)$, when are they algebraically dependent over $\varphi^*\mathfrak{K}(Y)$?

<u>Problem 5.</u> Let G be an open, irreducible[4)] subset of Y such

that $\tilde{G} = \varphi^{-1}(G) \neq \emptyset$. Let $\varphi_G = \varphi: \tilde{G} \to G$ be the restriction. Suppose that $f_1,\ldots,f_k$ are meromorphic functions on X such that $f_1|\tilde{G},\ldots,f_k|\tilde{G}$ are algebraically dependent over $\varphi^*_G\,\mathfrak{K}(G)$. When are $f_1,\ldots,f_k$ algebraically dependent over $\varphi^*\mathfrak{K}(Y)$?

Historically, only the case of a constant map, that is the case when Y is a point, has found much attention. In this case $\mathfrak{K}_\varphi(X) = \varphi^*\mathfrak{K}(Y) = \mathbb{C}$. Hence Problem 1 and 5 become trivial. φ-dependence is called analytic dependence. Problem 3 and 4 were solved by the following results.

Algebraic function field theorem (Problem 3). Let X be an irreducible, compact complex space of dimension m. Then $\mathfrak{K}(X)$ is an algebraic function field over $\mathbb{C}$ with a transcendence degree of almost m.

Dependence Theorem (Problem 4). Let X be an irreducible, compact complex space of dimension m. Let $f_1,\ldots,f_k,f$ be analytically dependent meromorphic functions on X, such that $f_1,\ldots,f_k$ are analytically independent. Then $k \leqq m$ and $f_1,\ldots,f_k,f$ are algebraically dependent over $\mathbb{C}$ and $f_1,\ldots,f_k$ are algebraically independent over $\mathbb{C}$.

Weierstrass (1869) formulated these theorems if X is a complex torus and if m = k. He never succeeded to give a complete and precise proof. Other well known mathematicians as Poincaré, Wirtinger and Osgood attempted correct proofs. In 1939, Thimm [22] succeeded in the case that m = k and that X is a compact complex manifold. Thimm's thesis remained unknown. In 1954 he proved the general case in [23] [24]. In 1956, Remmert [11] used his famous proper mapping theorem to obtain both theorems. Subsequently, Stein extended this method to a general theory of dependence of holomorphic and meromorphic maps which uses his construction of a complex basis. In 1955, Siegel [15] proved both theorems on manifolds by a completely different method, which originated in the theory of module functions and uses a generalization of Schwarz's Lemma. In [2], his method is used to extend both theorems to a pseudoconcave complex spaces X. For the dependence theorem X is assumed to be locally irreducible. For the algebraic function field theorem X is assumed to be normal.

For the solution of Problems 1-5, Remmert's proper mapping method and Siegel's Schwarz Lemma method will both be used. Kuhlmann [6] [7] introduced the concept of a semiproper map, which plays an essential role in these investigations. Here a whole "properness scale" will be introduced which bridges the gap between the extremes "proper" and "semiproper".

A holomorphic map $\varphi: X \to Y$ is said to be quasiproper if and only if for every compact subset K of Y a compact subset K' of X exists such that $B \cap K' \neq \emptyset$ for each branch B of $\varphi^{-1}(y)$ whenever $y \in K \cap \varphi(X)$. A quasiproper map is semiproper.

Theorem. If X and Y are irreducible complex spaces of dimension m and n respectively with $m \geqq n$, and if $\varphi: X \to Y$ is a quasiproper, holomorphic map of rank n, then $\mathfrak{K}_\varphi(X)$ is a finite algebraic extension of $\varphi^*\mathfrak{K}(Y)$. (Theorem 9.2).

A holomorphic map $\varphi: X \to Y$ is said to be quasiproper of codimension k, if and only if for every compact subset K of Y a compact subset K' of X exists satisfying the following condition:

(P) Take $y \in \varphi(X) \cap K$. Let B be a branch of $\varphi^{-1}(y)$. Let C be an irreducible analytic subset of B with $\dim B - \dim C \leqq k$. Then $C \cap K' \neq \emptyset$.

If φ is quasiproper of codimension k, and if $k' \leqq k$, then φ is quasiproper of codimension k'. A holomorphic map φ is quasiproper of codimension 0, if and only if φ is quasiproper. The map φ is proper, if and only if φ is quasiproper of codimension k for all $k \geqq 0$.

<u>Theorem.</u> Let X and Y be irreducible complex spaces of dimension m and n respectively with $m - n = q \geqq 0$. Let $\varphi: X \to Y$ be a holomorphic map of rank n which is quasiproper of codimension k. Let $f_1, \ldots, f_k$ be φ-independent meromorphic functions on X. Define $F = \{f_1, \ldots, f_k\}$. Then $\mathfrak{K}_\varphi(X;F)$ is an algebraic function field over $\varphi^* \mathfrak{K}(Y)$. The field $\varphi^* \mathfrak{K}(Y)(f_1, \ldots, f_k)$ is a pure transcendental extension of transcendence degree k of $\varphi^* \mathfrak{K}(Y)$ and $\mathfrak{K}_\varphi(X,F)$ is a finite algebraic extension of $\varphi^* \mathfrak{K}(Y)(f_1, \ldots, f_k)$. (Theorem 10.5).

<u>Corollary.</u> Let X and Y be irreducible complex spaces of dimension m and n respectively with $q = m - n \geqq 0$. Let $\varphi: X \to Y$ be a holomorphic map which is quasiproper of codimension q. Then $\mathfrak{K}(X)$ is an algebraic function field over $\varphi^* \mathfrak{K}(Y)$ with transcendence degree at most q. (Theorem 10.8).

<u>Corollary.</u> Let X and Y be irreducible complex spaces of dimension m and n respectively with $m - n = q > 0$. Let $\varphi: X \to Y$ be a proper holomorphic map of rank n. Then $\mathfrak{K}(X)$ is an algebraic function field over $\varphi^* \mathfrak{K}(Y)$ with transcendence degree at most q. (Theorem 10.9).

A holomorphic map $\varphi: X \to Y$ is said to be full if and only if every point $b \in Y$ has an open neighborhood V such that

$$\varphi^{-1}(V) = \bigcup_{\nu=1}^{\infty} U_\nu$$

where each U_ν is open and where $\varphi|U_\nu: U_\nu \to V$ is quasiproper whenever $U_\nu \neq \emptyset$. A quasiproper map is full. The concept of a full map generalizes the concept of a covering space.

<u>Theorem.</u> Let X and Y be irreducible complex spaces of dimension m and n respectively with $q = m - n \geqq 0$. Let $\varphi: X \to Y$ be a full holomorphic map of rank n. Let $G \neq \emptyset$ be an open, irreducible subset of Y. Then $\tilde{G} = \varphi^{-1}(G) \neq \emptyset$. Define $\varphi_G = \varphi: \tilde{G} \to G$. Let $f_1,\ldots,f_k,f$ be meromorphic functions on X such that $f_1|\tilde{G},\ldots,f_k|\tilde{G}$ are algebraically independent over $\varphi^*_G\mathfrak{K}(G)$ and such that $f_1|\tilde{G},\ldots,f_k|\tilde{G},f|\tilde{G}$ are algebraically dependent over $\varphi^*_G\mathfrak{K}(G)$ with

$$r = [(f|\tilde{G}): \varphi^*_G\mathfrak{K}(G)(f_1|\tilde{G},\ldots,f_k|\tilde{G})]$$

Then $f_1,\ldots,f_k$ are algebraically independent over $\varphi^*\mathfrak{K}(Y)$ and $f_1,\ldots,f_k,f$ are algebraically dependent over $\varphi^*\mathfrak{K}(Y)$ with

$$r = [f: \varphi^*\mathfrak{K}(Y)(f_1,\ldots,f_k)]$$

(Theorem 12.7 for $1 \leqq k \leqq q$, Theorem 12.6 for $k = 0$).

For full maps, a remarkable extension theorem holds:

Extension theorem. Let X and Y be irreducible complex spaces of dimension m and n respectively with $m - n = q \geqq 0$. Let $\varphi: X \to Y$ be a full holomorphic map of rank n. Let $G \neq \emptyset$ be an open subset of Y. Then $\tilde{G} = \varphi^{-1}(G) \neq \emptyset$. Define $\varphi_G = \varphi: \tilde{G} \to G$. Let f be a meromorphic function on X such that a meromorphic function g on G with $f|\tilde{G} = \varphi^*_G(g)$ exists. Then g is unique. One and only one meromorphic function h on Y exists such that $f = \varphi^*(h)$. Moreover, $g = h|G$.

The concept of a pseudoconcave map is introduced in §16. Based on the Schwarz Lemma method of Siegel [15] and its application to pseudoconcave spaces in [2], a complicated proof yields the following result:

Theorem. Let X and Y be irreducible complex spaces of dimension m and n respectively with $m < n$. Let $\varphi: X \to Y$ be a full holomorphic map of rank n, which is pseudoconcave over the open subset $G \neq \emptyset$ of Y. Let $f_1, \ldots, f_d$ be φ-dependent meromorphic functions on X. Then $f_1, \ldots, f_d$ are algebraically dependent over $\varphi^* \mathfrak{K}(Y)$. (Theorem 16.2).

Theorem. Let X be an irreducible, normal complex space of dimension m. Let Y be an irreducible complex space of dimension n with $m - n = q > 0$. Let $\varphi: X \to Y$ be a full holomorphic map of rank n, which is pseudoconcave over the open subset $G \neq \emptyset$ of Y. Then $\mathfrak{K}(X)$ is an algebraic function field over $\varphi^* \mathfrak{K}(Y)$ with transcendence degree almost q. (Theorem 16.6). [Actually, it suffices to require only that $\varphi^{-1}(G)$ is normal. Theorem 16.5].

Denote by $\mathfrak{H}(X)$ the ring of holomorphic functions on X. In §17, an example of Kodaira [5] and Kas is given. A two dimensional, connected complex manifold X and a full, surjective, regular, one-fibering holomorphic map $\varphi: X \to D$ onto the unit disc D is constructed such that $\varphi^* \mathfrak{K}(D) = \mathfrak{K}(X)$. Two φ-dependent meromorphic functions ξ, η on X are constructed which are algebraically independent over $\mathfrak{K}(X)$ and hence over $\varphi^* \mathfrak{K}(D)$.

§1 The rank of a holomorphic map

Remmert [12] proved a number of important theorems for holomorphic maps between normal complex spaces. Possible extensions to (reduced) complex spaces are needed here. They are used and even proved in the literature, but it seems to be impossible to locate one place where all these extensions are formulated and proved as needed here. Therefore the proofs shall be given here.

A subset T of a complex space X is said to be <u>thin of dimension p</u> (respectively <u>codimension q</u>) if and only if every point $x \in T$ has an open neighborhood U containing an analytic subset S of dimension p (respectively codimension q) with $T \cap U \subseteq S$. A thin subset of codimension 1 is also called <u>thin</u>. A subset T of X is said to be <u>almost thin of dimension p</u> (respectively <u>codimension q</u>) if and only if T is the atmost countable union of thin sets of dimension p (respectively codimension q). An almost thin subset of codimension 1 is also called <u>almost thin</u> in X.

Let X and Y be complex spaces. A holomorphic map $\varphi: X \to Y$ is said to be q-fibering, if and only if $\varphi^{-1}(\varphi(x))$ is pure q-dimensional for each $x \in X$. A o-fibering map is said to be <u>light</u>. A holomorphic map $\varphi: X \to Y$ is said to be <u>regular</u> at $a \in X$ if and only if a and $b = \varphi(a)$ are simple points of X and Y respectively and if the Jacobian of φ at a has rank n with $n = \dim_b Y$. A holomorphic map $\varphi: X \to Y$ is said to be <u>proper</u> if and only if $\varphi^{-1}(K)$ is compact for each compact subset K of Y.

Let X be a complex space. Let $\pi: \hat{X} \to X$ be the normalization of X. Then $\hat{\pi}$ is a proper, light, surjective, holomorphic map. The set N of normal points of X is open and dense in X and $\hat{N} = \pi^{-1}(N)$ is open and dense in $\hat{X}$. The restriction $\pi: \hat{N} \to N$ is biholomorphic. If U is open in X, then $\pi: \pi^{-1}(U) \to U$ is the normalization of U. If $a \in X$, an open, connected neighborhood U of a exists such that $\pi^{-1}(U) = U_1 \cup \ldots \cup U_p$ where $U_\nu \cap U_\mu = \emptyset$ if $\mu \neq \nu$, where each U_ν is open and connected, and where $\pi(U_1),\ldots,\pi(U_p)$ are the branches of U. Moreover, $\pi^{-1}(a) \cap U_\nu$ consists of exactly one point. If $\{\hat{X}_\lambda\}_{\lambda\in\Lambda}$ is the family of connectivity components of $\hat{X}$, then $\{\hat{X}_\lambda\}_{\lambda\in\Lambda}$ is the family of branches of $\hat{X}$ and $\{\pi(\hat{X}_\lambda)\}_{\lambda\in\Lambda}$ is the family of branches of X. Here $\pi\colon \hat{X}_\lambda \to \pi(\hat{X}_\lambda)$ is the normalization of X_λ. (See Abhyankar [1], Narasimhan [9]).

Theorem 1.1. (Proper mapping theorem of Remmert). Let X and Y be complex spaces. Let $\varphi\colon X \to Y$ be a proper, holomorphic map. Then $\varphi(X)$ is analytic.

Proof. If X and Y are both normal, see Remmert [12] Satz [24]. Suppose that only X is normal. Take $b \in Y$. An open neighborhood V of b and an injective, holomorphic map $\psi\colon V \to W$ into an open subset W of $\mathbb{C}^n$ exists such that $\psi(V)$ is an analytic subset of W and such that $\psi_0 = \psi\colon V \to \psi(V)$ is biholomorphic. Then $U = \varphi^{-1}(V)$ is open. Suppose that $U \neq \emptyset$. Define $\chi = \psi\circ\varphi\colon U \to W$. Let K be compact in W. Then $\chi^{-1}(K) = \varphi^{-1}(\psi_0^{-1}(\psi(V) \cap K))$ is compact. Hence χ is proper. Therefore $\psi(\varphi(U)) = \chi(U)$ is analytic in W. Hence

$$\varphi(X) \cap V = \varphi(U) = \psi_0^{-1}(\chi(U))$$

is analytic in V. Therefore $\varphi(X)$ is analytic in Y.

If X is not normal, let $\pi\colon \hat{X} \to X$ be the normalization of X. Then $\varphi\circ\pi\colon \hat{X} \to Y$ is proper. Therefore $\varphi(X) = \varphi(\pi(\hat{X}))$ is analytic in Y; q.e.d.

Lemma 1.2. Let X be a complex space of dimension m. Let Y

be a complex space of pure dimension $n > m$. Let $\varphi: X \to Y$ be a holomorphic map. Then $\varphi(X)$ is of first category, especially $\varphi(X)$ does not contain any non-empty open subset of Y.

<u>Proof.</u> Obviously, X can be assumed to be irreducible. The Lemma is true if $m = 0$, because $\varphi(X)$ is a point in this case. Suppose the Lemma is true for $\dim X = m - 1$. Assume, X with $\dim X = m < n$ is given. Let S_X and S_Y be the sets of non-simple points of X and Y respectively. Then $\dim S_X \leqq m - 1$ and $\varphi(S_X)$ has first category. The set $\tilde{S}_Y = \varphi^{-1}(S_Y)$ is analytic in X. If $X = \tilde{S}_Y$, then $\varphi(X) \subseteq S_Y$ has first category. Hence $X \neq \tilde{S}_Y$ can be assumed. Then $X_0 = X - (S_X \cup \tilde{S}_Y) \neq \emptyset$ is open, connected and dense in X, since X is irreducible. Define $Y_0 = Y - S_Y$ and $\varphi_0 = \varphi: X_0 \to Y_0$. Let p be the maximum of the rank of the Jacobian of φ_0. Then $p \leqq m < n$.

The set R of all points of X_0 where the rank of the Jacobian of φ_0 is smaller than p is analytic and thin in X_0, since X_0 is connected. Therefore, $\dim R \leqq m - 1$ and $\varphi(R)$ has first category. If $a \in X_1 = X_0 - R$, an open neighborhood U_a of a in X_1 exists such that $\varphi(U_a)$ is a pure p-dimensional complex manifold in Y_0 (Implicit function theorem). Therefore, $\varphi(U_a)$ has first category. A sequence $\{a_\nu\}_{\nu\in\mathbb{N}}$ of points $a_\nu \in X_1$ exist such that $X_1 = \bigcup_{\nu=1}^{\infty} U_{a_\nu}$. Then $\varphi(X_1) = \bigcup_{\nu=1}^{\infty} \varphi(U_{a_\nu})$ has first category. Now,

$$\varphi(X) \subseteq S_Y \cup \varphi(X_1) \cup \varphi(X_X) \cup \varphi(R).$$

Hence $\varphi(X)$ has first category; q.e.d.

Let X and Y be complex spaces. Let $\varphi: X \to Y$ be a holomorphic map. The <u>pseudo-rank</u> of φ at $a \in X$ is defined by

$$(1.1) \qquad \widetilde{\text{rank}}_a\varphi = \dim_a X - \dim_a \varphi^{-1}(\varphi(a)).$$

Let $\pi: \hat{X} \to X$ be the normalization of X. The <u>rank</u> of φ at $a \in X$ is defined by

$$(1.2) \qquad \text{rank}_a\varphi = \text{Min}\{\widetilde{\text{rank}}_z \varphi\circ\pi \mid z \in \pi^{-1}(a)\}.$$

If $\emptyset \neq M \subseteq X$, define

$$(1.3) \qquad \text{rank}_M\varphi = \sup\{\text{rank}_x\varphi \mid x \in M\}$$

$$(1.4) \qquad \text{rank}\ \varphi = \text{rank}_X\varphi .$$

The map φ has pure rank if $\text{rank}_x\varphi = \text{rank}\ \varphi$ for all $x \in X$. The map φ is said to have strict rank r if and only if $r = \text{rank}\ \varphi$ and if every branch B of X contains a point $x \in B$ with $\text{rank}_x\varphi = r$.

Obviously, $\text{rank}_x\varphi = \widetilde{\text{rank}}_x\varphi$ at every normal point x of X. Also the definition implies trivially

<u>Lemma 1.3.</u> Let X and Y be complex spaces. Let $\varphi: X \to Y$ be a holomorphic map. Let $U \neq \emptyset$ and V be open in X and Y respectively such that $\varphi(U) \subseteq V$. Define $\varphi_0 = \varphi: U \to V$. Then $\widetilde{\text{rank}}_x\varphi = \widetilde{\text{rank}}_x\varphi_0$ and $\text{rank}_x\varphi = \text{rank}_x\varphi_0$ for every $x \in U$.

<u>Lemma 1.4.</u> Let X, Y and Z be complex spaces. Let $\varphi: X \to Y$ and $\psi: Y \to Z$ be holomorphic maps. Suppose that ψ is injective. Then

$$\widetilde{\mathrm{rank}}_x\varphi = \widetilde{\mathrm{rank}}_x\psi\circ\varphi \quad \text{and} \quad \mathrm{rank}_x\varphi = \mathrm{rank}_x\psi\circ\varphi$$

for all $x \in X$.

Proof. Define $\chi = \psi\circ\varphi$. Then

$$\chi^{-1}(\chi(x)) = \varphi^{-1}(\psi^{-1}(\psi(\varphi(x)))) = \varphi^{-1}(\varphi(x))$$

if $x \in X$. Hence $\widetilde{\mathrm{rank}}_x\varphi = \widetilde{\mathrm{rank}}_x\chi$. Let $\pi\colon \hat{X} \to X$ be the normalization of X. Then $\widetilde{\mathrm{rank}}_z\chi\circ\pi = \widetilde{\mathrm{rank}}_z\varphi\circ\pi$ if $z \in \hat{X}$. Hence (1.2) implies $\mathrm{rank}_x\varphi = \mathrm{rank}_x\chi$ if $x \in X$; q.e.d.

Lemma 1.5. Let X and Y be complex spaces. Let $\varphi\colon X \to Y$ be holomorphic. Let S be analytic in Y. Suppose $\varphi(X) \subseteq S$. Define $\varphi_0 = \varphi\colon X \to S$. Then $\widetilde{\mathrm{rank}}_x\varphi = \widetilde{\mathrm{rank}}_x\varphi_0$ and $\mathrm{rank}_x\varphi = \mathrm{rank}_x\varphi_0$ for each $x \in X$.

Proof. Let $j\colon S \to Y$ be the inclusion. Then $\varphi = j\circ\varphi_0$. Apply Lemma 1.4; q.e.d.

Lemma 1.6. Let X and Y be complex spaces. Let $\varphi\colon X \to Y$ be a holomorphic map. Take $a \in X$. Then an open neighborhood U of a exists such that $\mathrm{rank}_x\varphi \geqq \mathrm{rank}_a\varphi$ for all $x \in U$.

Proof. If a and $\varphi(a) = b$ are normal points of X and Y respectively, see Remmert [12], Satz [15]. Suppose that only a is a normal point. An open neighborhood V of b and an injective holomorphic map $\psi\colon V \to \mathbb{C}^n$ exist. An open neighborhood U of a with $\varphi(U) \subseteq V$ exists such that $\mathrm{rank}_x\psi\circ\varphi \geqq \mathrm{rank}_a\psi\circ\varphi$ for all $x \in U$.

Lemma 1.4 implies $\operatorname{rank}_x \varphi \geqq \operatorname{rank}_a \varphi$ for all $x \in U$.

Now, consider the general case. Let $\pi\colon \hat{X} \to X$ be the normalization of X. Then $\{a_1,\ldots,a_p\} = \pi^{-1}(a)$. An open neighborhood U_ν of a_ν exists such that $\operatorname{rank}_z \varphi\circ\pi \geqq \operatorname{rank}_{a_\nu} \varphi\circ\pi$ if $z \in U_\nu$. Because π is proper, an open neighborhood U of a exists such that $\pi^{-1}(U) \subseteq U_1 \cup \ldots \cup U_p$. Take $x \in U$. Then $z \in \pi^{-1}(x)$ exists such $\operatorname{rank}_x \varphi = \operatorname{rank}_x \varphi\circ\pi$. Moreover, $z \in U_\nu$ for an index ν. Hence $\operatorname{rank}_x \varphi\circ\pi \geqq \operatorname{rank}_{a_\nu} \varphi\circ\pi \geqq \operatorname{rank}_a \varphi$. Therefore, $\operatorname{rank}_x \varphi \geqq \operatorname{rank}_a \varphi$ if $x \in U$; q.e.d.

Lemma 1.7. Let X and Y be complex spaces. Let $\varphi\colon X \to Y$ be a holomorphic map. Suppose that X consists of finitely many branches $X_1,\ldots,X_p$. Suppose that each X_μ contains a and is irreducible at a. Define $\varphi_\mu = \varphi\colon X_\mu \to Y$ for $\mu = 1,\ldots,p$. Then[5)]

$$\operatorname{rank}_a \varphi = \operatorname{Min}\{\operatorname{rank}_a \varphi_\mu \mid \mu = 1,\ldots,p\}. \tag{1.5}$$

Proof. Let $\pi\colon \hat{X} \to X$ be the normalization. Let $\hat{X}_1,\ldots,\hat{X}_p$ be the branches of $\hat{X}$. Then $\pi_\mu = \pi\colon\hat{X}_\mu \to X_\mu$ is the normalization of X_μ. One and only one point $a_\mu \in \hat{X}_\mu$ exists with $\pi(a_\mu) = a$. Then $\pi^{-1}(a) = \{a_1,\ldots,a_p\}$. Moreover $\operatorname{rank}_a \varphi_\mu = \operatorname{rank}_{a_\mu} \varphi_\mu\circ\pi_\mu$ by definition. Because $\hat{X}_\mu$ is open in $\hat{X}$, Lemma 1.3 implies $\operatorname{rank}_{a_\mu} \varphi_\mu\circ\pi_\mu = \operatorname{rank}_{a_\mu} \varphi\circ\pi$. Therefore, (1.2) implies (1.5); q.e.d.

If a is a point of a complex space X, the situation of Lemma 1.7 can always be achieved for an open neighborhood of a.

Lemma 1.8. Let X and Y be complex spaces. Let $\varphi: X \to Y$ be a holomorphic map. Take $a \in X$ and define $b = \varphi(a)$. Then

$$\widetilde{\operatorname{rank}}_a \varphi \leqq \dim_b Y \quad \text{and} \quad \operatorname{rank}_a \varphi \leqq \dim_b Y.$$

Especially, $\operatorname{rank} \varphi \leqq \dim Y$.

Proof. If A and B are analytic in an open subset of $\mathbb{C}^m$ and if $x \in A \cap B$, then

$$\dim_x A \cap B \geqq \dim A + \dim B - m.$$

By Lemma 1.3, it can be assumed that X is an analytic subset of an open subset U of $\mathbb{C}^m$ and that Y is an analytic subset of $V \times W$ where V and W are open in $\mathbb{C}^n$ and $\mathbb{C}^p$ respectively with $n = \dim_b Y$ and such that the projection $\pi: Y \to V$ is proper with $\pi^{-1}(\pi(b)) = \{b\}$. Define $\psi = \pi \circ \varphi: X \to V$. Then

$$\psi^{-1}(\psi(a)) = \varphi^{-1}(\pi^{-1}(\pi(b))) = \varphi^{-1}(b) = \varphi^{-1}(\varphi(a)).$$

Hence $\widetilde{\operatorname{rank}}_a \psi = \widetilde{\operatorname{rank}}_a \varphi$. Define $c = \psi(a) = \pi(b)$. The graph

$$B = \{(x, \psi(x)) \mid x \in X\}$$

is analytic in $U \times V$ and contained in $X \times V$. A biholomorphic map $\rho: X \to B$ is defined by $\rho(x) = (x, \psi(x))$ if $x \in X$. Hence $\dim_{(a,c)} B = \dim_a X$. Define $A = U \times \{c\}$. Then $\dim_{(a,c)} A = m$. Observe that

$$A \cap B = \psi^{-1}(\psi(a)) \cap \{c\}.$$

Hence,

$$\dim_a \psi^{-1}(\psi(a)) = \dim_{(a,c)} A \cap B \geqq \dim_{(a,c)} A + \dim_{(a,c)} B - (m+n)$$

$$= m + \dim_a X - m - n = \dim_a X - \dim_b Y$$

or

$$\widetilde{\operatorname{rank}}_a \varphi = \dim_a X - \dim_a \varphi^{-1}(\varphi(a)) \leqq \dim_b Y.$$

Let $\pi\colon \hat{X} \to X$ be the normalization. Then $b = \varphi(\pi(z))$ if $z \in \pi^{-1}(a)$ and $\widetilde{\operatorname{rank}}_z \varphi \circ \pi \leqq \dim_b Y$. By (1.2), $\operatorname{rank}_a \varphi \leqq \dim_b Y$; q.e.d.

<u>Lemma 1.9.</u> Let X and Y be complex spaces. Let $\varphi\colon X \to Y$ be a proper light, holomorphic, surjective map. Take $b \in Y$. Then

$$\dim_b Y = \operatorname{Max}\{\dim_a X \mid a \in \varphi^{-1}(b)\}.$$

<u>Proof.</u> By Lemma 1.8

$$\dim_b Y \geqq \widetilde{\operatorname{rank}}_a \varphi = \dim_a X - \dim_a \varphi^{-1}\varphi(a) = \dim_a X$$

if $a \in \varphi^{-1}(b)$.

Now, assume that Y is irreducible, hence pure n-dimensional. Let $X_1, \ldots, X_p$ be the branches of X with $\varphi^{-1}(b) \cap X_\mu \neq \emptyset$. Let $m_1, \ldots, m_p$ be the dimensions of $X_1, \ldots, X_p$ respectively with $m_1 \leqq \ldots \leqq m_p$. Then $m_p = \operatorname{Max}\{\dim_a X \mid a \in \varphi^{-1}(b)\}$. An open neighborhood U of b exists such that $V = \varphi^{-1}(U) \subseteq X_1 \cup \ldots \cup X_p$. The map $\varphi_0 = \varphi\colon V \to U$ is proper, light and surjective with $\dim V = m_p$. Lemma 1.2 implies $m_p \geqq n = \dim U = \dim Y = \dim_b Y$. By the first part of the proof, $\dim_b Y \geqq m_p$. Hence, $\dim_b Y = m_p$.

Now, consider the case where Y is reducible. Let Y_0 be a branch of Y such that $b \in Y_0$ and $\dim_b Y = \dim Y_0$. Then $Z = \varphi^{-1}(Y_0)$ is analytic and $\varphi: Z \to Y_0$ is proper light and surjective. Hence,

$$\begin{aligned}\dim_b Y = \dim Y_0 &= \mathrm{Max}\{\dim_a Z \mid a \in \varphi^{-1}(b) \cap Z\}\\ &\leqq \mathrm{Max}\{\dim_a X \mid a \in \pi^{-1}(b)\} = m_p\\ &\leqq \dim_b Y.\end{aligned}$$

Hence, $m_p = \dim_b Y$, q.e.d.

<u>Lemma 1.10.</u> Let X and Y be complex spaces. Let $\varphi: X \to Y$ be a proper, light, open, holomorphic map. Take $a \in X$ and define $b = \varphi(a)$. Then $\dim_b Y = \dim_a X$. Especially, if Y is pure n-dimensional, then X is pure n-dimensional. If X is pure n-dimensional, and if φ is surjective, then Y is pure n-dimensional.

<u>Proof.</u> Define $\{a_1, \ldots, a_p\} = \varphi^{-1}(b)$ with $a = a_1$. Let U_μ be an open neighborhood of a_μ such that $U_\mu \cap U_\nu = \emptyset$ if $\mu \neq \nu$. An open neighborhood U of b exists such that $V = \varphi^{-1}(U) \subseteq U_1 \cup \ldots \cup U_p$. The map $\varphi_0 = \varphi: V \to U$ is proper, light and open. Now $W = V \cap U_1 = V - (U_2 \cup \ldots \cup U_p)$ is open and closed in V. Hence, $\varphi_1 = \varphi_0: W \to U$ is proper light and open with $\varphi_1^{-1}(b) = \{a\}$. By Lemma 1.9, $\dim_b Y = \dim_b U = \dim_a W = \dim_a X$.

If Y is pure n-dimensional, then $\dim_x X = \dim_{\varphi(x)} Y = n$ for all $x \in X$. Suppose that X has pure dimension n and that φ is surjective. Take $y \in Y$. Then $x \in \varphi^{-1}(y)$ exists and $\dim_y Y = \dim_x X = n$; q.e.d.

Lemma 1.11. Let X and Y be complex spaces. Let $\pi: \hat{X} \to X$ be the normalization of X. Let $\varphi: X \to Y$ be a holomorphic map. Then φ is q-fibering if and only if $\varphi\circ\pi$ is q-fibering.

Proof. a) Suppose that φ is q-fibering. Take $a \in \hat{X}$. Let B be a branch of $(\varphi\circ\pi)^{-1}(\varphi(\pi(a))) = F$ with $\dim_a F = \dim B = s$. Define $b = \pi(a)$ and $G = \varphi^{-1}(\varphi(b))$. Then G is pure q-dimensional. By Theorem 1.1, $\pi(B)$ is an analytic subset of G. The map $\pi: B \to \pi(B)$ is proper, light and surjective. By Lemma 1.9, $\dim_b \pi(B) = s$. Because $\pi(B) \subseteq G$, this implies $s \leqq q$.

Let $\hat{C}$ be a branch of $\hat{X}$ with $\hat{C} \supseteq B$. Then $\pi(\hat{C}) = C$ is a branch of X. Let S_X be the set of non-simple points of X. Then $x \in \hat{C} - \pi^{-1}(S_X)$ exists. Because $\pi: \hat{X} - \pi^{-1}(S_X) \to X - S_X$ is biholomorphic

$$\text{rank}_x \varphi\circ\pi = \dim \hat{C} - \dim \varphi^{-1}\varphi(\pi(x)) = \dim \hat{C} - q.$$

By Lemma 1.6. an open neighborhood U of a exists such that $\text{rank}_x \varphi\circ\pi \geqq \text{rank}_a \varphi\circ\pi$ for all $a \in U$. The open set $\hat{C} - \pi^{-1}(S_X)$ is dense in $\hat{C}$ with $a \in \hat{C}$. Hence, $x \in U \cap (\hat{C} - \pi^{-1}(S_X))$ exists which implies

$$\text{rank}_a \varphi\circ\pi \leqq \text{rank}_x \varphi\circ\pi = \dim \hat{C} - q.$$

Because $\hat{C}$ is a neighborhood of $a \in \hat{X}$, this implies

$$q \geqq s = \dim_a F = \dim \hat{C} - \text{rank}_a \varphi\circ\pi \geqq q$$

Hence, $\dim_a F = q$. The map $\varphi\circ\pi$ is q-fibering.

b) Suppose that $\varphi\circ\pi$ is q-fibering. Take $a \in X$. Take $\hat{a} \in \pi^{-1}(a)$. Then

$$\pi_0 = \pi\colon (\varphi\circ\pi)^{-1}(\varphi\circ\pi(\hat{a})) \to \varphi^{-1}(\varphi(a))$$

is a proper, light, surjective, holomorphic map. By Lemma 1.9 $\dim_y\varphi^{-1}(\varphi(a)) = q$ because $(\varphi\circ\pi)^{-1}(\varphi\circ\pi(\hat{a}))$ is pure q-dimensional; q.e.d.

Lemma 1.12. Let X and Y be complex spaces. Let $\pi\colon \hat{X} \to X$ be the normalization of X. Let $\varphi\colon X \to Y$ be a holomorphic map. Then φ has pure rank r if and only if $\varphi\circ\pi$ has pure rank r.

Proof. a) Suppose that $\varphi\circ\pi$ has pure rank r. By (1.2), $\mathrm{rank}_x\varphi = r$ for all $x \in X$.

b) Suppose that φ has pure rank r. Take $a \in \hat{X}$. A sequence $\{a_\nu\}_{\nu\in\mathbb{N}}$ of points of $\hat{X}$ converges to a such that $\pi(a_\nu) = b_\nu$ is a simple point of X. Then $r = \mathrm{rank}_{b_\nu}\varphi = \mathrm{rank}_{a_\nu}\varphi\circ\pi$. By Lemma 1.6 an index ν_0 exists such that $\mathrm{rank}_{a_\nu}\varphi\circ\pi \geqq \mathrm{rank}_a\varphi\circ\pi$ for $\nu \geqq \nu_0$. Hence, $r \geqq \mathrm{rank}_a\varphi\circ\pi$. By 1.2, $\mathrm{rank}_a\varphi\circ\pi \geqq \mathrm{rank}_{\pi(a)}\varphi = r$. Together, this implies $r = \mathrm{rank}_a\varphi\circ\pi$. The map $\varphi\circ\pi$ has pure rank r; q.e.d.

Lemma 1.13. Let X and Y be complex spaces. Suppose that X is pure dimensional. Let $\varphi\colon X \to Y$ be a holomorphic map. Then $\mathrm{rank}_x\varphi = \widetilde{\mathrm{rank}}_x\varphi$ for all $x \in X$.

Proof. Let $\pi\colon \hat{X} \to X$ be the normalization of X. Then $\hat{X}$ and X are pure m-dimensional. Take $x \in X$. Define $F = \varphi^{-1}(\varphi(x))$. Then

$\pi^{-1}(x) = \{z_1,\ldots,z_p\}$ with $q_\mu = \dim_{z_\mu} \pi^{-1}(F)$ and $q_1 \leqq \ldots \leqq q_p$. Here $\pi^{-1}(F) = (\varphi\circ\pi)^{-1}(\varphi\circ\pi(z_\mu))$ for $\mu = 1,\ldots,p$. Hence, $\mathrm{rank}_x\varphi = m - q_p$. The map $\pi\colon \pi^{-1}(F) \to F$ is proper, light, surjective and holomorphic. Lemma 1.9 implies $q_p = \dim_x F$. Therefore

$$\widetilde{\mathrm{rank}}_x\varphi = \dim_x X - \dim_x F = m - q_p = \mathrm{rank}_x\varphi; \qquad \text{q.e.d.}$$

<u>Theorem 1.14.</u> Let X and Y be complex spaces. Let $\varphi\colon X \to Y$ be a holomorphic map. Let p be a non-negative integer. Let $\pi\colon \hat{X} \to X$ be the normalization of X. Then

$$\hat{E}(p) = \{x \in \hat{X} \mid \mathrm{rank}_x \varphi\circ\pi \leqq p\}$$

$$E(p) = \{x \in X \mid \mathrm{rank}_x \varphi \leqq p\}$$

are analytic sets in $\hat{X}$ and X respectively and $\pi(\hat{E}(p)) = E(p)$.

<u>Proof.</u> If X and Y are normal, then $\hat{X} = X$ and the set $\hat{E}(p) = E(p)$ is analytic by Remmert [12], Satz [17]. Suppose that only X is normal. Again $\hat{X} = X$ and $\hat{E}(p) = E(p)$. Take $a \in X$ and define $b = \varphi(a)$. An open neighborhood V of b and an injective holomorphic map $\psi\colon V \to \mathbb{C}^n$ exists. Define $U = \varphi^{-1}(V)$ and $\chi = \psi\circ\varphi\colon U \to V$. By Lemma 1.4 $\mathrm{rank}_x\chi = \mathrm{rank}_x\varphi$ if $x \in U$. Therefore

$$E(p) \cap U = \{x \in U \mid \mathrm{rank}_x\chi \leqq p\}$$

is analytic in U. Therefore E(p) is analytic.

Now, consider the general case. Because $\hat{X}$ is normal, the set $\hat{E}(p)$ is analytic. Take $x \in E(p)$. Then $z \in \pi^{-1}(x)$ exists such that $\operatorname{rank}_z \varphi\circ\pi = \operatorname{rank}_x \varphi \leqq p$. Hence $z \in \hat{E}(p)$ and $x = \pi(z) \in \pi(\hat{E}(p))$. If $x \in \pi(\hat{E}(p))$, then $z \in \hat{E}(p)$ with $\pi(z) = x$ exists. Then $\operatorname{rank}_x \varphi \leqq \operatorname{rank}_z \varphi\circ\pi \leqq p$. Hence $x \in E(p)$. Therefore $E(p) = \pi(\hat{E}(p))$. The restriction $\pi: \hat{E}(p) \to X$ is proper. By Theorem 1.1, $E(p) = \pi(\hat{E}(p))$ is analytic; q.e.d.

<u>Lemma 1.15.</u> Let X and Y be complex spaces. Let $\varphi: X \to Y$ be a holomorphic map. Suppose that X or Y have only finitely many branches. Then $\operatorname{rank} \varphi < \infty$.

<u>Proof.</u> If Y has only finitely many branches, then $\dim Y < \infty$. By Lemma 1.8, $\operatorname{rank} \varphi \leqq \dim Y < \infty$. Suppose that X has only finitely many branches $X_1, \ldots, X_p$. Take $a_\mu \in X_\mu$ for each $\mu = 1,\ldots,p$. Define $n = \operatorname{Max}\{\dim_{\varphi(a_\mu)} Y \mid \mu = 1,\ldots,p\} < \infty$. By Theorem 1.14, the set $E = \{x \in X \mid \operatorname{rank}_x \varphi \leqq n\}$ is analytic. By Lemma 1.6, a_μ is an interior point of E; hence $X_\mu \subseteq E$ for each $\mu = 1,\ldots,p$. Hence $X = E$ and $\operatorname{rank} \varphi \leqq n$; q.e.d.

<u>Lemma 1.16.</u> Let X and Y be complex spaces. Let $\varphi: X \to Y$ be a holomorphic map of $\operatorname{rank} \varphi \leqq r$. Then the following statements are equivalent:

1. The map φ has strict rank r.
2. The set $D = \{x \mid \operatorname{rank}_x \varphi < r\}$ is analytic and thin in X.
3. If B is a branch of X, a simple point x of X with $x \in B$ and $\operatorname{rank}_x \varphi = r$ exists.
4. If B is a branch of X, then $\operatorname{rank} \varphi|B = r$.

<u>Proof</u> <u>1. implies 2.:</u> D is analytic by Theorem 1.14. By assumption, D contains no branch of X. Hence D is thin; q.e.d.

<u>2. implies 3.:</u> Let B be a branch of X. Then $B = D \neq \emptyset$ is open and dense on B. Hence a simple point x of X with $x \in B - D$ exists. Hence $\operatorname{rank}_x \varphi = r$.

<u>3. implies 4.:</u> Let B be a branch of X. A simple point a of X exists such that $r = \operatorname{rank}_a \varphi$ with $a \in B$. Define

$$E = \{x \in B \mid \operatorname{rank}_x \varphi|B \leqq r\}.$$

By Theorem 1.14, E is analytic in B, at every point $x \in B$ which is a simple point of X, $\operatorname{rank}_x \varphi = \operatorname{rank}_x \varphi|B$. An open neighborhood U of a with $U \subseteq B$ exist such that U consists only of simple points of X and such that $r \geqq \operatorname{rank}_x \varphi = \operatorname{rank}_x \varphi|B$. Hence $U \subseteq E$. Therefore $E = B$. Since $r = \operatorname{rank}_a \varphi = \operatorname{rank}_a \varphi|B$, this implies $\operatorname{rank} \varphi|B = r$.

<u>4. implies 1.:</u> For each branch B, define

$$D_B = \{x \mid \operatorname{rank}_x \varphi|B < r\}.$$

Then D_B is analytic and thin in B. A simple point x_0 of X with $x \in B - D_B$ exists. Then $r = \operatorname{rank}_x \varphi|B = \operatorname{rank}_x \varphi$. Therefore φ has strict rank r; q.e.d.

<u>Lemma 1.17.</u> Let X and Y be complex spaces. Let $\{X_\lambda\}_{\lambda \in \Lambda}$ be the family of branches of X. Let $\varphi: X \to Y$ be a holomorphic map. Then

$$\operatorname{rank} \varphi = \sup\{\operatorname{rank} \varphi|X_\lambda \mid \lambda \in \Lambda\}.$$

Proof. Define $\varphi_\lambda = \varphi: X_\lambda \to Y$. Define $r_\lambda = \operatorname{rank} \varphi_\lambda$. The set

$$D_\lambda = \{x \in X_\lambda \mid \operatorname{rank} \varphi_\lambda < r_\lambda\}$$

is thin and analytic in X_λ. Take a simple point $x \in X$ with $x \in X_\lambda - D_\lambda$. Then $\operatorname{rank}_x \varphi = \operatorname{rank}_x \varphi_\lambda = r_\lambda$. Hence $\operatorname{rank} \varphi \geqq \sup\{r_\lambda \mid \lambda \in \Lambda\}$.

Define $r = \sup\{r_\lambda \mid \lambda \in \Lambda\}$. If $r = \infty$, then $r \geqq \operatorname{rank} \varphi$. Assume $r < \infty$. Suppose that $a \in X$ exists such that $r < \operatorname{rank}_a \varphi$. By Lemma 1.6 a simple point $x \in X$ with $\operatorname{rank}_x \varphi \geqq \operatorname{rank}_a \varphi$ exists. Now $x \in X_\lambda$ for some $\lambda \in \Lambda$ (and only one!). Hence $r \geqq r_\lambda \geqq \operatorname{rank}_x \varphi | X_\lambda = \operatorname{rank}_x \varphi > r$ which is impossible. Hence $\operatorname{rank}_x \varphi \leqq r$ for all $x \in X$; i.e., $\operatorname{rank} \varphi \leqq r$. Together, $r = \operatorname{rank} \varphi$ is obtained; q.e.d.

Lemma 1.18. Let C be an irreducible analytic subset of the complex space X. Define $s = \operatorname{Min}\{\dim_x X \mid x \in C\}$. Then $D = \{x \in C \mid \dim_x X > s\}$ is a thin analytic subset of X.

Proof. For each $r \in \mathbb{Z}$, the set $X_r = \{x \in X \mid \dim_x X \geqq r\}$ is analytic in X. Because C is irreducible, $C \subseteq X_r$ for some number r. Hence $t = \sup\{r \mid C \subseteq X_r\}$ is defined. Since each point of X is contained only in finitely many sets X_r, the supremum is finite and assumed. Then $C_1 = X_{t+1} \cap C$ is analytic and thin in C. Take $a \in C$ with $\dim_a X = s$. Then $a \in X_t$ and $s \geqq t$. If $x \in C - C_1$, then $t \geqq \dim_x X \geqq s$. Therefore, $s = t$ and $C - C_1 \subseteq C - D$. If $x \in C - D$, then $\dim_x X = s = t$, hence $x \in C - C_1$. Therefore $C - C_1 = C - D$ and $C_1 = D$; q.e.d.

Let X and Y be complex spaces. Let $\varphi\colon X \to Y$ be a holomorphic map. A subset S of X is said to be <u>φ-saturated</u> if and only if $\varphi^{-1}(\varphi(x)) \cap S$ is a union of branches of $\varphi^{-1}(\varphi(x))$ if $x \in S$. If S is analytic, then S is φ-saturated, if and only if each branch of $\varphi^{-1}(\varphi(x)) \cap S$ is a branch of $\varphi^{-1}(\varphi(x))$.

<u>Lemma 1.19.</u> Let X and Y be complex spaces. Let $\varphi\colon X \to Y$ be a holomorphic map. Let p be a non-negative integer. Assume that $\mathrm{rank}_x\varphi = \widetilde{\mathrm{rank}}_x\varphi$ for all $x \in X$. Then $E = \{x \in X \mid \widetilde{\mathrm{rank}}_x\varphi \leqq p\}$ is φ-saturated. Especially, this is the case if X is pure dimensional.

<u>Proof:</u> Take $x \in E$. Define $F = \varphi^{-1}(\varphi(x))$. Let B be a branch of $F \cap E$. Let $b \in B$ be a simple point of $F \cap E$. Let C be a branch of F with $b \in C$ and $\dim C = \dim_b F$. Then

$$\dim C = \dim_b F = \dim_b X - \mathrm{rank}_b\varphi \geqq \dim_b X - p.$$

Let S_F be the set of non-simple points of F. Define

$$s = \mathrm{Min}\{\dim_z X \mid z \in C\}$$

$$D = \{z \in C \mid \dim_z X > s\}.$$

By Lemma 1.18, D is a thin analytic subset of C. Take $z \in C - (D \cup S_F)$. Then

$$\mathrm{rank}_z\varphi = \dim_z X - \dim_z F = s - \dim_z C \leqq \dim_b X - \dim C \leqq p.$$

Therefore $C - (D \cup S_F) \subseteq E$. Because D and $S_F \cap C$ are thin in C, and because E is analytic, $C \subseteq E$. Therefore $C \subseteq F \cap E$, where C is irreducible. Hence $C \subseteq B^1$ where B^1 is a branch of $F \cap E$. But B is the only branch of $F \cap E$ containing b. Therefore $B = B^1$ and $C \subseteq B \subseteq F$. Because C is a branch of F and because B is irreducible, $C = B$; q.e.d.

Proposition 1.20. Let X and Y be complex space. Suppose that X has pure dimension m. Let $\varphi: X \to Y$ be a holomorphic map. Let p be a non-negative integer. Define

$$E = \{x \in X \mid \text{rank}_x \varphi \leqq p\}.$$

Suppose that E is thin. Then rank $\varphi | E \leqq p - 1$.

Proof. Define $D = \{x \in E \mid \text{rank}_x \varphi | E \leqq p - 1\}$. Then D is analytic in E. Let a be a simple point of E. Let B be a branch of $\varphi^{-1}(\varphi(a)) \cap E$ with $a \in B$ and

$$\dim B = \dim_a \varphi^{-1}(\varphi(a)) \cap E.$$

Now, $\dim_a E \leqq m - 1$, since E is thin in X. Hence

$$\text{rank}_a \varphi | E = \dim_a E - \dim B \leqq m - 1 - \dim B.$$

Now, B is also a branch of $\varphi^{-1}(\varphi(a))$ by Lemma 1.19. Take a simple point x of $\varphi^{-1}(\varphi(a))$ with $x \in B \subseteq E$. Then $\dim B = \dim_x \varphi^{-1}(\varphi(a))$ and $p \geqq \text{rank}_x \varphi = m - \dim B$. Hence $\text{rank}_a \varphi | E \leqq p - 1$. Therefore all simple points of E belong to D. Hence $E = D$ and $\text{rank}_x \varphi | E \leqq p - 1$ for all $x \in E$. Then rank $\varphi | E \leqq p - 1$; q.e.d.

Proposition 1.21. Let X and Y be complex spaces. Let $\varphi: X \to Y$ be a holomorphic map of pure rank r. Let U be an open neighborhood of $a \in X$. Then an open neighborhood V of a with $V \subseteq U$ and an open neighborhood W of $\varphi(a)$ exist such that $\varphi(V)$ is a pure r-dimensional analytic subset of W.

Proof. 1. Case: If X and Y are normal, this holds by Remmert [12] Satz 19.

2. Case: Suppose that X is normal. An open neighborhood G of $b = \varphi(a)$, an open subset H of $\mathbb{C}^n$ and an injective holomorphic map $\psi: G \to H$ exists such that $\psi(G)$ is analytic in H and such that $\psi_0 = \psi: G \to \psi(G)$ is biholomorphic. An open neighborhood V of a with $V \subseteq U \cap \varphi^{-1}(G)$ and an open neighborhood W_0 of $\psi(b)$ with $W_0 \subseteq H$ exists such that $\psi(\varphi(V))$ is a pure r-dimensional analytic subset of W_0, because $\psi \circ \varphi: \varphi^{-1}(G) \to H$ has pure rank r by Lemma 1.4. Then $\psi(\varphi(V)) \subseteq \psi(G) \cap W_0$. The set $W = \psi^{-1}(W_0 \cap \psi(G))$ is open in Y with $b \in W$ and $\varphi(V) = \psi^{-1}(\psi(\varphi(V)))$ is a pure r-dimensional analytic subset of W.

3. Case: Consider the general situation. Take $a \in X$. Define $\varphi(a) = b$. Let $\pi: \hat{X} \to X$ be the mornalization of X. Define $\pi^{-1}(a) = \{a_1, \ldots, a_p\}$ with $a_\mu \neq a_\nu$ if $\mu \neq \nu$. By Lemma 1.12, $\varphi \circ \pi$ has pure rank r.

By induction, open set $U_\lambda, W_{\mu\lambda}, V_{\mu\lambda}$ for $\mu = 1, \ldots, p$ and $\lambda \in \mathbb{N}$ shall be constructed such that

1. For each $\lambda \in \mathbb{N}$ and $\mu = 1, \ldots, p$

$$a \in U_{\lambda+1} \subseteq U_\lambda \subseteq U$$

$$a_\mu \in V_{\mu,\lambda+1} \subseteq V_{\mu\lambda}$$

$$b \in W_{\mu,\lambda+1} \subseteq W_{\mu\lambda}$$

$$V_{\lambda+1} = \bigcup_{\mu=1}^{p} V_{\mu,\lambda+1} \subseteq \pi^{-1}(U_\lambda) \subseteq \bigcup_{\mu=1}^{p} V_\mu = V_\lambda .$$

2. For each $\lambda \in \mathbb{N}$ and $\mu = 1,\ldots,p$, the set $F_{\mu\lambda} = \varphi(\pi(V_{\mu\lambda}))$ is a pure r-dimensional analytic subset of $W_{\mu\lambda}$ with $b \in F_{\mu\lambda}$ and

$$\bigcup_{\mu=1}^{p} F_{\mu\lambda+1} \subseteq \varphi(U_\lambda) \subseteq \bigcup_{\mu=1}^{p} F_{\mu,\lambda} .$$

At first, table $\lambda = 1$. An open neighborhood $V_{\mu 1}$ of a_μ and an open neighborhood $W_{\mu 1}$ of b exist such that $F_{\mu 1} = \varphi(\pi(V_{\mu 1}))$ is a pure r-dimensional analytic subset of $W_{\mu 1}$. An open neighborhood U_1 of a with $U_1 \subseteq U$ and $\pi^{-1}(U_1) \subseteq \bigcup_{\mu=1}^{p} V_{\mu 1} = V_1$ exists, because π is proper. Then

$$\bigcup_{\mu=1}^{p} F_{\mu 1} = \bigcup_{\mu=1}^{p} \varphi(\pi(V_{\mu 1})) \supseteq \varphi(\pi(\pi^{-1}(U_1))) = \varphi(U_1).$$

This completes the construction for $\lambda = 1$.

Suppose that the construction has proceeded to λ including λ. Then an open neighborhood $V_{\mu,\lambda+1}$ of a_μ, and an open neighborhood $W_{\mu,\lambda+1}$ of b exist such that

$$V_{\mu,\lambda+1} \subseteq V_{\mu\lambda} \cap \pi^{-1}(U_\lambda) \cap \pi^{-1}(\varphi^{-1}(W_{\mu\lambda}))$$

$$W_{\mu\lambda+1} \subseteq W_{\mu\lambda}$$

and such that $F_{\mu,\lambda+1} = \varphi(\pi(V_{\mu,\lambda+1}))$ is a pure r-dimensional analytic subset of $W_{\mu,\lambda+1}$. Then

$$\bigcup_{\mu=1}^{p} F_{\mu,\lambda+1} = \bigcup_{\mu=1}^{p} \varphi(\pi(V_{\mu,\lambda+1})) \subseteq \varphi(\pi(\pi^{-1}(U_\lambda)) = \varphi(U_\lambda).$$

Because π is proper, an open neighborhood $U_{\lambda+1}$ of a in U_λ exists such that $\pi^{-1}(U_{\lambda+1}) \subseteq \bigcup_{\mu=1}^{p} V_{\mu,\lambda+1} = V_{\lambda+1}$. Then

$$\bigcup_{\mu=1}^{p} F_{\mu,\lambda+1} = \bigcup_{\mu=1}^{p} \varphi(\pi(V_{\mu,\lambda+1})) \supseteq \varphi(\pi(\bigcup_{\mu=1}^{p} V_{\mu,\lambda+1}) \supseteq$$

$$\varphi(\pi(\pi^{-1}(U_{\lambda+1})) = \varphi(U_{\lambda+1}).$$

This completes the construction for $\lambda + 1$.

Define $W_\lambda = W_{1\lambda} \cap \ldots \cap W_{p\lambda} \supseteq W_{\lambda+1}$. Then $b \in W_\lambda$. The set $F_\lambda = (F_{1\lambda} \cup \ldots \cup F_{p\lambda}) \cap W_\lambda$ is a pure r-dimensional analytic subset of W_λ with $b \in F_\lambda$ and $F_{\lambda+1} \subseteq F_\lambda \cap W_{\lambda+1}$. Let $F_{\lambda b}$ be the germ of F_λ at b. Let $\mathfrak{B}_\lambda$ be the set of branches of $F_{\lambda b}$. Then $\mathfrak{B}_\lambda \supseteq \mathfrak{B}_{\lambda+1}$, because $F_\lambda \supseteq F_{\lambda+1}$ and because each F_λ is pure r-dimensional. Let $g_\lambda = \# \mathfrak{B}_\lambda$ be the number of elements of $\mathfrak{B}_\lambda$. Then $\infty > g_\lambda \geqq g_{\lambda+1}$. A number $\rho \in \mathbb{N}$ exists such that $g_\rho = g_{\rho+1}$. Hence an open neighborhood W of b with $W \subseteq W_{\rho+1}$ exists such that

$W \subseteq W_{\rho+1}$ and $F_\rho \cap W = F_{\rho+1} \cap W$. Now

$$F_\rho \cap W \supseteq \varphi(U_\rho) \cap W \supseteq F_{\rho+1} \cap W = F_\rho \cap W$$

implies $F_\rho \cap W = \varphi(U_\rho) \cap W$. Define $V = \varphi^{-1}(W) \cap U_\rho$. Then V is an open neighborhood of a with $V \subseteq U$ and $\varphi(V) = \varphi(U_\rho) \cap W = F_\rho \cap W$ is a pure r-dimensional analytic subset of W; q.e.d.

Theorem 1.22. (Open mapping theorem of Remmert [12], Satz 28). Let X be a complex space. Let Y be a locally irreducible, complex space of pure dimension n. Let $\varphi\colon X \to Y$ be a holomorphic map of pure rank n. Then φ is open.

Proof. Let $U \neq \emptyset$ be open in X. Take $b \in \varphi(U)$. Take $a \in U$ with $b = \varphi(a)$. Open neighborhoods V of a with $V \subseteq U$ and W of b in Y exist such that $\varphi(V)$ is a pure n-dimensional analytic subset of W (Proposition 1.21). Because Y is locally irreducible and pure n-dimensional, $\varphi(V)$ is a neighborhood of b with $\varphi(V) \subseteq \varphi(U)$, hence $\varphi(U)$ is open; q.e.d.

An open holomorphic map $\varphi\colon X \to Y$ into a locally irreducible complex space Y exists such that φ does not have pure rank. (Compare Theorem 2.6).

Proposition 1.23. Let X and Y be complex spaces. Let $\varphi\colon X \to Y$ be a holomorphic map of rank $n < \infty$. Then $\varphi(X)$ is almost thin of dimension n.

Proof. If φ has rank 0, then φ is constant on each branch of X. Hence $\varphi(X)$ is almost countable, i.e., almost thin of dimension 0. Suppose the proposition is proved if $\operatorname{rank} \varphi < n$. Suppose that a map with $\operatorname{rank} \varphi = n$ is given. Let $\{X_\lambda\}_{\lambda \in \Lambda}$ be the family of branches of X. Define $\varphi_\lambda = \varphi: X_\lambda \to Y$. Then $\operatorname{rank} \varphi_\lambda = n_\lambda \leqq n$. If $n_\lambda < n$, then $\varphi_\lambda(X_\lambda)$ is almost thin of dimension n_λ, hence of dimension n. If $n_\lambda = n$, define

$$E_\lambda = \{x \in X_\lambda \,|\, \operatorname{rank}_x \varphi_\lambda \leqq n = 1\}.$$

By Lemma 1.16 and Lemma 1.20, $\operatorname{rank} \varphi_\lambda | E_\lambda \leqq n - 2$. Hence $\varphi_\lambda(E_\lambda) = \varphi(E_\lambda)$ is thin of dimension $n - 2$, hence thin of dimension n. The map $\varphi_\lambda: X_\lambda - E_\lambda \to Y$ has pure rank n. For every point $a \in X_\lambda - E_\lambda$, an open neighborhood $V(a)$ of a and an open neighborhood $W(a)$ of $\varphi(a)$ exists such that $\varphi_\lambda(V(a))$ is an analytic set of pure dimension n in $W(a)$. A sequence $\{a_\rho\}_{\rho \in \mathbb{N}}$ of points $a_\rho \in X_\lambda - E_\lambda$ exists such that $X_\lambda - E_\lambda = \bigcup_{\rho \in \mathbb{N}} V(a_\rho)$. Hence

$$\varphi(X_\lambda - E_\lambda) = \bigcup_{\rho \in \mathbb{N}} \varphi(V(a_\rho))$$

is almost thin of dimension n in Y. Hence $\varphi(X_\lambda) = \varphi(E_\lambda) \cup \varphi(X_\lambda - E_\lambda)$ is almost thin of dimension n in Y. Therefore

$$\varphi(X) = \bigcup_{\lambda \in \Lambda} \varphi(X_\lambda)$$

is almost thin of dimension n in Y; q.e.d.

Proposition 1.24. Let X and Y be complex spaces. Suppose that X has pure dimension m. Let $\varphi: X \to Y$ be a holomorphic map of strict rank n. Define $D = \{x \in X \mid \text{rank}_x \varphi < n\}$. Then D is a thin analytic subset of X with rank $\varphi|D \leqq n - 2$. Also $D' = \varphi(D)$ is almost thin of dimension n-2. If Y is pure dimensional, then D' is almost thin.

Proof. By Lemma 1.16, D is analytic and thin. By Lemma 1.20 rank $\varphi|D \leqq n - 2$. By Proposition 1.23, $\varphi(D) = D'$ is almost thin of dimension n - 2. By Lemma 1.8, dim $Y \geqq$ rank $\varphi = n$. Hence, if Y is pure dimensional, D' is almost thin; q.e.d.

Lemma 1.25. Let X and Y be complex spaces. Suppose that Y has pure dimension n. Let $\varphi: X \to Y$ be a holomorphic map of strict rank n. Let S be an almost thin subset of Y. Let $U \neq \emptyset$ be an open subset of X. Then $\varphi(U)$ is not contained in S.

Proof. The set $D = \{x \in X \mid \text{rank}_x \varphi < n\}$ is thin and analytic in X by Lemma 1.16. The map $\varphi_0 = \varphi: U - D \to Y$ has pure rank n. Take $a \in U - D$. By Proposition 1.21 an open neighborhood V of a with $V \subseteq U$ and an open subset W of Y exists such that $\varphi(V)$ is a pure n-dimensional analytic subset of W. Hence $\varphi(V) \not\subseteq S$ which implies $\varphi(U) \not\subseteq S$, q.e.d.

Lemma 1.26. Let X and Y be complex spaces of pure dimension m and n respectively. Let S be a thin analytic subset of X. Let $\varphi: X \to Y$ be a holomorphic map. Let T be the set of all points $y \in \varphi(X)$ such that a branch of $\varphi^{-1}(y)$ is contained in S. Then T is almost thin in Y.

Proof. Let $\{S_\lambda\}_{\lambda \in \Lambda}$ be the family of branches of S. Define

$$N_\lambda = \{x \in S_\lambda \mid \operatorname{rank}_x \varphi | S_\lambda \leqq n - 1\}.$$

Then N_x is analytic in S_λ and in X. Take $y \in T$. Let B be a branch of $\varphi^{-1}(y)$ with $B \subseteq S$. Then $B \subseteq S_\lambda$ for some $\lambda \in \Lambda$. Let $x \in B$ be a simple point of $\varphi^{-1}(y)$. Then

$$\dim_x \varphi^{-1}(\varphi(x)) = \dim_x B \leqq \dim_x \varphi^{-1}(\varphi(x)) \cap S_\lambda \leqq \dim_x \varphi^{-1}(\varphi(x)).$$

Hence,

$$\dim_x \varphi^{-1}(\varphi(x)) = \dim_x \varphi^{-1}(\varphi(x)) \cap S_\lambda.$$

Since S_λ is pure dimensional, Lemma 1.13 implies

$$\operatorname{rank}_x \varphi | S_\lambda = \widetilde{\operatorname{rank}}_x \varphi | S_\lambda = \dim_x S_\lambda - \dim_x \varphi^{-1}(\varphi(x)) \cap S_\lambda$$

$$< \dim_x X - \dim_x \varphi^{-1}(\varphi(x)) = \operatorname{rank}_x \varphi \leqq n.$$

Therefore, $x \in N_\lambda$ and $y = \varphi(x) \in \varphi(N_\lambda)$. Hence

$$T \subseteq \bigcup_{\lambda \in \Lambda} \varphi(N_\lambda).$$

If $N_\lambda \neq S_\lambda$, then N_λ is thin in S_λ and Proposition 1.20 implies $\operatorname{rank} \varphi | N_\lambda \leqq n - 1$. By Proposition 1.23, $\varphi(N_\lambda)$ is almost thin of dimension n - 1. If $N_\lambda = S_\lambda$, then $\operatorname{rank} \varphi | N_\lambda = \operatorname{rank} \varphi | S_\lambda \leqq n - 1$. By Proposition 1.23 $\varphi(N_\lambda)$ is almost thin of dimension n - 1. Because Λ is at most countable, and because Y has pure dimension n, the union T is almost thin.

In a certain sense, this Lemma can be considered as a substitute for Sard's theorem for complex spaces.

<u>Lemma 1.27.</u> Let X be an irreducible, complex space. Let $\varphi: X \to Y$ be a surjective, holomorphic map. Then Y is irreducible. If φ is also q-fibering, then $\dim X = q + \dim Y$.

<u>Proof.</u> Let S_X and S_Y be the sets of non-simple points of X and Y respectively. Then $\tilde{S}_Y = \varphi^{-1}(S_Y) \neq X$ is a thin analytic subset of X. Hence $X - \tilde{S}_Y \neq \emptyset$ is connected. Therefore $Y - S_Y = \varphi(X-\tilde{S}_Y)$ is connected. Hence Y is irreducible. Define $n = \dim Y$.

Suppose that φ is also q-fibering. By Lemma 1.8, $\operatorname{rank} \varphi \leqq n$. By Proposition 1.23, $\operatorname{rank} \varphi \geqq n$, because φ is surjective. Hence $\operatorname{rank} \varphi = n$. By Lemma 1.16, $E = \{x \mid \operatorname{rank}_x \varphi < n\}$ is thin and analytic. Take $a \in X - (S_X \cup E)$. Then

$$n = \operatorname{rank}_a \varphi = \dim X - \dim_a \varphi^{-1}(\varphi(a)) = \dim_X - q, \qquad \text{q.e.d.}$$

If S is a set, define $S^p = S \times \ldots \times S$ (p-times). Let $\Delta_S = \Delta_S(p) = \{(x,\ldots,x) \in S^p \mid x \in S\}$ be the diagonal of S^p. A bijective map $\delta_S: \Delta_S \to S$ is defined by $\delta_S(x,\ldots,x) = x$. If X is a complex space, Δ_X is an analytic subset of X^p. The map δ_X is biholomorphic. If $\varphi: X \to Y$ is a holomorphic map into the complex space Y, a holomorphic map

$$\varphi_p: X^p \to Y^p$$

is defined by

$$\varphi_p(x_1,\ldots,x_p) = (\varphi(x_1),\ldots,\varphi(x_p)).$$

Define $X^p_\varphi = \varphi_p^{-1}(\Delta_Y)$ as the cartesian product relative to φ. Then X^p_φ is an analytic subset in X^p, which defines X^p_φ as a complex space. Observe that Δ_X is an analytic subset of X^p_φ. Let $\pi_\nu\colon X^p_\varphi \to X$ be the projection onto the νth factor: $\pi_\nu(x_1,\ldots,x_p) = x_\nu$. The map $\delta_\varphi = \delta_Y \circ \varphi_p\colon X^p_\varphi \to Y$ is holomorphic. Moreover $\delta_\varphi = \varphi \circ \pi_\nu$ for $\nu = 1,\ldots,p$. If $y \in Y$, then

$$(1.6) \qquad \delta_\varphi^{-1}(y) = \varphi^{-1}(y) \times \ldots \times \varphi^{-1}(y) = (\varphi^{-1}(y))^q.$$

If $x \in X$ and $y = \varphi(x)$, then

$$(1.7) \qquad \pi_\nu^{-1}(x) = (\varphi^{-1}(y))^{\nu-1} \times \{x\} \times (\varphi^{-1}(y))^{p-\nu}.$$

<u>Lemma 1.28.</u> Let X and Y be complex spaces. Let $\varphi\colon X \to Y$ be a light holomorphic map. Let $p > 1$ be an integer. Let B be a branch of the diagonal Δ_X in X^p_φ. Then B is a branch of X^p_φ. If X is irreducible, Δ_X is a branch of X^p_φ.

<u>Proof.</u> Let C be a branch of X^p_φ with $B \subseteq C$. Take a simple $a = (b,\ldots,b)$ of Δ_X with $a \in B$. Then $\dim B = \dim_a \Delta_X$. Because $\delta_X\colon \Delta_X \to X$ is biholomorphic, $\dim_a \Delta_X = \dim_b X$. By (1.7) $\dim_a \pi_1^{-1}(\pi_1(a)) = 0$. Hence $\widetilde{\operatorname{rank}}_a \pi_1 = \dim_a X^p_\varphi$. By Lemma 1.8, $\dim_b X \geqq \widetilde{\operatorname{rank}}_a \pi_1$. Therefore

$$\dim B = \dim_a \Delta_X = \dim_b X \geqq \widetilde{\operatorname{rank}}_a \pi_1 = \dim_a X^p_\varphi \geqq \dim C$$

Therefore, $B = C$, q.e.d.

Lemma 1.29. Let X and Y be complex spaces. Let $\varphi: X \to Y$ be a holomorphic map of strict rank n. Define $D = \{x \in X \mid \operatorname{rank}_x \varphi < n\}$. Then $D' = \varphi(D)$ is almost thin of dimension $n - 2$.

Proof. Let $\pi: \hat{X} \to X$ be the normalization of X. Define $\hat{D} = \{x \in X \mid \operatorname{rank}_x \varphi\circ\pi < n\}$. By Theorem 1.14, D and $\hat{D}$ are analytic with $\pi(\hat{D}) = D$. Let $\mathfrak{B}$ be the set of branches of $\hat{X}$. Take $B \in \mathfrak{B}$. Take $a \in B$. By Lemma 1.6 an open neighborhood U of a in B exists such that $\operatorname{rank}_a \varphi\circ\pi|B \leqq \operatorname{rank}_x \varphi\circ\pi|B$ for all $x \in U$. A point $x \in U$ exists such that $\pi(x)$ is a simple point of $\hat{X}$. Then $\operatorname{rank}_x \varphi\circ\pi|B = \operatorname{rank}_x \varphi\circ\pi = \operatorname{rank}_{\pi(X)} \varphi \leqq n$. Hence $\operatorname{rank}_a \varphi\circ\pi|B \leqq n$. Now $\pi(B)$ is a branch of X. By Lemma 1.16 a point $x \in B$ exists such that $\pi(X)$ is a simple point of X with $\operatorname{rank}_{\pi(x)} \varphi = n$. As before $\operatorname{rank}_x \varphi\circ\pi|B = \operatorname{rank}_{\pi(x)} \varphi = n$. Therefore $\operatorname{rank} \varphi\circ\pi|B = n$ for each $B \in \mathfrak{B}$. By Lemma 1.16 $\varphi\circ\pi$ has strict rank n and $\hat{D}$ is thin in $\hat{X}$. Also $D_B = \{x \in B \mid \operatorname{rank}_x \varphi\circ\pi|B < n\}$ is thin and analytic in B. Because B is open in $\hat{X}$, $\operatorname{rank}_x \varphi\circ\pi|B = \operatorname{rank}_x \varphi\circ\pi$ for each $x \in B$. Hence $D_B = B \cap D$ and $D = \bigcup_{B\in\mathfrak{B}} D_B$. By Proposition 1.24, $\varphi\circ\pi(D_B) = D_B'$ is almost thin of dimension $n - 2$ in Y. Because $\mathfrak{B}$ is at most countable, also $D' = \bigcup_{B\in\mathfrak{B}} D_B'$ is almost thin of dimension $n - 2$; q.e.d.

Lemma 1.30. Let X and Y be complex spaces. Let $\varphi: X \to Y$ be a holomorphic map. Let p be a non-negative integer. Define

$$E = \{x \in X \mid \operatorname{rank}_x \varphi \leqq p\}$$

Then $E' = \varphi(E)$ is almost thin of dimension p in Y and

$$\operatorname{rank} \varphi | E \leqq p.$$

<u>Proof.</u> Let $\pi: \hat{X} \to X$ be the normalization of X. Define

$$\hat{E} = \{x \in X | \operatorname{rank}_x \varphi \circ \pi \leqq p\}.$$

By Theorem 1.14, E and $\hat{E}$ are analytic with $\pi(\hat{E}) = E$. Hence $E' = \varphi(\pi(\hat{E}))$. Let $\mathcal{L}$ be the set of branches of $\hat{X}$. Each branch $B \in \mathcal{L}$ is open in the normal space $\hat{X}$. Hence $\operatorname{rank}_x \varphi \circ \pi = \operatorname{rank}_x \varphi \circ \pi | B$ for every $x \in B$. Therefore

$$E_B = \{x \in B | \operatorname{rank}_x \varphi \circ \pi | B \leqq p\} = \hat{E} \cap B.$$

If $B = E_B$, then $\operatorname{rank} \varphi \circ \pi | B \leqq p$ and $\varphi(\pi(E_B)) = E'_B$ is almost thin of dimension p by Proposition 1.23. If $B \neq E_B$, then E_B is a thin analytic subset of the pure dimensional complex space B. By Lemma 1.20, $\operatorname{rank} \varphi \circ \pi | E_B \leqq p - 1$. By Proposition 1.23, $\varphi(\pi(E_B)) = E'_B$ is almost thin of dimension p - 1. Because $\mathcal{L}$ is almost countable

$$E' = \varphi(\pi(\hat{E})) = \varphi(\pi(\bigcup_{B \in} E_B)) = \bigcup_{B \in} E'_B$$

is almost thin of dimension p.

Take $a \in E$. Then $\operatorname{rank}_a \varphi | E \leqq p$ has to be proved. An open neighborhood U of a exists such that $U \cap E$ consists only of finitely many branches. By Lemma 1.15, $\operatorname{rank} \varphi | U \cap E = m < \infty$. A point

$b \in U \cap E$ exists such that $\operatorname{rank}_b \varphi | U \cap E = m$. By Lemma 1.6, an open neighborhood U_1 of b with $U_1 \subseteq U$ exists such that

$$m = \operatorname{rank}_b \varphi | U \cap E \leqq \operatorname{rank}_x \varphi | U \cap E \leqq \operatorname{rank} \varphi | U \cap E = m$$

for all $x \in U_1 \cap E$. Hence $\operatorname{rank}_x \varphi | U \cap E = m$ if $x \in U_1$. By Proposition 1.21, an open neighborhood U_2 of b with $U_2 \subseteq U_1$ exists such that $\varphi(U_2 \cap E)$ is a pure m-dimensional analytic subset of some open subset of Y. The set $E' = \varphi(E)$ is almost thin of dimension p in Y with $\varphi(U_2 \cap E) \subseteq E'$. Then $m \leqq p$. Now,

$$\operatorname{rank}_a \varphi | E = \operatorname{rank}_a \varphi | U \cap E \leqq m \leqq p, \qquad \text{q.e.d.}$$

§2. Product representations

Let X and Y be complex spaces. Let $\varphi: X \to Y$ be a light, holomorphic map. Take $a \in X$. An open neighborhood U of a is said to be distinguished if and only if $\overline{U}$ is compact and $\{a\} = \varphi^{-1}(\varphi(a)) \cap \overline{U}$. The distinguished neighborhoods of a form a neighborhood base of a. Let U be a distinguished neighborhood of a, then

$$(2.1) \qquad \tilde{\nu}_\varphi(a) = \limsup_{z \to a} \# \varphi^{-1}(\varphi(z)) \cap U$$

is finite, integral, independent of U and called the order of φ at a. Let $\pi: \hat{X} \to X$ be the normalization of X. Let $b_1, \ldots, b_r$ be the different points of $\pi^{-1}(a)$. The map $\psi = \varphi \circ \pi$ is light. The multiplicity of φ at a is defined by

$$(2.2) \qquad \nu_\varphi(a) = \sum_{\lambda=1}^{r} \tilde{\nu}_\psi(b_\lambda).$$

If X and Y are pure m-dimensional and if Y is locally irreducible, then $\nu_\varphi = \tilde{\nu}_\varphi$.

Proposition 2.1. Let $\varphi: X \to Y$ be a light, holomorphic map. Take an integer $p > 1$. Then

$$(2.3) \qquad N_p = \{x \in X \mid \tilde{\nu}_\varphi(x) \geqq p\}$$

is a thin analytic subset of X.

Proof. Define X^p_φ, δ_φ, π_ν, Δ_X and Δ_Y as in Lemma 1.28. If $1 \leqq \mu < \nu \leqq p$, the set

$$E_{\mu\nu} = \{x \in X^p_\varphi | \pi_\nu(x) = \pi_\mu(x)\}$$

is analytic. Also

$$E = \bigcup_{1 \leqq \mu < \nu \leqq p} E_{\mu\nu}$$

is analytic with $\Delta_X \subseteq E$. Let $\{B_\lambda\}_{\lambda \in \Lambda}$ be the family of branches of X^p_φ. Let $\Lambda_0 = \{\lambda \in \Lambda | B_\lambda \nsubseteq E\}$. Define $S = \bigcup_{\lambda \in \Lambda_0} B_\lambda$. The set S is analytic and $S \cap E$ is thin in S. If B is a branch of Δ_X, then $B = B_\lambda$ for some $\lambda \in \Lambda$ by Lemma 1.28. Now, $B_\lambda \subseteq \Delta_X \subseteq E$ implies $\lambda \in \Lambda - \Lambda_0$. Therefore, $\Delta_X \cap S$ is a thin analytic subset of Δ_X. Because $\delta_X\colon \Delta_X \to X$ is biholomorphic, $F = \delta_X(\Delta_X \cap S)$ is a thin analytic subset of X. Now, $F = N_p$ is claimed.

Take $a \in F$. Let U be a distinguished neighborhood of a. Let V be any open neighborhood of a with $V \subseteq U$. Because $(a,\ldots,a) \in \Delta_X \cap \overline{S - E}$, a point $(x_1,\ldots,x_p) \in V^p \cap (S-E)$ exists. Hence $x_\mu \neq x_\nu$ if $\mu \neq \nu$ and $x_\mu \in \varphi^{-1}(\varphi(x_1)) \cap U$ for $\mu = 1,\ldots,p$, which implies $p \leqq \#\varphi^{-1}(\varphi(x_1)) \cap U$. Because $x_1 \in V$ and because V was arbitrary small in U, this implies that $\tilde{\nu}_\varphi(a) \geqq p$. Hence $a \in N_p$.

Take $a \in N_p$. Let W be any neighborhood of $(a,\ldots,a) \in X^p_\varphi$. A distinguished neighborhood U of a with $U^p \cap X^p_\varphi \subseteq W$ exists. A point $x \in U$ exists such that $\varphi^{-1}(\varphi(x)) \cap U$ contains at least p different points $x_1 = x, x_2, \ldots, x_p$. Then $(x_1,\ldots,x_p) \in B_\lambda - E$ for some $\lambda \in \Lambda$. Obviously, $\lambda \in \Lambda_0$. Hence $(x_1,\ldots,x_p) \in (S-E) \cap W$. Therefore

$$(a,\dots,a) \in \overline{S - E} \cap \Delta_X = S \cap \Delta_X,$$

which implies $a \in \delta_X(S \cap \Delta_X) = F$. Hence, $N_p = F$ in a thin analytic subset of X; q.e.d.

Let $\varphi: X \to Y$ be a light holomorphic map. Take $a \in X$. For $b \in Y$, define

$$(2.4) \qquad \nu_\varphi(a;b) = \begin{cases} 0 & \text{if } \varphi(a) \neq b \\ \nu_\varphi(a) & \text{if } \varphi(a) = b \end{cases}$$

as the <u>b-multiplicity</u> of a. If $G \subseteq X$, define

$$(2.5) \qquad n_\varphi(G;b) = \sum_{x \in G} \nu_\varphi(x;b)$$

as the <u>valence</u> of φ for b in G. Obviously

$$(2.6) \qquad \# \varphi^{-1}(b) \cap G \leqq n_\varphi(G,b) \leqq \infty .$$

If $\overline{G}$ is compact, $n_\varphi(G;b) < \infty$. If X and Y are pure m-dimensional complex spaces, if Y is locally irreducible and irreducible, and if $\varphi: X \to Y$ is light, proper and holomorphic, then $n_\varphi(X;y)$ is a positive, finite, constant function of $y \in Y$, whose constant function value is called the sheet number of φ. (See [21] Lemma 2.3.)

<u>Lemma 2.2.</u> Let X and Y be pure m-dimensional complex spaces. Suppose that Y is irreducible and locally irreducible. Let $\varphi: X \to Y$ be a light, proper and holomorphic map with sheet number s.

Then

1. The map φ is open and surjective.

2. The space X consists of finitely many branches $B_1, \ldots, B_r$ with $r \leqq s$. Moreover, $\varphi(B_\lambda) = Y$ for $\lambda = 1, \ldots, r$.

3. If $\varphi^{-1}(\varphi(a)) = \{a\}$ for some point $a \in X$, then X is connected. If in addition, X is locally irreducible at a, then X is irreducible.

4. The set $N_2 = \{x \in X \mid \nu_\varphi(x) > 1\}$ is thin and analytic in X. Then $S' = \varphi(N_2)$ and $S = \varphi^{-1}(S')$ are thin analytic subsets of Y and X respectively.

5. The restriction $\varphi_0 = \varphi: X - S \to Y - S'$ is a locally topological, holomorphic, surjective, proper map. If Y is normal, then φ_0 is locally biholomorphic.

6. If $y \in Y - S'$, then $\#\varphi^{-1}(y) = s$. If $y \in S'$, then $\#\varphi^{-1}(y) < s$.

7. If $s = 1$, then $\varphi: X \to Y$ is topological. If, in addition, Y is normal, then φ is biholomorphic.

<u>Proof.</u> 1. By Lemma 1.13, φ has pure rank m. By Theorem 1.22, φ is open. By Theorem 1.1 $\varphi(X)$ is analytic in Y. Also $\varphi(X)$ is open. Because Y is irreducible, $Y = \varphi(X)$. The map φ is surjective.

2. If B is a branch of X, then B is pure m-dimensional and $\varphi: B \to Y$ is light, proper and holomorphic. Hence $\varphi(B) = Y$ by 1. Take $y \in Y$. Then $\varphi^{-1}(y)$ is compact and intersects each branch of X. Hence X consists of finitely many branches $B_1, \ldots, B_r$. Later $r \leqq s$ will be shown.

3. If $\varphi^{-1}(\varphi(a)) = \{a\}$, then $a \in B_1 \cap \ldots \cap B_r$.

Hence $X = B_1 \cup \ldots \cup B_r$ is connected. If X is locally irreducible at a, then a is contained in exactly one branch, which implies $r = 1$. Hence $X = B_1$ is irreducible.

4. Because X and Y are pure m-dimensional, and because Y is locally irreducible $\nu_\varphi = \tilde{\nu}_\varphi$. By Proposition 2.1, N_2 is analytic and thin. By Theorem 1.1, $S' = \varphi(N_2)$ is analytic. By Lemma 1.2, S' is thin, because $\dim Y = m > \dim N_2$. Because φ is open, $\varphi^{-1}(S') = S$ is a thin analytic subset of X. The map

$$\varphi_0 = \varphi: X - S \to Y - S'$$

is surjective, open, light and proper. Because Y is irreducible, Y - S' is irreducible. If $y \in S'$, then $x_0 \in N_2 \cap \varphi^{-1}(y)$ exists. Hence $\nu_\varphi(x_0;y) = \nu_\varphi(x_0) > 1$ and

$$s = \sum_{x \in X} \nu_\varphi(x;y) > \# \varphi^{-1}(y).$$

If $y \in Y - S'$, then $\varphi^{-1}(y) \subseteq X - N_2$ and $\nu_\varphi(x) = 1$ if $x \in \varphi^{-1}(y)$. Hence

$$s = \sum_{x \in X} \nu_\varphi(x;y) = \# \varphi^{-1}(y).$$

Take $x \in X - S$. Let U be a distinguished neighborhood of x with $U \subseteq X - S$. Suppose that $\varphi|V$ is not injective for every open neighborhood V of x in U. Then $\nu_\varphi(x) = \tilde{\nu}_\varphi(x) \geqq 2$, which is wrong. Hence, an open neighborhood V of x in U exists such that $\varphi: V \to Y$ if injective. Here, $\varphi(V) = W$ is open and $\varphi_1 = \varphi: V \to W$ is open, bijective and holomorphic, hence topological. Let S_V and S_W be the sets of non-simple points of V and W respectively. The map

$\varphi_1^{-1}: W \to V$ is continuous and holomorphic on $W - (S_W \cup \varphi_1(S_V))$. Here $\varphi_1(S_V)$ is analytic, because φ_1 is proper, and thin because φ_1 is topological. Hence, if Y is normal, then φ_1^{-1} is holomorphic and φ_1 is biholomorphic. This proves 4, 5 and 6.

7. If $s = 1$, then N_2; S' and S are empty by 1 and 6. Also by 6, φ is injective. Hence, 5 implies that φ is topological. If in addition Y is normal, φ is topological and locally biholomorphic. Hence, φ is biholomorphic.

2. Let $b \in Y - S'$ be a simple point of Y. Let U be an open, connected, simply connected neighborhood of b in $Y - S'$ consisting of simple points only. Let $a_1, \ldots, a_s$ be the points of $\varphi^{-1}(b)$. Define $V = \varphi^{-1}(U)$. By 5, $\varphi: V \to U$ is an s-sheeted covering map, which is locally biholomorphic. Because U is simply connected, $V = \bigcup_{\lambda=1}^{s} V_\lambda$, where $a_\lambda \in V_\lambda$ and $V_\lambda \cap V_\mu = \emptyset$ if $\lambda \neq \mu$ such that $\varphi: V_\lambda \to U$ is biholomorphic. Then $V_\lambda \subseteq B_{\rho_\lambda}$ for one and only one number ρ_λ with $1 \leqq \rho_\lambda \leqq r$. If B_ρ is a branch of X, then $B_\rho \cap \varphi^{-1}(a) \neq \emptyset$. Hence $a_\lambda \in B_\rho$ for some λ with $1 \leqq \lambda \leqq s$, which implies $\rho = \rho_\lambda$. Therefore $r \leqq s$; q.e.d.

Let X be a complex space. A proper, light, open and holomorphic map $\alpha: U_\alpha \to U'_\alpha$ of an open subset U_α into an open, connected subset U'_α of $\mathbb{C}^m$ is called a <u>chart</u> of X. By Lemma 1.10, U_α is pure m-dimensional. Hence, by Lemma 2.2, U_α consists of finitely many branches, and α is surjective and has a constant sheet number. The chart $\alpha: U_\alpha \to U'_\alpha$ is said to be a <u>chart at $a \in X$</u> if $a \in U_\alpha$. The chart α is said to be <u>centered at a</u> if and only if $\alpha^{-1}(\alpha(a)) = \{a\}$. In this case, U_α is connected by Lemma 2.2. The chart $\alpha: U_\alpha \to U'_\alpha$

is said to be schlicht if and only if α is injective. If α is a schlicht chart, α has sheet number 1. By Lemma 2.2, α is biholomorphic. Hence a schlicht chart provides local coordinates.

Proposition 2.3. Let X be a complex space of pure dimension m. Let U be an open neighborhood of $a \in X$. Then a chart $\alpha: U_\alpha \to U'_\alpha$ centered at a exists with $U_\alpha \subseteq U$.

Proof. A biholomorphic map $\beta: U_\beta \to U'_\beta$ of an open neighborhood U_β of a with $U_\beta \subseteq U$ onto an analytic subset U'_β of an open connected subset G of a complex vector space V of dimension n exists with $\beta(a) = 0$. If $m = n$, then $G = U'_\beta$ and the proof is finished. If $n > m$, V can be identified with $\mathbb{C}^q \times \mathbb{C}^m$ such that $(\mathbb{C}^q \times \{0\}) \cap U'_\beta$ contains 0 as an isolated point. Let $\pi: \mathbb{C}^q \times \mathbb{C}^m \to \mathbb{C}^m$ be the projection. An open connected neighborhood U'_α of $0 \in \mathbb{C}^m$ and an open connected neighborhood W of $0 \in \mathbb{C}^q$ exist such that $\tilde{U}_\alpha = (U'_\alpha \times W) \cap U'_\beta$ and such that $\pi_0 = \pi: \tilde{U}_\alpha \to U'_\alpha$ is a proper, light holomorphic map with $\pi_0^{-1}(0) = \{0\}$. Because U'_β is pure m-dimensional, π_0 is surjective and open by Lemma 2.2. Define $U_\alpha = \beta^{-1}(\tilde{U}_\alpha)$. The map $\alpha = \pi_0 \circ \beta: U_\alpha \to U'_\alpha$ is holomorphic, light, proper, open and surjective with $\alpha^{-1}(\alpha(a)) = \{a\}$; q.e.d.

Let X and Y be complex spaces. Consider a holomorphic map $\varphi: X \to Y$. Then (α, β) or (α, β, π) are said to be a product representation of φ if and only if

1. The maps $\alpha: U_\alpha \to U'_\alpha$ and $\beta: U_\beta \to U'_\beta$ are charts on X and Y respectively with $\varphi(U_\alpha) \subseteq U_\beta$.

2. An open connected subset U''_α of $\mathbb{C}^q$ exists such that $U'_\alpha = U''_\alpha \times U'_\beta$.

3. If $\pi\colon U''_\alpha \times U'_\beta \to U'_\beta$ denotes the projection, then $\pi\circ\alpha = \beta\circ\varphi$ on U_α.

If α is a chart at a (respectively centered at a), then (α,β) is a <u>product representation of φ at a</u> (respectively centered at a). If β is a chart at b (respectively centered at b), then (α,β) is a <u>product representation of φ over b</u> (respectively centered over b). The product representation is said to be <u>smooth</u> if and only if α and β are biholomorphic. A smooth product representation of φ at a exists if and only if φ is regular at a.

If (α,β) is a product representation of φ, then U_α is pure m-dimensional, U_β is pure n-dimensional with $q = m - n$. If $q = 0$, then U''_α is a point and $U''_\alpha \times U'_\beta$ can be identified with U'_β such that π becomes the identity.

<u>Proposition 2.4.</u> Let X and Y be complex spaces of pure dimension m and n respectively with $m - n = q \geqq 0$. Let $\varphi\colon X \to Y$ be a q-fibering, holomorphic map. Take $a \in X$ and define $b = \varphi(a)$. Let U be an open neighborhood of a. Let $\delta\colon U_\delta \to U'_\delta$ be a chart of Y which is centered at b. Then a product representation (α,β,π) of φ exists which is centered at a and over b, such that $U_\alpha \subseteq U$ and $U_\beta \subseteq U_\delta$ with $\beta = \delta|U_\beta$.

<u>Proof.</u> Let $\gamma\colon U_\gamma \to U'_\gamma$ be a chart of X which is centered at a with $\gamma(a) = 0$ and $U_\gamma \subseteq U \cap \varphi^{-1}(U_\delta)$. Then $F = \varphi^{-1}(b) \cap U_\gamma$ is a pure q-dimensional analytic subset of U_γ with $a \in F$. The set $F' = \gamma(F)$ is analytic and pure q-dimensional by Theorem 1.1 and Lemma 1.9.

Without loss of generality $\delta(b) = 0 \in \mathbb{C}^n$ can be assumed. Also it can be assumed that $F' \cap (\mathbb{C}^n \times \{0\})$ contains $0 \in \mathbb{C}^m$ as an isolated point. Bounded, open, connected neighborhoods Z of $0 \in \mathbb{C}^q$ and W of $0 \in \mathbb{C}^n$ exist such that $\overline{W} \times \overline{Z} \subset U'_\gamma$ and $(\overline{W} \times \{0\}) \cap F' = \{0\}$. Define $\tilde{U}_\gamma = \gamma^{-1}(W \times Z)$. Then

$$\Gamma = \{(x, \varphi(x)) | x \in \tilde{U}_\gamma\}$$

is a pure m-dimensional analytic subset of $\tilde{U}_\gamma \times U_\delta$. A biholomorphic map $\mu\colon \tilde{U}_\gamma \to \Gamma$ is defined by $\mu(x) = (x, \varphi(x))$ for $x \in \tilde{U}_\gamma$. Because the map

$$\eta = \gamma \times \delta\colon \tilde{U}_\gamma \times U_\delta \to W \times Z \times U'_\delta$$

is proper, light, surjective, open and holomorphic, $\Gamma' = \eta(\Gamma)$ is a pure m-dimensional analytic subset of $W \times Z \times U'_\delta$ and $\eta_0 = \eta\colon \Gamma \to \Gamma'$ is a proper, light, surjective and holomorphic map.

Take $w \in \overline{W}$ with $(w,0,0) \in \Gamma'$. Then $(x, \varphi(x)) \in \Gamma$ exists with $\gamma(x) = (w,0)$ and $\delta(\varphi(x)) = 0$. Hence $\varphi(x) = b$ and $x \in F$. Therefore $\gamma(x) \in F' \cap (\overline{W} \times \{0\}) = \{0\}$, which implies $w = 0$. Consequently, $(\overline{W} \times \{0\} \times \{0\}) \cap \Gamma' = \{0\}$. Let $\psi\colon \Gamma' \to Z \times U'_\delta$ be the projection. Bounded, open, connected neighborhoods U''_α of $0 \in \mathbb{C}^q$ and U'_β of $0 \in \mathbb{C}^n$ with $\overline{U}''_\alpha \subset Z$ and $\overline{U}'_\beta \subset U'_\delta$ exist such that

$$G = \Gamma' \cap (W \times U''_\alpha \times U'_\beta)$$

$$\psi_0 = \psi | G\colon G \to U''_\alpha \times U'_\beta$$

is proper. Of course, ψ_0 is holomorphic. Because G is an analytic subset of the open subset $W \times U''_\alpha \times U'_\beta$ of $\mathbb{C}^{m+n}$, the proper map ψ_0 is light. Define $\tilde{G} = \eta_0^{-1}(G) \subseteq \Gamma$ and $U_\alpha = \mu^{-1}(\tilde{G})$. Then U_α is an open neighborhood of a in X with $U_\alpha \subseteq \tilde{U}_\gamma \subseteq U_\gamma \subseteq U$ and $\varphi(U_\alpha) \subseteq U_\delta$.

The map $\alpha = \psi_0 \circ \eta_0 \circ \mu: U_\alpha \to U'_\alpha = U''_\alpha \times U'_\beta$ is defined, proper, light and holomorphic. By Lemma 2.2, α is open and surjective. Hence, α is a chart of X at a. Observe that $\alpha(a) = 0 \in \mathbb{C}^m$. If $x \in \alpha^{-1}(0)$, then $\psi_0(\eta_0(\mu(x))) = 0$ or $\eta_0(\mu(x)) \in \Gamma' \cap (W \times \{0\} \times \{0\}) = \{0\}$. Hence $\eta(\mu(x)) = 0$, which implies $\gamma(x) = 0$ and $\delta(\varphi(x)) = 0$. Because the chart γ is centered at a with $\gamma(a) = 0$, this implies $x = a$. Therefore, the chart α is centered at a.

Define $U_\beta = \delta^{-1}(U'_\beta)$ and $\delta = \beta | U_\beta: U_\beta \to U'_\beta$. Then U_β is an open neighborhood of b with $U_\beta \subseteq U_\delta$ and β is a proper, light, open and surjective map with $\beta^{-1}(\beta(b)) = \{b\}$. Hence, β is a chart centered at b.

Let $\pi: U''_\alpha \times U'_\beta \to U'_\beta$ be the projection. Take $x \in U_\alpha$. Then

$$\pi(\alpha(x)) = \pi(\psi(\eta(x,\varphi(x)))) = \pi(\psi(\gamma(x),\delta(\varphi(x)))) = \delta(\varphi(x))$$

Hence $\delta(\varphi(x)) \in U'_\beta$ and $\varphi(x) \in U_\beta$. Therefore $\varphi(U_\alpha) \subseteq U_\beta$ and $\beta \circ \varphi = \pi \circ \alpha$ on U_α.

Hence, (α,β,π) is a product representation of φ centered at a and over b; q.e.d.

Let $\varphi: X \to Y$ be a holomorphic map. An analytic subset S of an open subset U of X is said to be a <u>section</u> of φ over the open subset V of Y if and only if $\varphi(S) \subseteq V$ and if $\varphi: S \to V$ is proper and light. The section S is said to be <u>central in U</u> if and only if $\varphi^{-1}(y) \cap S \neq \emptyset$ for every $y \in \varphi(U)$. The section S is said to be

<u>strictly central in U</u> if and only if each branch of $\varphi^{-1}(y) \cap U$ intersects S if $y \in \varphi(U)$. The section S <u>covers</u> V if and only if $\varphi\colon S \to V$ is also open and surjective. The section S is said to be <u>schlicht</u> if $\varphi\colon S \to V$ is biholomorphic.

<u>Lemma 2.5.</u> Let X and Y be complex spaces. Let $\varphi\colon X \quad Y$ be a holomorphic map. Let (α,β,π) be a product representation of φ. Let s be the sheet number of α. Then

1. U_α and U_β have pure dimension m and n respectively with $m - n = q \geqq 0$.

2. If $z \in U'_\beta$ then $\alpha^{-1}(U''_\alpha \times \{z\})$ is pure q-dimensional and consists of at most s branches. If B is a branch of $\alpha^{-1}(U''_\alpha \times \{z\})$, then

$$\alpha\colon \alpha^{-1}(U''_\alpha \times \{z\}) \to U''_\alpha \times \{z\}$$

$$\alpha\colon \quad B \longrightarrow U''_\alpha \times \{z\}$$

are proper, light, open, surjective and holomorphic. Moreover, $\varphi|B$ is constant and $\varphi(B) \subseteq \beta^{-1}(z)$.

3. If $y \in \varphi(U_\alpha)$, then each branch of $\varphi^{-1}(y) \cap U_\alpha$ is a branch of $\alpha^{-1}(U''_\alpha \times \{\beta(y)\})$, and $\varphi^{-1}(y) \cap U_\alpha$ is a union of at most s branches of $\alpha^{-1}(U''_\alpha \times \{\beta(y)\})$.

4. The restriction $\varphi\colon U_\alpha \to U_\beta$ is q-fibering and has pure rank n.

5. If $c \in U''_\alpha$, then $S = \alpha^{-1}(\{c\} \times U'_\beta)$ is a strictly central section of φ in U_α over U_β. The complex space S is pure n-dimensional and consists of at most s branches. If $y \in U_\beta$, then the fiber $\varphi^{-1}(y) \cap S$ consists of at most s points. If U_β is connected

and locally irreducible, then S covers U_β. If α and β are biholomorphic, then S is schlicht.

6. The image set $\varphi(U_\alpha)$ is an analytic set of pure dimension n in U_β and is a union of branches of U_β.

7. If (α, β, π) is centered over b, then $b \in \varphi(U_\alpha)$.

8. If (α, β, π) is centered over b and if Y is locally irreducible at b, then U_β is irreducible and $\varphi(U_\alpha) = U_\beta$.

Proof. 1. The definition and Lemma 1.10 imply 1.

2. Take $z \in U'_\beta$. Define $F' = U''_\alpha \times \{z\}$ and $F = \alpha^{-1}(F')$. Then F' is a q-dimensional, connected complex submanifold of U'_α and F is analytic in U_α. The restriction

$$\alpha_z = \alpha | F\colon F \to F'$$

if proper, light, surjective and holomorphic. Let A be open in F. An open subset $\tilde{A}$ of U_α with $A = F \cap \tilde{A}$ exists. Then $\alpha_z(A) = \alpha(\tilde{A}) \cap F'$ is open, because $\alpha(\tilde{A})$ is open. Hence α_z is an open map. By Lemma 1.10, F is pure q-dimensional. Clearly, the sheet number of α_z is at most s. By Lemma 2.2 F consists of at most s branches.

If B is branch of F, then B has pure dimension q. Then $\alpha | B\colon B \to F'$ is proper, light and holomorphic. By Lemma 2.2, $\alpha | B$ is open and surjective. If $x \in B$, then $\beta(\varphi(x)) = \pi(\alpha(x)) = z$. Hence $\varphi(x) \in \beta^{-1}(z)$. Since $\beta^{-1}(z)$ is finite and B irreducible, $\varphi | B$ is constant and contained in $\beta^{-1}(z)$.

3. Take $y \in \varphi(U_\alpha)$. Define $z = \beta(y)$ and define F, F', α_z as in 2. Let B be a branch of $\varphi^{-1}(y) \cap U_\alpha$. If $x \in \varphi^{-1}(y) \cap U_\alpha$, then $\pi(\alpha(x)) = \beta(\varphi(y)) = z$ and $x \in \alpha^{-1}(\pi^{-1}(x)) = F$. Hence

$\varphi^{-1}(y) \cap U_\alpha \subseteq F$. A branch C of F contains B. Because $\varphi|C$ is constant and $\varphi(B) = \{y\}$, the branch C is contained in $\varphi^{-1}(y) \cap U_\alpha$. Hence B = C. Therefore, each branch of $\varphi^{-1}(y) \cap U_\alpha$ is a branch of F, which has at most s branches. Hence $\varphi^{-1}(y) \cap U_\alpha$ is an union of at most s branches of F. Especially $\varphi^{-1}(y) \cap U_\alpha$ is pure q-dimensional. Hence, $\varphi|U_\alpha$ is q-fibering with $q = m - n$. Because U_α is pure dimensional, Lemma 1.13 shows that $\varphi|U_\alpha$ has pure rank n. This proves 3. and 4.

5. Take $c \in U''_\alpha$. Then $S = \alpha^{-1}(\{c\} \times U'_\beta)$ is an analytic subset of U_α. Again $\alpha|S\colon S \to \{c\} \times U'_\beta$ is proper, light, open, surjective and holomorphic. By Lemma 1.10, S is pure n-dimensional. Since $\alpha|S$ has at most s sheets, S consists of at most s branches. Take $x \in S$. If $z \in \varphi^{-1}(\varphi(x)) \cap S$, then

$$\alpha(z) = (c, \beta(\varphi(z))) = (c, \beta(\varphi(x))) = \alpha(x).$$

Hence, $\varphi^{-1}(\varphi(x)) \cap S \subseteq \alpha^{-1}(\alpha(x))$. Therefore, $\varphi^{-1}(\varphi(x)) \cap S$ consists of at most s points. The map $\varphi\colon S \to U_\beta$ is light.

Let K be a compact subset of U_β. Then $K' = \alpha^{-1}(\{c\} \times \beta(K))$ is a compact subset of S. Take $x \in \varphi^{-1}(K) \cap S$. Then $\alpha(x) = (c, \beta(\varphi(x)))$. Hence, $x \in K'$. Therefore, $\varphi^{-1}(K) \cap S \subseteq K'$. The map $\varphi\colon S \to U_\beta$ is proper. Take $y \in \varphi(U_\alpha)$. Let B be a branch of $\varphi^{-1}(y) \cap U_\alpha$. Because $\alpha(B) = U''_\alpha \times \{\beta(y)\}$, a point $x \in B$ with $\alpha(x) = (c, \beta(y))$ exists. Hence $x \in B \cap S$. Therefore, S is a strictly central section of φ in U_α over U_β.

If Y is locally irreducible at every point of U_β and if U_β is connected, the proper, light holomorphic map $\varphi\colon S \to U_\beta$ is surjective by Lemma 2.2. Hence S covers U_β. Trivially, if α and β

are biholomorphic, S is schlicht.

6. By Theorem 1.1, $\varphi(U_\alpha) = \varphi(S)$ is analytic. By Lemma 1.9, $\varphi(S)$ has pure dimension n. Hence $\varphi(U_\alpha)$ is an union of branches of U_β.

7. Let (α, β, π) be centered over b. Because $\beta \circ \varphi = \pi \circ \alpha$ is surjective $x \in U_\alpha$ exists with $\beta(\varphi(x)) = \beta(b)$. Now $\beta^{-1}(\beta(b)) = \{b\}$ implies $\varphi(x) = b$. Hence $b \in \varphi(U_\alpha)$.

8. Let (α, β, π) be centered over b and let Y be locally irreducible at b. By Lemma 2.2.3, U_β is irreducible. Because $\varphi(U_\alpha)$ is an union of branches of U_β, this implies $U_\beta = \varphi(U_\alpha)$; q.e.d.

Now, the inverse of the open mapping theorem of Remmert can be proved (Theorem 1.22.).

<u>Theorem 2.6.</u> (Open mapping theorem of Remmert). Let X and Y be complex spaces of pure dimension m and n respectively. Suppose that Y is locally irreducible. Let $\varphi: X \to Y$ be holomorphic. Then the following three statements are equivalent.

a) The map φ is q-fibering with $m - n = q$.

b) The map φ has pure rank n.

c) The map φ is open.

<u>Proof.</u> By Lemma 1.13, a) and b) are equivalent. By Theorem 1.22, b) implies c). Now, it shall be shown that c) implies a). If $n = 0$, the statement if true. Suppose that it is true for $0, 1, \ldots, n - 1$. Then it shall be proved for n.

Pick $b \in \varphi(X)$. Let $\beta: U_\beta \to U'_\beta$ be a chart of Y centered at b with $\beta(b) = 0 \in \mathbb{C}^n$. Define $U = \varphi^{-1}(U'_\beta)$. Then $\psi = \beta \circ \varphi \colon U \to U'_\beta$ is open and holomorphic with $\psi^{-1}(0) = \varphi^{-1}(\beta^{-1}(0)) = \varphi^{-1}(b)$. Let $\lambda \colon U'_\beta \to \mathbb{C}$ be a linear function which is not the zero function. Then $L = \lambda^{-1}(0) \cap U'_\beta$ is a complex manifold of pure dimension $n - 1$. Define $M = \psi^{-1}(L) = (\lambda \circ \psi)^{-1}(0)$. Because $\lambda \circ \psi \colon U \to \mathbb{C}$ is an open map and a holomorphic function, $(\lambda \circ \psi)^{-1}(0)$ has pure codimension 1 in U. Hence M has pure dimension $m - 1$. Let A be open in M, then $A = M \cap \tilde{A}$ where $\tilde{A}$ is open in U. Now, $\psi(\tilde{A}) = A'$ is open in U'_β. Hence $\psi(A \cap M) = A' \cap L$ is open in L. The map $\psi|M \colon M \to L$ is open. By induction $\psi^{-1}(0) \cap M = \psi^{-1}(0) = \varphi^{-1}(b)$ has pure dimension

$$(m-1) - (n-1) = M - n = q \qquad \text{q.e.d.}$$

Let X and Y be complex spaces. Let $\varphi \colon X \to Y$ be a holomorphic map. For $y \in Y$, let $\tau_\varphi(y)$ be the number of branches of $\varphi^{-1}(y)$. Then $0 \leqq \tau_\varphi(y) \leqq \infty$.

Lemma 2.7. Let X and Y be complex spaces of pure dimension m and n respectively with $q = m - n \geqq 0$. Let $\varphi \colon X \to Y$ be a proper, q-fibering, holomorphic map. Then τ_φ is locally bounded.

Proof. Take $b \in Y$. Because φ is proper, $\varphi(X)$ is closed in Y. Hence $\tau_\varphi(y) = 0$ if $y \in Y - \varphi(X)$. Therefore, only $b \in \varphi(X)$ has to be considered. Define $F = \varphi^{-1}(b)$. A finite number of points $a_1, \ldots, a_p$ in F and product representations $(\alpha_i, \beta_i, \pi_i)$ of φ centered at a_i and over b exist such that $F \subseteq U_{\alpha_1} \cup \ldots \cup U_{\alpha_p}$. Because φ is proper, an open neighborhood V of b with $V \subseteq \bigcap_{i=1}^{p} U_{\beta_i}$ and $\varphi^{-1}(V) \subseteq \bigcup_{i=1}^{p} U_{\alpha_i}$ exists.

Let s_i be the sheet number of α_i. Define $s = s_1 + \dots + s_p$. Then $s \geqq \tau_\varphi(y)$ for all $y \in V$ is claimed. Pick $y \in V$. Let $\mathfrak{B}$ be the set of branches of $\varphi^{-1}(y)$. Let $\mathfrak{B}_i$ be the set of branches of $\varphi^{-1}(y) \cap U_{\alpha_i}$. By definition $\tau_\varphi(y) = \#\mathfrak{B}$. By Lemma 2.5, $s_i \geqq \#\mathfrak{B}_i$. If $H \in \mathfrak{B}_i$, one and only one $\lambda_i(H) \in \mathfrak{B}$ with $\lambda_i(H) \supseteq H$ exists. A map $\lambda_i : \mathfrak{B}_i \to \mathfrak{B}$ is defined. If $\tilde{H} \in \mathfrak{B}$, then $\tilde{H} \subseteq \bigcup_{i=1}^{p} U_{\alpha_i}$. Hence an index i exists with $\tilde{H} \cap U_{\alpha_i} \neq \emptyset$. Since $\tilde{H} \cap U_{\alpha_i}$ is a union of branches of $\varphi^{-1}(y) \cap U_{\alpha_i}$ a branch $H \in \mathfrak{B}_i$ with $H \subseteq \tilde{H} \cap U_{\alpha_i}$ exists. Hence $\lambda_i(H) = H$. Therefore $\mathfrak{B} = \bigcup_{i=1}^{p} \lambda_i(\mathfrak{B}_i)$ and

$$\tau_\varphi(y) = \#\mathfrak{B} \leqq \sum_{i=1}^{p} \#\lambda_i(\mathfrak{B}_i) \leqq \sum_{i=1}^{p} \#\mathfrak{B}_i \leqq \sum_{i=1}^{p} s_i = s$$

for each $y \in V$; q.e.d.

§3. Meromorphic functions

Since this paper deals extensively with meromorphic functions an outline of their definition and well-known properties shall be given.

Let X be a complex space. If $U \neq \emptyset$ is open in X, let $\mathfrak{H}(U)$ be the ring of holomorphic functions on U. If U is irreducible, $\mathfrak{H}(U)$ is an integral domain. Let $\mathfrak{N}(U)$ be the set of non-zero divisors of $\mathfrak{H}(U)$, i.e., the set of elements $f \in \mathfrak{H}(U)$ such that $f|B \not\equiv 0$ for each branch B of U. Let $\mathfrak{Q}(U)$ be the total quotient ring of $\mathfrak{H}(U)$. If U is irreducible, then $\mathfrak{N}(U) = \mathfrak{H}(U) - \{0\}$ and $\mathfrak{Q}(U)$ is a field called the quotient field of $\mathfrak{H}(U)$. If $\emptyset \neq V \subseteq U$ and if V and U are open, the restriction map $r^U_V\colon \mathfrak{H}(U) \to \mathfrak{H}(V)$ maps $\mathfrak{N}(U)$ into $\mathfrak{N}(V)$ and induces a restriction map $r^U_V\colon \mathfrak{Q}(U) \to \mathfrak{Q}(V)$. Then $\mathfrak{H} = \{\mathfrak{H}(U), r^U_V\}$ and $\mathfrak{Q} = \{\mathfrak{Q}(U), r^U_V\}$ are presheaves of rings on X. The associated sheaf $\mathcal{O}$ to $\mathfrak{H}$ is the sheaf $\mathcal{O} = \mathcal{O}_X$ of <u>germs of holomorphic functions</u> on X, also called the <u>structure</u> sheaf of X. The stalk $\mathcal{O}_x$ is an integral domain, if and only if X is locally irreducible at x. The presheaf $\mathfrak{H}$ is canonical. Hence, $\mathfrak{H}(U) = \Gamma(U, \mathcal{O})$ can be identified. If $f \in \mathfrak{H}(U)$, then $f(x) \in \mathbb{C}$ denotes the function value of f at x and $r^U_x(f) = f_x \in \mathcal{O}_x$ denotes the germ of f at x.

The associated sheaf $\mathfrak{M}$ to $\mathfrak{Q}$ is called the <u>sheaf $\mathfrak{M} = \mathfrak{M}_X$ of germs of meromorphic functions on X</u>. It contains $\mathcal{O}$ as a subsheaf. The ring $\mathfrak{M}_x$ is the total quotient ring of $\mathcal{O}_x$. The ring $\mathfrak{M}_x$ is a field if and only if X is locally irreducible at x. In general, $\mathfrak{Q}$ is not canonical. However, the natural map $\mathfrak{Q}(U) \to \Gamma(U, \mathfrak{M})$ is injective and defines $\mathfrak{Q}(U)$ as a subring of $\Gamma(U, \mathfrak{M})$. Hence $\mathfrak{Q}$ is a subpresheaf of $\Gamma(\mathfrak{M})$. A section in $\mathfrak{M}$ over the open set $U \neq \emptyset$ is called a <u>meromorphic function</u> on U and $\mathfrak{K}(U) = \Gamma(U, \mathfrak{M})$ is the ring of meromorphic functions on U. If and only if U is irreducible, $\mathfrak{K}(U)$ is a field. Clearly, $\mathfrak{H}(U) \subseteq \mathfrak{A}(U) \subseteq \mathfrak{K}(U)$. Let $\mathfrak{K}^*(U)$ be the set of units in $\mathfrak{K}(U)$. Then $\mathfrak{K}^*(U) = \{f \in \mathfrak{K}(U) \mid f_x \neq 0$ if $x \in U\}$. (Observe, that $f \in \mathfrak{M}_x$ with $f \neq 0$ may not be an unit in $\mathfrak{M}_x$.). Moreover, $\mathfrak{T}(U) = \mathfrak{H}(U) \cap \mathfrak{K}^*(U)$. Especially, $\mathfrak{T}(U) = \{f \in \mathfrak{H}(U) \mid f_x \neq 0$ if $x \in U\}$.

Let $f \in \mathfrak{K}(U)$ be a meromorphic function on the open subset $U \neq \emptyset$ of X. Take $a \in U$. Then $f_a \in \mathfrak{M}_a$. Hence an open neighborhood V of a in U exists such that $f_a = r_a^V(g/h)$ with $g \in \mathfrak{H}(V)$ and $h \in \mathfrak{T}(V)$. Moreover, $f_a = g_a/h_a$ where $g_a \in \mathcal{O}_a$ and where h_a is a non-zero divisor in $\mathcal{O}_a$. Define $s \in \Gamma(V, \mathfrak{M})$ by $s_x = r_x^V(g/h) = g_x/h_x$. Then $s_a = f_a$. An open neighborhood W of a in V exists such that $s|W = f|W$. Observe that $s \in \mathfrak{A}(W)$ and $s = f = g/h$ on W. Define

$$\mathfrak{J} = \mathfrak{J}_f = \bigcup_{x \in U} \{h \in \mathcal{O}_x \mid hf_x \in \mathcal{O}_x\}$$

Clearly, $\mathfrak{J}_x \neq \emptyset$ for each x, and $\mathfrak{J}$ is a sheaf of ideals in $\mathcal{O}|U$.

<u>Lemma 3.1.</u> The sheaf $\mathfrak{J}$ is coherent.

<u>Proof.</u> Pick $a \in U$. An open neighborhood V of a in U exists such that $f = g/h$ on V where $g \in \mathfrak{H}(V)$ and $h \in \mathfrak{T}(V)$. Define $\mathcal{O}^2 = \mathcal{O} \oplus \mathcal{O}$. Define $\beta\colon \mathcal{O}^2 \to \mathcal{O}$ over V by $\beta(u,v) = ug_x - vh_x$ if $(u,v) \in \mathcal{O}_x^2$ and $x \in V$. Then $\mathfrak{R} = \ker \beta$ is coherent. Define $\gamma\colon \mathfrak{R} \to \mathcal{O}$ over V by $\gamma(u,v) = u$. If $u \in (\operatorname{Im} \gamma)_x$, then $v \in \mathcal{O}_x$ with $(u,v) \in \mathfrak{R}_x$ exists. Hence $uf_x = ug_x/h_x = v_x \in \mathcal{O}_x$. Therefore

$u \in \mathfrak{J}_x$. If $u \in \mathfrak{J}_x$, then $uf_x = v \in \mathcal{O}_x$ and $ug_x - vh_x = 0$. Hence $(u,v) \in \mathfrak{R}_x$ and $u \in (\mathrm{Im}\ \gamma)_x$. Therefore $\mathfrak{J}\,|V = \mathrm{Im}\ \gamma$ is coherent. Hence $\mathfrak{J}$ is coherent; q.e.d.

Let $f \in \mathfrak{K}(X)$ be a meromorphic function on the complex space X. A point $a \in X$ is said to be a point of holomorphy if and only if $f \in \mathcal{O}_a$. A point $a \in X$ is said to be a pole if $f_a \in \mathfrak{M}_a - \mathcal{O}_a$. A point $a \in X$ is a pole if and only if $1_a \notin \mathfrak{J}_a$, i.e., if and only if $\mathfrak{J}_a \neq \mathcal{O}_a$. Hence, the set P_f of poles is the support of $\mathcal{O}/\mathfrak{J}$. Therefore, P_f is analytic. The local quotient representation shows that P_f is thin.

Lemma 3.2[6]. Let X be a Stein space. Let $U \neq \emptyset$ be open and suppose that $\overline{U}$ is compact. Let $f \in \mathfrak{K}(X)$ be a meromorphic function on X. Let D be an at most countable subset of $U - P_f$. Then a holomorphic function $h \in \mathfrak{H}(X)$ exists such that $h \cdot f$ is holomorphic on X, such that $h|U \in \Gamma(U)$ is not a zero divisor in $\mathfrak{H}(U)$ and such that $h(x) \neq 0$ if $x \in D$.

Proof. Because P_f is thin, an at most countable subset E of $U - P_f$ exists such that at least each branch of U contains a point of E. Define $F = E \cup D$. Because $\overline{U}$ is compact, and because $\mathfrak{J} = \mathfrak{J}_f$ is coherent, finitely may sections $h_\mu \in \Gamma(X, \mathfrak{J})$ $(\mu = 1, \ldots, p)$ generate $\mathfrak{J}$ over U:

$$\mathfrak{J} = \mathcal{O} h_1 + \ldots + \mathcal{O} h_p$$

Then

$$P_f \cap U = \{x \in U | h_1(x) = \ldots = h_p(x) = 0\}.$$

Hence $(h_1(x),\ldots,h_p(x)) \neq 0 \in \mathbb{C}^p$ if $x \in F$. For $x \in F$, the set

$$L_x = \{z_1,\ldots,z_p) \in \mathbb{C}^p \mid \sum_{\mu=1}^{p} z_\mu h_\mu(x) = 0\}$$

is a $(p-1)$-dimensional subspace of C^p. Because F is at most countable,

$$(a_1,\ldots,a_p) \in \mathbb{C}^p - \bigcup_{x \in F} L_x$$

exists. Then $h = a_1h_1 + \ldots + a_ph_p \in \Gamma(X, \mathfrak{J})$ and $h(x) \neq 0$ if $x \in D \cup E$. Especially, $h \cdot f$ is holomorphic on X and $h(x) \neq 0$ if $x \in D$. Because each branch B of U contains a point of E, the restriction $h|B$ is not the zero function. Hence $h \in$ (U); q.e.d.

Let X and Y be complex spaces. Let $\varphi: X \to Y$ be a holomorphic map. A homomorphism $\varphi^*: \mathfrak{H}(Y) \to \mathfrak{H}(X)$ is defined by $\varphi^*(f) = f \circ \varphi$. Obviously, φ^* is injective, if φ is surjective. If for every open subset U of X, the image set $\varphi(U)$ is not thin in Y, then $\varphi^*(\mathfrak{T}(Y)) \subseteq \mathfrak{T}(X)$. If $\psi: Y \to Z$ is a holomorphic map of Y into the complex space Z, then $(\psi \circ \varphi)^* = \varphi^* \circ \psi^*$. If U and $V \neq \emptyset$ are open in X and Y respectively and if $\varphi(U) \subseteq V$, define $\varphi_0 = \varphi: U \to V$, then $\varphi^*(f)|U = \varphi_0^*(f|V)$ if $f \in \mathfrak{H}(X)$. Write $\varphi^*(\mathfrak{H}(Y)) = \varphi^*\mathfrak{H}(Y)$.

If $U \neq \emptyset$ is open in Y, define $\tilde{U} = \varphi^{-1}(U)$ and

$$\mathcal{L}_\varphi(U) = \{f \in \mathfrak{K}(U) \mid \varphi^{-1}(P_f) \text{ is thin on } \tilde{U}\}.$$

If $f \in \mathfrak{K}(U)$ and $g \in \mathfrak{K}(U)$, then $P_{f-g} \subseteq P_f \cup P_g$ and $P_{fg} \subseteq P_f \cup P_g$. Hence $f \in \mathcal{L}_\varphi(U)$ and $g \in \mathcal{L}_\varphi(U)$ implies $f - g \in \mathcal{L}_\varphi(U)$ and

$f \cdot g \in \mathcal{L}_\varphi(U)$. Therefore $\mathcal{L}_\varphi(U)$ is a subring of $\mathfrak{K}(U)$. Obviously $\mathfrak{H}(U)$ is a subring of $\mathcal{L}_\varphi(U)$.

Let $\emptyset \neq V \subseteq U \subseteq Y$ and suppose that U and V are open. Then the restriction may r_V^U: $\mathfrak{K}(U) \to \mathfrak{K}(V)$ is defined. If $\mathfrak{A} \subseteq \mathfrak{K}(U)$, define $\mathfrak{A}|V = r_V^U(\mathfrak{A})$. Now, $\varphi^{-1}(P_{f|V}) = \varphi^{-1}(P_f \cap V) = \varphi^{-1}(P_f) \cap \tilde{V}$ if $\tilde{V} = \varphi^{-1}(V)$. Hence, $\mathcal{L}_\varphi(U)|V \subseteq \mathcal{L}_\varphi(V)$. Therefore $\mathcal{L}_\varphi = \{\mathcal{L}_\varphi(U), r_V^U\}$ is a subpresheaf of $\Gamma(\mathfrak{M}_Y)$.

<u>Lemma 3.3.</u> Let X and Y be complex spaces. Suppose that Y is a Stein space. Let $\varphi: X \to Y$ be a holomorphic map. Let V be an open subset of Y such that $\overline{V}$ is compact and $U = \varphi^{-1}(V) \neq \emptyset$. Take $f \in \mathcal{L}_\varphi(Y)$. Then $g \in \mathfrak{H}(Y)$ and $h \in \mathfrak{H}(Y)$ exist such that $h \cdot f = g$ and $h|V \in \mathfrak{T}(V)$ and $h \circ \varphi|U \in \mathfrak{T}(U)$.

<u>Proof.</u> Let $\mathfrak{B}$ be the set of branches of U. If $B \in \mathfrak{B}$, then $\varphi^{-1}(P_f) \cap B$ is thin on B. Take $x_B \in B - \varphi^{-1}(P_f)$. Then $y_B = \varphi(x_B) \in V - P_f$. Hence, $D = \{y_B | B \in \mathfrak{B}\}$ is an at most countable subset of $V - P_f$. By Lemma 3.2, a holomorphic function $h \in \mathfrak{H}(Y)$ exists such that $h \cdot f = g \in \mathfrak{H}(Y)$ such that $h|V \in \mathfrak{T}(V)$ and such that $h(y_B) = h(\varphi(x_B)) = \varphi^*(h)(x_B) \neq 0$ if $B \in \mathfrak{B}$. Hence, $h \circ \varphi|B \not\equiv 0$ for each $B \in \mathfrak{B}$, i.e., $h \circ \varphi|U \in \mathfrak{T}(U)$; q.e.d.

<u>Lemma 3.4.</u> Let X and Y be complex spaces. Let $\varphi: X \to Y$ be an holomorphic map. Then $\varphi^*: \mathfrak{H}(Y) \to \mathfrak{H}(X)$ extends to one and only one homomorphism $\varphi^*: \mathcal{L}_\varphi(Y) \to \mathfrak{K}(X)$ satisfying the following condition:

1. If V is open in Y, if $U = \varphi^{-1}(V) \neq \emptyset$, if $f \in \mathcal{L}_\varphi(Y)$, if

$f = g/_h$ on V with $g \in \mathfrak{H}(V)$, $h \in \mathfrak{T}(V)$ and $h\circ\varphi \in \mathfrak{T}(U)$, then

$$\varphi^*(f)|U = \frac{g\circ\varphi|U}{h\circ\varphi|U} = \frac{\varphi^*(y)|U}{\varphi^*(h)|U} .$$

Moreover, the following properties hold:

2. If $U \neq \emptyset$ and V are open in X and Y respectively, with $\varphi(U) \subseteq V$, if $\varphi_0 = \varphi: U \to V$, then $\mathcal{L}_{\varphi_0}(V) \supseteq \mathcal{L}_\varphi(Y)|U$ and

$$\varphi_0{}^*(f|V) = \varphi^*(f)|U.$$

3. If for every branch B of X, the image $\varphi(B)$ is not contained in a thin analytic subset of Y, then $\mathcal{L}_\varphi(Y) = \mathfrak{K}(Y)$.

4. If for every open subset $U \neq \emptyset$ of X, the image $\varphi(U)$ is not thin in Y, and if $\varphi(X)$ intersects each branch of Y in at least one point where Y is locally irreducible, then $\mathcal{L}_\varphi(Y) = \mathfrak{K}(Y)$ and $\varphi^*: \mathfrak{K}(Y) \to \mathfrak{K}(X)$ is injective.

5. If Z is another complex space, if $\psi: Y \to Z$ is holomorphic, if $f \in \mathcal{L}_\psi(Z)$, if $\psi^*(f) \in \mathcal{L}_\varphi(Y)$ and if $f \in \mathcal{L}_{\psi\circ\varphi}(Z)$, then $\varphi^*(\psi^*(f)) = (\psi\circ\varphi)^*(f)$.

<u>Proof 1.</u> Let $\{W_i\}_{i\in I}$ be a locally finite covering of Y by open Stein sets W_i. Let $\{V_i\}_{i\in I}$ be an open covering of Y such that $\overline{V}_i$ is compact and contained in W_i if $i \in I$. Define $U_i = \varphi^{-1}(V_i)$ and $I_0 = \{i \in I | U_i \neq \emptyset\}$ and $I_1 = \{(i,j) \in I^2 | U_i \cap U_j \neq \emptyset\}$. If $i \in I_0$, then $g_i \in \mathfrak{H}(W_i)$ and $h_i \in \mathfrak{H}(W_i)$ exist such that $h_i f = g_i$ on W_i, such that $h_i|V_i = \mathfrak{T}(V_i)$ and $h\circ\varphi|U_i \in \mathfrak{T}(U_i)$. Then $f = g_i/h_i$ on V_i. A meromorphic function $F_i = {}^{g_i\circ\varphi}/h_i\circ\varphi \in \mathfrak{K}(U_i)$ is

defined. If $(i,j) \in I_1$, then $V_i \cap V_j \neq \emptyset$ and $g_i h_j = g_j h_i$ on $V_i \cap V_j$. Hence $(g_i \circ \varphi)(h_j \circ \varphi) = (g_j \circ \varphi)(h_i \circ \varphi)$ on $U_i \cap U_j$ which implies $F_i | U_i \cap U_j = F_j | U_j \cap U_i$. Therefore, one and only one meromorphic function $\varphi^*(f) \in \mathfrak{K}(X)$ exists such that $\varphi^*(f)|U_i = F_i$ for each $i \in I_0$. Hence $\varphi^*: \mathcal{L}_\varphi(X) \to \mathfrak{K}(X)$ is defined.

Let V be open in Y. Suppose that $U = \varphi^{-1}(V) \neq \emptyset$. Take $f \in \mathcal{L}_\varphi(Y)$. Suppose that $g \in \mathfrak{H}(V)$ and $h \in \mathfrak{T}(V)$ with $h \circ \varphi \in \mathfrak{T}(U)$ and $f = {}^g/h$ are given. Then $h_i g = h \cdot g_i$ on $V_i \cap V$ if $U_i \cap U \neq \emptyset$. Hence $(h_i \circ \varphi) \cdot (g \circ \varphi) = (h \circ \varphi)(g_i \circ \varphi)$ on $U_i \cap U$ which implies

$$\varphi^*(f)|U \cap U_i = \frac{g_i \circ \varphi}{h_i \circ \varphi} \mid U \cap U_i = \frac{g \circ \varphi}{h \circ \varphi} \mid U \cap U_i .$$

Hence $\varphi^*(f) = {}^{g \circ \varphi}/h \circ \varphi$ on U. This proves property 1.

If $f \in \mathfrak{H}(Y)$, take $V = Y$, $g = f$ and $h = 1$, then $U = \varphi^{-1}(V) = X$ and $\varphi^*(f) = {}^{g \circ \varphi}/h \circ \varphi = f \circ \varphi$ on X. Hence $\varphi^*: \mathcal{L}_\varphi(Y) \to \mathfrak{K}(X)$ extends $\varphi^*: \mathfrak{H}(Y) \to \mathfrak{H}(X)$.

Take $f_\lambda \in \mathcal{L}_\varphi(Y)$ for $\lambda = 1,2$. Then $f_\lambda = g_{\lambda i}/h_{\lambda i}$ on V_i with $g_{\lambda i} \in \mathfrak{H}(W_i)$ and $h_{\lambda i} \in \mathfrak{H}(W_i)$ and $h_{\lambda i}|V_i \in \mathfrak{T}(V_i)$ and $h_{\lambda i} \circ \varphi | U_i \in \mathfrak{T}(U_i)$ if $i \in I_0$. Moreover, $h_{2i} g_{1i} - g_{2i} h_{1i} \in \mathfrak{H}(V_i)$ and $g_{1i} g_{2i} \in \mathfrak{H}(V_i)$. Also $h_{1i} h_{1i} \in \mathfrak{T}(V_i)$ and $(h_{1i} \circ \varphi) \cdot (h_{2i} \circ \varphi) \in \mathfrak{T}(U_i)$. Also,

$$f_1 - f_2 = \frac{h_{2i} g_{1i} - g_{2i} h_{1i}}{h_{1i} \cdot h_{2i}}$$

$$f_1 \cdot f_2 = \frac{g_{1i} \cdot g_{2i}}{h_{1i} \cdot h_{2i}} .$$

Therefore, by Property 1:

$$\varphi^*(f_1 - f_2) = \frac{(h_{2i} \circ \varphi)(g_{1i} \circ \varphi) - (g_{2i} \circ \varphi)(h_{1i} \circ \varphi)}{(h_{1i} \circ \varphi) \cdot (h_{2i} \circ \varphi)}$$

$$= \frac{g_{1i} \circ \varphi}{h_{1i} \circ \varphi} - \frac{g_{2i} \circ \varphi}{h_{2i} \circ \varphi} = \varphi^*(f_1) - \varphi^*(f_2)$$

$$\varphi^*(f_1 \cdot f_2) = \frac{(g_{1i} \circ \varphi)(g_{2i} \circ \varphi)}{(h_{1i} \circ \varphi)(h_{2i} \circ \varphi)} = \frac{g_{1i} \circ \varphi}{h_{1i} \circ \varphi} \, \frac{(g_{2i} \circ \varphi)}{(h_{2i} \circ \varphi)} = \varphi^*(f_1) \cdot \varphi^*(f_2)$$

on U_i. Hence $\varphi^*(f_1 - f_2) = \varphi^*(f_1) - \varphi^*(f_2)$ and $\varphi^*(f_1 f_2) = \varphi^*(f_1)\varphi^*(f_2)$ on X. The map $\varphi^*: \; {}_{\varphi}(Y) \to \; (X)$ is a ring homomorphism.

Suppose that $\hat{\varphi}^*: \mathcal{L}_{\varphi}(Y) \to \mathcal{M}(X)$ is another map satisfying Property 1. Take $f \in \mathcal{L}_{\varphi}(Y)$. Take $i \in I_0$. Then $\hat{\varphi}^*(f)|U_i = {}^{g_i \circ \varphi}/_{h_i \circ \varphi}$ by Property 1. Hence $\hat{\varphi}^*(f)|U_i = F_i = \varphi^*(f)|U_i$ which implies $\hat{\varphi}^*(f) = \varphi^*(f)$ for each $f \in \mathcal{L}_{\varphi}(Y)$. Therefore $\hat{\varphi}^* = \varphi^*$.

2. Let $U \neq \emptyset$ and V be open in X and Y respectively with $\varphi(U) \subseteq V$. Define $\varphi_0 = \varphi: U \to V$. Now $\mathcal{L}_{\varphi}(Y)|V \subseteq \mathcal{L}_{\varphi_0}(V)$ has already been proven. Take $f \in \mathcal{L}_{\varphi}(Y)$. Take $a \in U$ and define $b = \varphi(a)$. An open neighborhood W of b with $W \subseteq V$ and holomorphic functions g and h on W with $h \in \mathfrak{T}(W)$ and $h \circ \varphi \in \mathfrak{T}(\varphi^{-1}(W))$ exist by Lemma 3.3. Then $\varphi^*(f) = g \circ \varphi / h \circ \varphi$ on $\varphi^{-1}(W)$. Also

$$h \circ \varphi_0 | \varphi_0^{-1}(W) = h \circ \varphi | U \cap \varphi^{-1}(W) \in \mathfrak{T}(U \cap \varphi^{-1}(W)) = \mathfrak{T}(\varphi_0^{-1}(W))$$

Hence $\varphi_0^*(f|V) = {}^{g \circ \varphi}/_{h \circ \varphi} = \varphi^*(f)$ on $\varphi^{-1}(W) \cap U$. Hence $\varphi_0^*(f|V) = \varphi^*(f)|U$.

3. Suppose that $\varphi(B)$ is not contained in any thin analytic subset of Y, if B is a branch of X. Take $f \in \mathfrak{K}(Y)$. Then $\varphi^{-1}(P_f)$ does not contain a branch of X. Hence $\varphi^{-1}(P_f)$ is a thin analytic subset of X. Therefore $f \in \mathfrak{L}_\varphi(Y)$. Hence $\mathfrak{L}_\varphi(Y) = \mathfrak{K}(Y)$; q.e.d.

4. Suppose that $\varphi(U)$ is not thin in Y if $U \neq \emptyset$ is open in X. Suppose that $\varphi(X)$ intersects each branch of Y in at least one point where Y is locally irreducible. Then $\varphi(B)$ is not contained in any thin analytic subset of Y, if B is a branch of X, because B contains a non-empty open subset of X. Hence $\mathfrak{L}_\varphi(Y) = \mathfrak{K}(Y)$ by 3. Take $f \in \mathfrak{K}(Y)$ with $\varphi^*(f) = 0$. Take a branch C of Y. Take $x \in X$ with $\varphi(x) = y \in C$ such that Y is irreducible at y. An open, irreducible neighborhood N of y exists such that $N \cap C = N \cap Y$. An open neighborhood V of y with $V \subseteq N$ and holomorphic functions g and h on V with $h \in \mathfrak{T}(V)$ and $h \circ \varphi \in \mathfrak{T}(U)$ with $U = \varphi^{-1}(V)$ exist such that $f|V = g/_h$. Then $\varphi^*(f)|U = {}^{g\circ\varphi}/h\circ\varphi$ on U. Hence $g \circ \varphi = 0$ on U. Therefore $\varphi(U) \subseteq g^{-1}(0)$. Because $\varphi(U)$ is not thin and because N is irreducible $g = 0$ on N. Hence $f = 0$ on N. Therefore $f = 0$ on C for each branch C of Y. Hence $f = 0$. The map φ^* is injective.

5. Let Z be another complex space. Let $\psi\colon Y \to Z$ be a holomorphic map. Take $f \in \mathfrak{L}_\psi(Z)$ with $\psi^*(f) \in \mathfrak{L}_\varphi(Y)$. Then $\varphi^*(\psi^*(f))$ is defined. Assume that $f \in \mathfrak{L}_{\psi\circ\varphi}(Z)$. Then $(\psi\circ\varphi)^*(f)$ is defined.

Pick $a \in X$. Define $b = \varphi(a)$ and $c = \psi(b)$. Take an open Stein neighborhood W of c and an open neighborhood V of c such that $\overline{V}$ is compact and contained in W. Define $U = \psi^{-1}(V)$ and $T = \varphi^{-1}(U)$. Let $\mathfrak{B}$ and $\mathfrak{C}$ be the sets of branches of U and T respectively. If $B \in \mathfrak{B}$, then $\psi(B) \not\subseteq P_f$. Hence $y_B \in B$ with $\psi(y_B) \in V - P_f$ exists. If $C \in \mathfrak{C}$, then $\psi(\varphi(C)) \not\subseteq P_f$. Hence $x_C \in C$ with $\psi(\varphi(x_C)) \in V - P_f$

exists. Define $D_1 = \{\psi(y_B) \mid B \in \mathcal{B}\}$ and $D_2 = \{\psi(\varphi(x_C)) \mid C \in \mathcal{C}\}$. Then $D = D_1 \cup D_2 \subseteq V - P_f$ is at most countable. By Lemma 3.2, $h \in \mathfrak{g}(W)$ exists such that $g = h \cdot f \in \mathfrak{g}(W)$ and such that $h|V \in \mathfrak{T}(V)$ and such that $h(\psi(y_B)) \not\equiv 0$ if $B \in \mathcal{B}$ and such that $h(\varphi(\psi(x_C))) \not\equiv 0$ if $C \in \mathcal{C}$. Hence $h \circ \psi | B \not\equiv 0$ if $B \in \mathcal{B}$ and $h \circ \psi \circ \varphi | C \not\equiv 0$ if $C \in \mathcal{C}$. Therefore $h \circ \psi \in \mathfrak{T}(U)$ and $(h \circ \psi \circ \varphi) \in \mathfrak{T}(T)$. Hence

$$(\psi \circ \varphi)^*(f)|T = \frac{g \circ \psi \circ \varphi | T}{h \circ \psi \circ \varphi | T} \qquad \psi^*(f)|U = \frac{g \circ \psi | U}{h \circ \psi | U} .$$

Here $T = \varphi^{-1}(U)$ and $g \circ \psi \in \quad (U)$ and $h \circ \varphi \in \quad (U)$ with $h \circ \psi \circ \varphi \in \quad (T)$. Hence

$$\varphi^*(\psi^*(f))|T = \frac{g \circ \psi \circ \varphi | T}{g \circ \psi \circ \varphi | T} = (\psi \circ \varphi)^*(f)|T$$

where T is an open neighborhood of a. Hence $\varphi^* \circ \psi^* = (\psi \circ \varphi)^*$; q.e.d.

There are examples where $f \in \mathcal{L}_\psi(Z)$ and $\psi^*(f) \in \mathcal{L}_\varphi(Y)$ and $f \notin \mathcal{L}_{\psi \circ \varphi}(Z)$.

Remark 3.5. By Lemma 1.25, the conditions of Lemma 3.4.4 are satisfied in the following situations.

1. The complex space Y has pure dimension n. The map $\varphi: X \to Y$ is surjective, holomorphic and has strict rank n.

2. The complex space Y is irreducible with $n = \dim Y$. The map $\varphi: X \to Y$ is holomorphic and has strict rank n.

3. The complex spaces X and Y are irreducible with $\dim Y = n$. The holomorphic map $\varphi: X \to Y$ has rank n.

4. The holomorphic map $\varphi: X \to Y$ is open and surjective.

5. The complex space Y is irreducible and the holomorphic map $\varphi: X \to Y$ is open.

In all these cases, $\mathcal{L}_\varphi(Y) = \mathcal{K}(Y)$ and $\varphi^*: \mathcal{K}(Y) \to \mathcal{K}(X)$ is injective.

Lemma 3.6. Let X be a complex space. Let A be an open, dense subset of X. Define $S = X - A$. Let $\{N_\lambda\}_{\lambda\in\Lambda}$ be a family of open subsets of X such that $S \subseteq \bigcup_{\lambda\in\Lambda} N_\lambda$. Suppose that f is a meromorphic function on A. Suppose that $\{f_\lambda\}_{\lambda\in\Lambda}$ is a family of meromorphic functions $f_\lambda \in \mathcal{K}(N_\lambda)$ with $f_\lambda|N_\lambda \cap A = f|N_\lambda \cap A$ for each $\lambda\in\Lambda$. Then one and only one meromorphic function F on X exists with $F|A = f$. Moreover, $F|N_\lambda = f_\lambda$.

Proof. Define $\Lambda_0 = \Lambda \cup \{0\}$ and $N_0 = A$ and $f_0 = f$. Then $\{N_\lambda\}_{\lambda\in\Lambda_0}$ is an open covering of X. If $N_\lambda \cap N_\mu \neq \emptyset$, then $A \cap N_\lambda \cap N_\mu \neq \emptyset$ is open and dense in $N_\lambda \cap N_\mu$. Moreover

$$f_\lambda|A \cap N_\lambda \cap N_\mu = f|A \cap N_\lambda \cap N_\mu = f\ |A \cap N_\lambda \cap N_\mu.$$

Therefore, $f_\lambda|N_\lambda \cap N_\mu = f_\mu|N_\lambda \cap N_\mu$ for all $\lambda\in\Lambda_0$ and $\mu\in\Lambda_0$ with $N_\lambda \cap N_\mu \neq \emptyset$. Hence $F \in \mathcal{K}(X)$ exists such that $F|N_\lambda = f_\lambda$ for each $\lambda\in\Lambda_0$, especially for each $\lambda\in\Lambda$. Also $F|A = f_0 = f$ holds. Because A is open and dense in X the function F is unique; q.e.d.

Let X be a complex space. A holomorphic function $u \in \mathcal{O}(U)$, which is a non-zero divisor is called an universal denominator on the open subset $U \neq \emptyset$ of X if and only if the following property $(\mathfrak{U})$ holds:

$(\mathfrak{U})$. Let A and B be open subsets of U such that $\emptyset \neq A \subseteq B \subseteq U$

and such that B - A is a thin analytic subset of B. Let f be a holomorphic function on A. Suppose that for every point $z \in B - A$ an open neighborhood W of z in B exists such that $|f|$ is bounded on $W \cap A$. Then a holomorphic function F on B exists such that $u \cdot f = F$ on B.

If $a \in X$, an open neighborhood U of a and a holomorphic function $u \in \mathfrak{O}(U)$ exists such that u is an universal denominator on U. (See Narasimhan [9], pp. 50, 56, 58, 110. There B is the set of simple points of A. But the definitions are equivalent, since in a neighborhood of each simple point the Riemann extension theorem holds.)

Lemma 3.7. Let A be an open subset of a complex space X such that $S = X - A$ is a thin analytic subset of X. Let f be a meromorphic function on A. Suppose that for every point $x \in S$ an open neighborhood W_x and bounded holomorphic functions g_x and h_x on $W_x \cap A_x$ exist, such that $h_x \in \mathfrak{O}(W_x \cap A)$ is not a zero-divisor and such that $h_x \cdot f = g_x$ on $W_x \cap A$. Then one and only one meromorphic function F on X exists such that $f = F|A$.

Proof. For every $x \in S$, an open neighborhood U_x of x and an universal denominator $u_x \in \mathfrak{O}(U_x)$ exist. Then $N_x = U_x \cap W_x$ is an open neighborhood of x. Now, $g_x|N_x \cap A$ and $h_x|N_x \cap A$ are bounded holomorphic functions and $N_x - A$ is a thin analytic subset of N_x. Therefore, holomorphic functions G_x and H_x exist on N_x such that $G_x = u_x g_x$ and $H_x = u_x h_x$ on $N_x \cap A$. Because u_x and h_x are non-zero divisors on U_x and W_x respectively, H_x is a non-zero divisor on N_x.

A meromorphic function $f_x = {}^{G_x}/H_x$ is defined on N_x such that

$$f_x|N_x \cap A = \frac{u_x g_x}{u_x h_x}|N_x \cap A = {}^{g_x}/h_x|N_x \cap A = f|N_x \cap A.$$

By Lemma 3.6, a meromorphic function F on X exists such that $F/A = f$. Obviously, F is unique; q.e.d.

Lemma 3.8. Let X and Y be complex spaces. Let $\varphi: X \to Y$ be a light, proper, surjective, holomorphic map. Suppose that a thin analytic subset T of Y exists such that $S = \varphi^{-1}(T)$ is a thin analytic subset of X and such that $\varphi_0 = \varphi: X - S \to Y - T$ is biholomorphic. Then $\varphi^*: \mathfrak{K}(Y) \to \mathfrak{K}(X)$ is an isomorphism.

Proof. Define $A = X - S$ and $B = Y - T$. Lemma 3.4.3 (or 4) implies $\mathcal{L}_\varphi(Y) = \mathfrak{K}(Y)$. Moreover, if $\varphi^*(f) = 0$, then $0 = \varphi^*(f)|A = \varphi_0^*(f|B)$. Because φ_0 is biholomorphic, φ_0^* is an isomorphism. Therefore $f|B = 0$, which implies $f = 0$. Hence φ^* is injective.

Take $\hat{f} \in \mathfrak{K}(X)$. A function $f \in \mathfrak{K}(Y)$ has to be constructed such that $\hat{\varphi}^*(f) = \hat{f}$. Define $\hat{f}_0 = \hat{f}|A$ and $f_0 = (\varphi_0^{-1})^*(\hat{f}_0) \in \mathfrak{K}(B)$. Take any $y \in T$. Let $x_1,\ldots,x_p$ be the different points of $\varphi^{-1}(y)$. Open neighborhoods V_μ of x_μ exist such that $V_\mu \cap V_\nu = \emptyset$ if $\mu \neq \nu$ and such that bounded holomorphic functions $\hat{g}_\mu \in \mathfrak{H}(V_\mu)$ and $\hat{h}_\mu \in \mathfrak{T}(V_\mu)$ exist with $\hat{h}_\mu \cdot f = \hat{g}_\mu$ on V_μ. The union $V = V_1 \cup \ldots \cup V_p$ if open. Bounded, holomorphic functions $\hat{g} \in \mathfrak{H}(V)$ and $\hat{h} \in \mathfrak{T}(V)$ are defined by $\hat{g}|V_\mu = \hat{g}_\mu$ and $\hat{h}|V_\mu = \hat{h}_\mu$ for $\mu = 1,\ldots,p$. Then $\hat{h}\cdot\hat{f} = \hat{g}$ on V. Because φ is proper, an open neighborhood W_y of y in Y exists such that $\varphi^{-1}(W_y) \subseteq V$. Then $g_y = \hat{g}\circ\varphi_0^{-1}$ and $h_y = \hat{h}\circ\varphi_0^{-1}$

are bounded holomorphic functions on $W_y \cap B$. Because φ_0^{-1} is biholomorphic, h_y is not a zero-divisor on $W_y \cap B$, i.e., $h_y \in \mathfrak{T}(W_y \cap B)$. By Lemma 3.7, a meromorphic function f exists on Y such that $f|B = f_0$. Then

$$\varphi^*(f)|A = \varphi_0^*(f|B) = \varphi_0^*(f_0) = \varphi_0^*((\varphi_0^{-1})^*(f_0)) = \hat{f}_0 = \hat{f}|A.$$

Because A is dense, $\varphi^*(f) = \hat{f}$ on X. The map φ^* is also surjective;

q.e.d.

Let X be a complex space and let S_X be the set of non-simple points of X. Then S_X is a thin analytic subset of X. Let $\pi\colon \hat{X} \to X$ be the normalization of X. Then $\hat{S} = \pi^{-1}(S_X)$ is a thin analytic subset of $\hat{X}$ and $\pi_0 = \pi\colon \hat{X} - \hat{S} \to X - S_X$ is biholomorphic. Also the map π is light, proper, surjective and holomorphic. Therefore, Lemma 3.8 implies:

Theorem 3.9. Let $\pi\colon \hat{X} \to X$ be the normalization of the complex space X, then $\pi^*\colon \mathfrak{K}(X) \to \mathfrak{K}(\hat{X})$ is an isomorphism.

Although Theorem 3.9 is "well known", it seems to be impossible to find it explicitly stated and proved in the existing textbook literature.

Lemma 3.10. Let $\mathfrak{B}$ be the set of branches of the complex space X. Suppose that for every $B \in \mathfrak{B}$ a meromorphic function $f_B \in \mathfrak{K}(B)$ is given. Then one and only one meromorphic function $f \in \mathfrak{K}(X)$ exists such that $j_B^*(f) = f_B$ for all $B \in \mathfrak{B}$, where $j_B\colon B \to X$ is the inclusion map. (Also write $j_B^*(f) = f|B$.)

Proof. Let $\pi: \hat{X} \to X$ be the normalization of X. One and only one branch $\hat{B}$ of $\hat{X}$ exists for each $B \in \mathfrak{B}$ such that $\pi(\hat{B}) = B$. Then $\pi_B = \pi: \hat{B} \to B$ is the normalization of B. Let $\hat{j}_B: \hat{B} \to \hat{X}$ be the inclusion map. Then $j_B \circ \pi_B = \pi \circ \hat{j}_B$. Then $\hat{f}_B = \pi_B^*(f_B)$ is meromorphic on $\hat{B}$. Since each $\hat{B}$ is open in $\hat{X}$ and since $\bigcup_{\hat{B} \in \mathfrak{B}} \hat{B} = \hat{X}$ is a disjoint union, one and only one meromorphic function $\hat{f} \in \mathfrak{K}(\hat{X})$ exists such $\hat{j}_B^*(f) = \hat{f}|\hat{B} = \hat{f}_B = \pi_B^*(f_B)$ for $B \in \mathfrak{B}$. Now, $f \in \mathfrak{K}(X)$ with $\pi^*(f) = \hat{f}$ exists. Hence $\hat{j}_B^*(\pi^*(f)) = \pi_B^*(f_B)$ if $B \in \mathfrak{B}$. By Lemma 3.4.3, $f \in \mathfrak{K}(X) = \mathfrak{L}_{\pi \circ \hat{j}_B}(X)$. Hence, Lemma 3.4.5 implies $\pi_B^*(f_B) = (\pi \circ \hat{j}_B)^*(f) = (j_B \circ \pi_B)^*(f)$. Here $\mathfrak{L}_{j_B}(X) = \mathfrak{K}(X)$ and $\mathfrak{L}_{\pi_B}(B) = \mathfrak{K}(B)$. Hence $(j_B \circ \pi_B)^*(f) = \pi_B^*(j_B^*(f))$. Because π_B^* is an isomorphism, $f_B = j_B^*(f)$; q.e.d.

Lemma 3.11. Let X be a complex space of pure dimension m. Let S be an analytic subset of X with $\dim S \leqq m - 2$. Define $A = X - S$. Suppose that $f \in \mathfrak{K}(A)$ is meromorphic on A. Then one and only one meromorphic function $F \in \mathfrak{K}(X)$ exists with $F|A = f$.

Proof. If X is normal, see Narasimhan [9], p 133, Theorem 4. Suppose that X is not normal. Let $\hat{\pi}: \hat{X} \to X$ be the normalization. Then $\hat{S} = \pi^{-1}(S)$ is analytic in $\hat{X}$. By Lemma 1.9 $\dim \hat{S} \leqq m - 2$. Moreover $\hat{X}$ is pure m-dimensional. Define $\hat{A} = \hat{X} - \hat{S} = \pi^{-1}(A)$ and $\pi_0 = \pi: \hat{A} \to A$. Then $\pi_0^*(f) = \hat{f} \in \mathfrak{K}(\hat{A})$. Hence $\hat{F} \in \mathfrak{K}(\hat{X})$ exists with $\hat{F}|\hat{A} = \hat{f}$. Now, $F \in \mathfrak{K}(X)$ exists with $\pi^*(F) = \hat{F}$. By Lemma 3.4.2, $\pi_0^*(f) = \pi^*(F)|\hat{A} = \pi_0^*(F|A)$. Hence $f = F|A$, because π_0^* is injective; q.e.d.

Later, it will be shown, that Lemma 3.11 remains true if S is a closed subset of X, which is almost thin of dimension n - 2.

Let V be a complex vector space of dimension n + 1 over C. An equivalence relation $\mathfrak{R}$ is defined on V - {0} by x ~ y if and only if $x = \lambda y$ for some complex number $\lambda \neq 0$. Let $\mathbb{P}(V) = (V - \{0\})/\mathfrak{R}$ be the quotient space with the quotient topology. Let $\rho: V - \{0\} \to \mathbb{P}(V)$ be the residual map. Then $\mathbb{P}(V)$ has one and only one complex structure such that ρ is holomorphic. The complex manifold $\mathbb{P}(V)$ is compact, connected, n-dimensional and homogeneous. The holomorphic map ρ is regular, 1-fibering and surjective. The space $\mathbb{P}(V)$ is called the <u>complex projective space associated</u> to V. If W is a linear subspace of dimension $p + 1 > 0$ of V, then $\mathbb{P}(W)$ is a smooth, p-dimensional, compact submanifold of $\mathbb{P}(V)$, called a <u>p-dimensional projective plane</u>.

Denote $\mathbb{P}^n = \mathbb{P}(C^{n+1})$. The map $j: \mathbb{C}^n \to \mathbb{P}^n$ which is defined by $j(z_1,\ldots,z_n) = \rho(1,z_1,\ldots,z_n)$ maps $\mathbb{C}^n$ biholomorphically onto an open subset $j(\mathbb{C}^n)$ of $\mathbb{P}^n$. The complement $\mathbb{P}^n - j(\mathbb{C}^n)$ is a projective plane of dimension n - 1. A biholomorphic map $\lambda: \mathbb{P}^{n-1} \to \mathbb{P}^n - j(\mathbb{C}^n)$ is defined by

$$\lambda(\rho(z_1,\ldots,z_n)) = \rho(0,z_1,\ldots,z_n)$$

if $(z_1,\ldots,z_n) \in \mathbb{C}^n - \{0\}$. If $\mathbb{C}^n$ and $j(\mathbb{C}^n)$ are identified by j and if $\mathbb{P}^{n-1}$ and $\mathbb{P}^n - j(\mathbb{C}^n)$ are identified by λ. Then

$$\mathbb{P}^n = \mathbb{C}^n \cup \mathbb{P}^{n-1} \qquad \mathbb{C}^n \cap \mathbb{P}^{n-1} = \emptyset.$$

In this sense, $\mathbb{P}^n$ is called the projective closure of $\mathbb{C}^n$ and $\mathbb{P}^{n-1}$ is called the infinite plane in $\mathbb{P}^n$. The space P^0 consist of exactly one point denoted by ∞. Hence

$$\mathbb{P} = \mathbb{P}^1 = \mathbb{C} \cup \{\infty\} \qquad \{\infty\} = \mathbb{P} - \mathbb{C}.$$

is the <u>closed plane</u>, <u>the Riemann sphere</u>, <u>the projective line</u>.

Let X and Y be complex spaces. Let A be open in X such that X - A is a thin analytic subset of X. Consider a holomorphic map $\tau: A \to Y$. Let $A^* = \{(x,\tau(x)) | x \in A\}$ be the graph of τ. Then A^* is analytic in $A \times Y$. Let $\Gamma = \Gamma_\tau$ be the closure of A^* in $X \times Y$. The map τ is said to be <u>meromorphic</u>[7)] if and only if Γ_τ is an analytic subset of $X \times Y$ and if the projection $\pi: \Gamma \to X$ is proper. Let τ be meromorphic. Then Γ is called the graph of τ over X. The map $\pi | A^*: A^* \to A$ is biholomorphic and A^* is dense in Γ. The set $\pi^{-1}(X-A) = \Gamma - A^*$ is thin and analytic in Γ. The map $\pi: \Gamma \to X$ is surjective and maps each branch of Γ onto one and only branch of X and each branch of X is obtained this way. If X is pure dimensional, so is Γ; if X is irreducible, so is Γ. A largest open subset $A_0 \supseteq A$ of X exists such that τ can be continued to a holomorphic map $\tau_0: A_0 \to Y$. Then $\tau_0 | A = \tau$. Moreover, $I_\tau = X - A_0$ is analytic and thin and is called the <u>indeterminacy</u> of τ. The set $\pi^{-1}(A_0) = A_0^*$ is open and dense in Γ with $A^* \subseteq A_0^*$ and $\Gamma - A_0^*$ is analytic and thin in Γ. The restriction $\pi | A_0^*: A_0^* \to A_0$ is biholomorphic. If X is normal, then I_τ has at least codimension 2 and $\dim_z \pi^{-1}(x) > 0$ if $z \in \pi^{-1}(x)$ and $x \in I_\tau$. If X is not normal, $\pi^{-1}(x)$ may consist of finitely many points, even only one if $x \in I_\tau$. Because π is proper, $\pi^{-1}(x)$ is compact for each $x \in X$.

<u>Proposition 3.12.</u> Let X be a complex space. Let A be an

open subset of X such that $D = X - A$ is a thin analytic subset of X. Let $f_1,\ldots,f_k$ be holomorphic functions on A. Define $\tau: A \to \mathbb{P}^k$ by

$$\tau(x) = (f_1(x),\ldots,f_k(x)) \in \mathbb{C}^k \subseteq \mathbb{P}^k$$

if $x \in A$. Define $A^* = \{(x,\tau(x)) \mid x \in A\}$. Let $\Gamma = \overline{A}^*$ be the closure of A^* in $X \times \mathbb{P}^k$. Then τ is meromorphic if and only if $f_1,\ldots,f_k$ continue to meromorphic functions $f_1,\ldots,f_k$ on X.

Proof. The Proposition is true if X is normal by Remmert [12] Satz 13 and [20] Satz 4.3. Suppose that X is not normal. Let $\rho: \hat{X} \to X$ be the normalization of X. Then $\sigma = \rho \times \mathrm{Id}: \hat{X} \times \mathbb{P}^k \to X \times \mathbb{P}^k$ is the normalization of $X \times \mathbb{P}^k$ especially, σ is light and proper. The projection $\pi: \Gamma \to X$ is proper and $\pi_0 = \pi: A^* \to A$ is biholomorphic. Define $\hat{A} = \rho^{-1}(A)$ and $\hat{\tau} = \tau\circ\pi: \hat{A} \to \mathbb{P}^k$. Then

$$\hat{\tau}(x) = \{f_1(\pi(x)),\ldots,f_k(\pi(x))) \qquad \text{if } x \in \hat{A}$$

Define $\hat{A}^* = \{(x,\hat{\tau}(x)) \mid x \in \hat{A}\}$ and let $\hat{\Gamma}$ be the closure of $\hat{A}^*$ in $\hat{X} \times \mathbb{P}^k$. The projection $\hat{\pi}: \hat{\Gamma} \to \hat{X}$ is proper and $\hat{\pi}_0 = \hat{\pi}: \hat{A}^* \to \hat{A}$ is biholomorphic. If $(x,z) \in \hat{A}^*$, then $z = \hat{\tau}(x) = \tau(\rho(x))$ and $\sigma(x,z) = (\rho(x),z) = (\rho(x),\tau(\rho(x)) \in A^*$. If $(x,z) \in A^*$, then $z = \tau(x)$ and $x = \rho(\hat{x})$ for some $\hat{x} \in \hat{A}$. Hence $\sigma(\hat{x},\hat{\tau}(\hat{x})) = (\rho(\hat{x}),\tau(\rho(\hat{x})) = (x,\tau(x)) = (x,z)$. Therefore $\sigma(\hat{A}^*) = A^*$. Define $\sigma_0 = \sigma: \hat{A}^* \to A^*$. Then $\pi_0 \circ \sigma_0 = \rho\circ\hat{\pi}_0$. The maps π_0 and $\hat{\pi}_0$ are biholomorphic. Hence $\sigma_0: \hat{A}^* \to A^*$ is the normalization of A^*.

Now, $A^* = \sigma(\hat{A}^*)$ implies $\Gamma \supseteq \sigma(\hat{\Gamma})$. If $(a,b) \in \Gamma$, a sequence $\{(x_\nu,z_\nu)\}_{\nu\in\mathbb{N}}$ of points of A^* converges to (a,b). Because ρ is proper a sequence $\{\hat{x}_{\nu_\lambda}\}_{\lambda\in\mathbb{N}}$ converges to $\hat{a}$ where $\nu_\lambda \to \infty$ for $\lambda \to \infty$ and where $\rho(\hat{x}_{\nu_\lambda}) = x_{\nu_\lambda}$. Then $z_{\nu_\lambda} = \tau(x_{\nu_\lambda}) = \tau(\rho(\hat{x}_{\nu_\lambda})) = \hat{\tau}(\hat{x}_{\nu_\lambda})$ and $(\hat{x}_{\nu_\lambda}, z_{\nu_\lambda}) \in \hat{A}_{\nu_\lambda}$ with $\sigma(\hat{x}_{\nu_\lambda}, z_{\nu_\lambda}) = (x_{\nu_\lambda}, z_{\nu_\lambda})$. Hence $(\hat{a}, b) \in \hat{\Gamma}$ and $\sigma(\hat{a}, b) = (a, b)$. Therefore $\Gamma \subseteq \sigma(\hat{\Gamma})$. This implies $\Gamma = \sigma(\hat{\Gamma})$. The restriction $\sigma_1 = \sigma\colon \hat{\Gamma} \to \Gamma$ is proper. Moreover $\sigma_0 = \sigma_1 : \hat{A}^* \to A^*$.

If $f_1, \ldots, f_p$ are meromorphic on X, then $\rho^*(f_1), \ldots, \rho^*(f_k)$ are meromorphic on $\hat{X}$ and extend the holomorphic functions $f_1 \circ \rho, \ldots, f_k \circ \rho$ on $\hat{A}$. Hence $\hat{\Gamma}$ is analytic. Because σ is proper, $\Gamma = \sigma(\hat{\Gamma})$ is analytic. Hence τ is meromorphic on X.

Now, assume, that τ is meromorphic. Then Γ is analytic. Select a metric η on $\hat{X}$ and a metric δ on $\mathbb{P}^k$. Let S_X be the set of simple points of X. Then $N = X - S_X$ is open and dense in X. Also $\hat{N} = \rho^{-1}(N)$ is open and dense in $\hat{X}$ and $\hat{A} = \hat{X} - \hat{N}$ is analytic in $\hat{X}$. Then $M = \hat{N} \times \mathbb{P}^k$ is open and dense in $\hat{X} \times \mathbb{P}^k$ and $T = \hat{S} \times \mathbb{P}^k$ is analytic and thin in $\hat{X} \times \mathbb{P}^k$. Define $\hat{\Gamma} \cap M = E$. Then $\overline{E} = \hat{\Gamma}$ is claimed. Because $\hat{\Gamma}$ is closed, $\overline{E} \subseteq \hat{\Gamma}$ is trivially true. Take $(a,b) \in \hat{\Gamma}$. A sequence $\{(x_\nu, z_\nu)\}_{\nu\in\mathbb{N}}$ in $\hat{A}^*$ converges to (a,b) with $z_\nu = \hat{\tau}(x_\nu)$ and $x_\nu \in \hat{A}$. Points $x'_\nu \in \hat{A} \cap \hat{N}$ with $\eta(x'_\nu, x_\nu) < \frac{1}{\nu}$ and $\delta(\hat{\tau}(x'_\nu), z_\nu) = \delta(\hat{\tau}(x'_\nu), \hat{\tau}(x_\nu)) < \frac{1}{\nu}$ exist. Then $(x'_\nu, \hat{\tau}(x'_\nu)) \to (a,b)$ for $\nu \to \infty$ with $(x'_\nu, \hat{\tau}(x'_\nu)) \in \hat{A} \cap M \subseteq \hat{\Gamma} \cap M = E$. Hence, $(a,b) \in \overline{E}$. Therefore $\hat{\Gamma} \subseteq \overline{E}$ which implies $\overline{E} = \hat{\Gamma}$.

The set $F = \sigma^{-1}(\Gamma)$ is analytic in $\hat{X} \times P^k$ with $F \supseteq \hat{\Gamma}$. Now, $F \cap M = \hat{\Gamma} \cap M = E$ is claimed. Take $(\hat{a}, b) \in F \cap M$. Then $(a,b) = (\rho(\hat{a}), b) = \sigma(\hat{a}, b) \in \Gamma$ and $a \in N$. A sequence $\{(x_\nu, z_\nu\}_{\nu\in\mathbb{N}}$ in A^* converges to (a,b). Because N is open, $x_\nu \in A \cap N$ can be assumed.

Because $\rho: \hat{N} \to N$ is biholomorphic, one and only one $\hat{x}_\nu \in \hat{A} \cap \hat{N}$ exists such that $\rho(\hat{x}_\nu) = x_\nu$. Moreover $\hat{x}_\nu \to \hat{a}$ for $\nu \to \infty$. Then $z_\nu = \tau(x_\nu) = \tau(\rho(x_\nu)) = \hat{\tau}(x_\nu) \to b$ for $\nu \to \infty$. Hence $(\hat{x}_\nu, \hat{\tau}(x_\nu)) \to (\hat{a}, b) \in \hat{\Gamma} \cap M$. Therefore $F \cap M \subseteq \hat{\Gamma} \cap M \subseteq F \cap M$, which implies $F \cap M = \hat{\Gamma} \cap M = E$. Let F_1 be the union of all branches of F which are not contained in $T = \hat{X} \times \mathbb{P}^k - M$. Then $F_1 \cap M = E$ and $F_1 \cap T$ is thin on F_1. Therefore, $F_1 = \overline{F_1 \cap M} = \overline{E} = \hat{\Gamma}$. The set $\hat{\Gamma}$ is analytic. Obviously, $\hat{\pi}: \hat{\Gamma} \to \hat{X}$ is proper. Hence, $\hat{\tau}$ is meromorphic. Therefore $f_1 \circ \rho | \hat{A}, \ldots, f_p \circ \rho | \hat{A}$ extend to meromorphic functions $\hat{F}_1, \ldots, \hat{F}_k$ on $\hat{X}$. Meromorphic function $F_1, \ldots, F_k$ on X exist such that $\rho^*(F_\mu) = \hat{F}_\mu$ for $\mu = 1, \ldots, k$. Define $\rho_0 = \rho: \hat{A} \to A$. Then $\rho_0^*(F_\mu | A) = \rho^*(F_\mu)|A = \hat{F}_\mu | A = \rho^*(f_\mu)$. Therefore $F_\mu | A = f_\mu$ for $\mu = 1, \ldots, k$; q.e.d.

Let X be a complex space. Let $f_1, \ldots, f_k$ be meromorphic functions on X. Consider $\mathbb{P}^k = \mathbb{C}^k \cup \mathbb{P}^{k-1}$ as the disjoint union of $\mathbb{C}^k$ and $\mathbb{P}^{k-1}$. Let A be the largest open subset of X such that $f_1, \ldots, f_k$ are holomorphic on X. Then $X = A = P_{f_1} \cup \ldots \cup P_{f_k}$ is a thin analytic subset of X. A holomorphic map $\tau: A \to \mathbb{P}^k$ is defined by

$$\tau(x) = (f_1(x), \ldots, f_k(x)) \in \mathbb{C}^k \subseteq \mathbb{P}^k$$

if $x \in A$. By Proposition 3.8, the map τ is meromorphic. Define $A^* = \{(x, \tau(x)) | x \in A\}$ and $\Gamma = \overline{A}^*$ as the closure of A^* in $X \times \mathbb{P}^k$. Then Γ is an analytic subset of $X \times \mathbb{P}^k$. The projections $\pi: \Gamma \to X$ and $\eta: \Gamma \to \mathbb{P}^k$ are holomorphic. The restriction $\pi_0 = \pi: A^* \to A$ is biholomorphic. The map π is proper, surjective and holomorphic. Let $U \neq \emptyset$ be open in Γ, then $U \cap A^* \neq \emptyset$. Hence $\pi(U)$ is not thin in X. By Lemma 3.4.4, $\mathfrak{L}_\pi(X) = \mathfrak{K}(X)$ and $\pi^*: \mathfrak{K}(X) \to \mathfrak{K}(\Gamma)$ is injective. Therefore $\pi^*(f_1),\ldots,\pi^*(f_k)$ exist with $\pi^*(f_\lambda)|A^* = \pi_0(f_\lambda|A) = f_\lambda \circ \pi_\varphi$ for $\lambda = 1,\ldots,k$. Define $g_\lambda: \mathbb{C}^k \to \mathbb{C}$ by

$$g_\lambda(z_1,\ldots,z_k) = z_\lambda.$$

Then g_λ extends to a meromorphic function $g_\lambda \in \mathfrak{K}(\mathbb{P}^k)$ with $P_{g_\lambda} = \mathbb{P}^{k-1}$. Now, $\eta(A^*) = \tau(A^*) \subseteq \mathbb{C}^k$. Hence $\eta^{-1}(\mathbb{P}^{k-1}) \subseteq \Gamma - A^*$. Therefore, $\eta^{-1}(P_{g_\lambda})$ is a thin analytic subset of Γ and $g_\lambda \in \mathfrak{L}_\eta(\mathbb{P}^k)$. Hence $\eta^*(g_\lambda) \in \mathfrak{K}(\Gamma)$ is defined with $\eta^*(g_\lambda)|A^* = g_\lambda \circ \eta$. If $(x,\tau(x)) \in A^*$, then $g_\lambda \circ \eta(x,\tau(x)) = g_\lambda(\tau(x)) = f_\lambda(x) = f_\lambda \circ \pi(x,\tau(x))$. Hence, $\eta^*(g_\lambda)|A^* = \pi^*(f_\lambda)|A^*$ which implies $\pi^*(f_\lambda) = \eta^*(g_\lambda)$ for $\lambda = 1,\ldots,k$.

The functions $g_\lambda \in \mathfrak{K}(\mathbb{P}^k)$ can be obtained by a different method. Let $\rho: \mathbb{C}^{k+1} - \{0\} \to \mathbb{P}^k$ be the residual map. By Lemma 3.4.4, $\mathfrak{L}_\rho(\mathbb{P}^k) = \mathfrak{K}(\mathbb{P}^k)$ and

$$\rho^*: \mathfrak{K}(\mathbb{P}^k) \to \mathfrak{K}(\mathbb{C}^{k+1} - \{0\}) = \mathfrak{K}(\mathbb{C}^{k+1})$$

is injective. Let $h_\lambda: \mathbb{C}^{k+1} \to \mathbb{C}$ be defined by

$$h_\lambda(z_0,\ldots,z_k) = z_\lambda$$

for $\lambda = 0,\ldots,k$. Then $\rho^*(g_\lambda) = h_\lambda/h_0$ for $\lambda = 1,\ldots,k$.

Now, consider the case $k = 1$: Let $f \in \mathfrak{K}(X)$ be a meromorphic function on the complex space X. Define $A = X - P_f$ and

$A^* = \{(x,f(x)) | x \in A\}$. Then $\Gamma = \overline{A}^*$ is analytic in $X \times \mathbb{P}$. The projections $\pi: \Gamma \to X$ and $\eta: \Gamma \to \mathbb{P}$ are holomorphic. The map $\pi: \Gamma \to X$ is proper and maps each branch onto a branch of X; different branches are mapped onto different branches, and all branches of X are obtained. The holomorphic map $f: A \to \mathbb{C} \subseteq \mathbb{P}$ can be continued to a holomorphic map $\tau_f: A_0 \to \mathbb{P}$ where $A_0 \supseteq A$ is the largest open subset for which such a continuation exists. A point $x \in X$ belongs to A_0 if and only if $f_x \in \mathcal{O}_x$ or $1/f_x \in \mathcal{O}_x$. The set $I_{\tau_f} = I_f = X - A_0$ is said to be the set of <u>indeterminacy of f</u>. A point $x \in P_f - I_f$ is said to be a <u>proper pole.</u> A point $a \in I_f$ is said to be a <u>point of indeterminacy</u>. If $a \in I_f$, then $\pi^{-1}(a)$ is an analytic subset of $\{a\} \times \mathbb{P}$. Hence $\pi^{-1}(a)$ is either finite, then a is called a <u>weak indeterminacy</u> of f, or $\pi^{-1}(a) = \{a\} \times \mathbb{P}$, then a is called a <u>strong indeterminacy</u> of f. Observe that in the case of a weak indeterminacy, $\pi^{-1}(a)$ may consist of one and only one point (a,b). If $b \in \mathbb{C}$, then f is said to be <u>weakly holomorphic</u> at b, if $b = \infty$, then f has a <u>weak pole</u> at a. In both cases, the map τ_f is continuous at a. If X is locally irreducible at a, then $\pi^{-1}(a)$ is connected. Hence f is either holomorphic at a, or weakly holomorphic at a, or has a proper pole at a or has a weak pole at a, or a is a strong indeterminacy of f. If X is normal at a, then f is either holomorphic at a, or has a proper pole at a or f has a strong indeterminacy at a. Define

$$I_f^s = \{x \in X | x \text{ is a strong determinacy of } f\}.$$

Then $I_f^s \subseteq I_f \subseteq \mathbb{P}_f$.

The function f lifts to $\pi^*(f) = \eta$. If $a \in \mathbb{P}$, the set $\eta^{-1}(a)$

is analytic in Γ. Then $N_f(a) = \pi(\eta^{-1}(a))$ is an analytic subset of X with $N_f(a) - I_f = \tau_f^{-1}(a)$. If $a \neq \infty$, then $N_f(a) - P_f = f^{-1}(a)$ where $f\colon A \to \mathbb{C}$ is regarded as a holomorphic map. If $a = \infty$, then $N_f(\infty) \subseteq P_f$ and $P_f = N_f(\infty) \cup I_f$. Obviously,

$$I_f^s = \bigcap_{a \in \mathbb{P}} N_f(a).$$

Hence I_f^s is an analytic subset of X. If X is normal, then $I_f^s = I_f$.

Lemma 3.9. Let X be a pure m-dimensional complex space. Let $f \in \mathcal{K}(X)$. Let $\Gamma \subseteq X \times \mathbb{P}$ be the graph of f. Then $\dim I_f^s \leqq m - 2$. Moreover, define

$$E = \{z \in \Gamma \mid \operatorname{rank}_z \pi \leqq m-1\}$$

Then $\pi(E) = I_f^s$.

Proof. If $a \in \pi(E)$, then $(a,b) \in E$ with

$$1 \geqq \dim_{(a,b)} \pi^{-1}(a) = m - \operatorname{rank}_{(a,b)} \pi \geqq m - (m-1) = 1$$

since Γ has pure dimension m. Hence $\pi^{-1}(a) = \{a\} \times \mathbb{P}$ and $a \in I_f^s$. If $a \in I_f^s$, then $\pi^{-1}(a) = \{a\} \times \mathbb{P}$. If $b \in \mathbb{P}$, then

$$\operatorname{rank}_{(a,b)} \pi = m - \dim_{(a,b)} \pi^{-1}(a) = m - 1.$$

Hence $(a,b) \in E$ and $a \in \pi(E)$. Therefore $\pi(E) = I_f^s$.

Let B be a branch of Γ. Then $a \in B \cap A^*$ exists and $\operatorname{rank}_a \pi =$

$\dim_a B = m$ because $\pi\colon A^* \to A$ is biholomorphic. Hence π has strict rank m. By Proposition 1.24 $\pi(E) = I_f^s$ is almost thin of dimension n-2. By Theorem 1.14 $\pi(E) = I_f^s$ is analytic. Hence $\dim I_f^s \leqq n - 2$;

q.e.d.

§4. Dependence

Let X and Y be complex spaces of pure dimension m and n respectively with $m - n = q \geqq 0$. Let $\varphi: X \to Y$ be a holomorphic map. Let $f_1,\ldots,f_k$ be meromorphic functions on X. Define $A = X - (P_{f_1} \cup \ldots \cup P_{f_k})$ as the largest open subset of X where all functions $f_1,\ldots,f_k$ are holomorphic. Define $f: A \to \mathbb{P}^k$ by

$$(4.1) \qquad f(x) = (f_1(x),\ldots,f_k(x)) \in \mathbb{C}^k \subseteq \mathbb{P}^k.$$

Then $f: A \to \mathbb{P}^k$ is holomorphic and meromorphic on X. The set $A^* = \{(x,f(x)) \mid x \in A\}$ is analytic and pure m-dimensional. Then $\Gamma = \overline{A}^*$ is a pure m-dimensional analytic subset of $X \times \mathbb{P}^k$. The projection $\pi: \Gamma \to X$ is holomorphic, proper and surjective. The restriction $\pi_0 = \pi: A^* \to A$ is biholomorphic. A holomorphic map $\psi: \Gamma \to Y \times \mathbb{P}^k$ is defined by $\psi(x,z) = (\varphi(x),z)$ if $(x,z) \in \Gamma$. Then rank $\varphi \leqq n + k$. The functions $f_1,\ldots,f_k$ are said to be <u>φ-dependent</u> (or dependent over φ) if and only if rank $\psi < n + k$. The functions $f_1,\ldots,f_k$ are said to be <u>φ-independent</u> (or independent over φ) if and only if ψ has strict rank $n + k$. If X is irreducible, then Γ is irreducible; then $f_1,\ldots,f_k$ are φ-dependent, if and only if $f_1,\ldots,f_k$ are not φ-dependent. If X is reducible, this alternative may be wrong.

By Lemma 1.16, the following statements are equivalent.

1. The functions $f_1,\ldots,f_k$ are φ-independent.
2. The map ψ has strict rank $n + k$.

3. The set $D = \{z \in \Gamma | \text{rank}_z \psi < n + k\}$ is thin and analytic.

4. Each branch B of Γ contains a simple point x of Γ such that $\text{rank}_x \psi = n + k$.

5. If B is a branch of Γ, then $\text{rank } \varphi | B = n + k$.

<u>Lemma 4.1.</u> Let X and Y be complex spaces of pure dimension m and n respectively with $q = m - n \geqq 0$. Let $\varphi: X \to Y$ be a holomorphic map. Let $f_1, \ldots, f_k$ be meromorphic functions on X with $k > q$, then $f_1, \ldots, f_k$ are φ-dependent.

<u>Proof.</u> By Lemma 1.13, $\text{rank } \psi \leqq \dim \Gamma = \dim X = m = n + q < n + k$; q.e.d.

Let X and Y be complex spaces of pure dimension m and n respectively with $q = m - n \geqq 0$. Let $\varphi: X \to Y$ be a holomorphic map. Let $f_1, \ldots, f_k$ be meromorphic functions on X. Then $f_1, \ldots, f_k$ satisfy the <u>Jacobian test</u> $J = J(U, \beta, \omega)$ on U if and only if the following conditions are satisfied.

J_1: The set $U \neq \emptyset$ is open on X and consist of simple points of X only.

J_2: The functions $f_1, \ldots, f_k$ are holomorphic on U.

J_3: A schlicht chart $\beta: U_\beta \to U'_\beta \subseteq \mathbb{C}^n$ of Y is given with $\varphi(U) \subseteq U_\beta$ and $\beta \circ \varphi = (g_1, \ldots, g_n)$.

J_4: Define

$$\omega = dg_1 \wedge \cdots \wedge dg_n \wedge df_1 \wedge \cdots \wedge df_k. \tag{4.2}$$

Lemma 4.2. Let X and Y be complex spaces of pure dimension m and n respectively with $m - n = q \geqq 0$. Let $\varphi: X \to Y$ be a holomorphic map. Let $f_1, \ldots, f_k$ be meromorphic functions on X. Then $f_1, \ldots, f_k$ are φ-independent if and only if each branch B of X contains an open set $U \neq \emptyset$, such that $f_1, \ldots, f_k$ satisfy the Jacobian test $J(U, \beta, \omega)$ for some schlicht chart β of Y with $\omega(x) \neq 0$ for all $x \in U$.

Proof. a) Suppose that $J(U, \beta, \omega)$ with $\omega(x) \neq 0$ for all $x \in U$ is satisfied. Then it will be shown that $f_1, \ldots, f_k$ are φ-independent.

Let $A = X - (P_{f_1} \cup \ldots \cup P_{f_k})$. Define $f: A \to \mathbb{P}^k$ by (4.1). Define $A^* = \{(x, f(x)) | x \in A\}$ and $\Gamma = \overline{A}^*$. Then Γ is a pure m-dimensional analytic subset of $X \times \mathbb{P}^k$. The projection $\pi: \Gamma \to X$ is holomorphic, proper and surjective. Define $\psi: \Gamma \to Y \times \mathbb{P}^k$ by $\psi(x,z) = (\varphi(x), z)$ if $(x,z) \in \Gamma$.

Take any branch B^* of Γ. Then $B = \pi(B^*)$ is a branch of X. Now, $J(U, \beta, \omega)$ is given with $U \subseteq B$. Then $U \subseteq A \cap B$ and $U^* = \pi^{-1}(U)$ is an open subset of B^* and of Γ. The restriction $\pi_1 = \pi: U^* \to U$ is biholomorphic. Now, ω is the Jacobian of the map $\gamma: U \to Y \times \mathbb{C}^k \subseteq Y \times \mathbb{P}^k$ where

$$(4.3) \qquad \gamma(x) = (\beta \circ \varphi(x), f(x)) = (g_1(x), \ldots, g_n(x), f_1(x), \ldots, f_k(x)).$$

Now, $\omega(x) \neq 0$ for all $x \in U$ is assumed. Hence γ is a regular map. Therefore γ has pure fiber dimension $m - n - k$. Define $\psi_1 = \psi: U^* \to U_\beta \times \mathbb{P}^k$. Then $(\beta \times \mathrm{Id}) \circ \psi_1 = \gamma \circ \pi_1$ has pure fiber dimension $m - n - k$ on U^*. Because $(\beta \times \mathrm{Id})$ is biholomorphic, ψ_1 has pure fiber dimension m - n - k on the pure m-dimensional open subset U^* of Γ and B. Hence $\text{rank}_z\psi = \widetilde{\text{rank}}_z\psi = n + k$ if $z \in U$. Therefore ψ has strict rank n + k. The functions $f_1,\ldots,f_k$ are φ-independent.

b) Suppose that $f_1,\ldots,f_k$ are φ-independent.

Let B be a branch of X. One and only one branch B^* of Γ exists such that $\pi(B^*) = B$. The analytic set

$$D = \{z \in \Gamma \mid \text{rank}_z\psi \leqq n + k - 1\}$$

is thin on Γ. Let S_X and S_Y be the sets of non-simple points of X and Y respectively. Then $S_Y \times \mathbb{P}^k$ is the set of non-simple points of $Y \times \mathbb{P}^k$. By Lemma 1.25, the inverse image $S_Y^* = \psi^{-1}(S_Y \times \mathbb{P}^k)$ is a thin analytic subset of Γ. Again by Lemma 1.25, $S_X^* = \pi^{-1}(S_X)$ is a thin analytic subset of Γ. Then $A^* - S_X^* = \pi^{-1}(A - S_X)$ is open and dense in Γ. Because $\pi_2 = \pi: A^* - S_X^* \to A - S_X$ is biholomorphic, $A^* - S_X^*$ is a complex manifold of pure dimension m. Then $A' = A^* = (S_X^* \cup S_Y^* \cup D)$ is open and dense in Γ and contains simple points of Γ only. If $z \in A'$, then $\text{rank}_z\psi = n + k$. Hence $\psi_2 = \psi: A' \to (Y - S_Y) \times \mathbb{P}^k$ is a (q-k)-fibering map into a pure (n+k)-dimensional complex manifold. By Theorem 2.6 (or Theorem 1.22), ψ_2 is open. Let T be the set of non-regular points of ψ_2. Then T is analytic in A'. By Sard's theorem, $\psi_2(T)$ has measure zero in $(Y - S_Y) \times \mathbb{P}^k$. Because ψ_2 is open, T is a thin analytic subset of A'. Hence $A' - T$ is dense in Γ. A point $d = (a,c) \in (A' - T) \cap B^*$ exists and $b = \varphi(a)$ is a simple point of Y. Moreover, $a \in B$ is a simple point of X. An open neighborhood U^* of d with $U^* \subseteq (A' - T) \cap B^*$ and a schlicht chart $\beta: U_\beta \to U'_\beta$ of Y at b with $\varphi(\pi(U^*)) \subseteq U_\beta$. Then $\pi(U^*) = U$ is open and $\pi_1 = \pi: U^* \to U$ is biholomorphic. The map $\psi_1 = \psi: U^* \to U_\beta \times \mathbb{P}^k$ is regular. The map $\beta \times \mathrm{Id}: U_\beta \times \mathbb{P}^k \to U'_\beta \times \mathbb{P}^k$ is biholomorphic. Define $\gamma: U \to U'_\beta \times \mathbb{C}^k \subseteq U'_\beta \times \mathbb{P}^k$ by (4.3) where $(g_1,\ldots,g_n) = \beta\circ\varphi$. Define ω by 4.2. Then ω is the Jacobian of the regular map $\gamma = (\beta \times \mathrm{Id})\circ\psi_1\circ\pi_1^{-1}$. Hence $\omega(x) \neq 0$ for $x \in U$. Then $J(U,\beta,\omega)$ is satisfied with $U \subseteq B$ and $\omega(x) \neq 0$ for all $x \in U$; q.e.d.

<u>Lemma 4.3.</u> Let X and Y be complex spaces of pure dimension m and n respectively with $m - n = q \geqq 0$. Let $\varphi: X \to Y$ be a holomorphic map. Let $f_1,\ldots,f_k$ be meromorphic functions on X. Then $f_1,\ldots,f_k$ are φ-independent if and only if each branch B of X contains an open subset $U \neq \emptyset$, such that $f_1,\ldots,f_k$ satisfy the Jacobian test $J(U,\beta,\omega)$ for some schlicht chart β of Y with $\omega \not\equiv 0$ on U.

<u>Proof.</u> If $f_1,\ldots,f_k$ are φ-independent, then $J(U,\beta,\omega)$ is satisfied with $U \subseteq B$ and $\omega(x) \neq 0$ for all $x \in U$, especially with $\omega \not\equiv 0$ on U. If $J(U,\beta,\omega)$ is satisfied with $U \subseteq B$ and $\omega \not\equiv 0$ on U, an open subset U_0 of U exists such that $\omega(x) \neq 0$ if $x \in U_0 \neq \emptyset$. Then $J(U_0,\beta,\omega)$ is satisfied with $\omega(x) \neq 0$ if $x \in U_0$ and $U_0 \subseteq B$. Hence $f_1,\ldots,f_k$ are φ-independent; q.e.d.

<u>Lemma 4.4.</u> Let X and Y be complex spaces of pure dimension m and n respectively with $m - n = q \geqq 0$. Let $\varphi: X \to Y$ be a holomorphic map. Let $f_1,\ldots,f_k$ be meromorphic functions on X. Then $f_1,\ldots,f_k$ are φ-independent (respectively φ-dependent) if and only if $f_1|B,\ldots,f_k|B$ are φ_B-independent (respectively φ_B-dependent) for

each branch B of X if $\varphi_B = \varphi\colon B \to Y$ is the restriction.

Proof. The case of φ-independence follows immediately from Lemma 4.2.

Let $A = X - (P_{f_1} \cup \ldots \cup P_{f_k})$. Define $f\colon A \to \mathbb{P}^k$ by (4.1). Define $A^* = \{(x,f(x))\,|\,x \in A\}$ and $\Gamma = \overline{A}^*$. The projection $\pi\colon \Gamma \to X$ is proper, surjective and holomorphic. The map $\psi\colon \Gamma \to Y \times \mathbb{P}^k$ is defined by $\psi(x,z) = (\varphi(x),z)$. The subset Γ of $X \times \mathbb{P}^k$ is analytic and pure m-dimensional. Let B be any branch of X. Then $B \cap A$ is open and dense in B. Now, $B' = \{(x,f(x))\,|\,x \in B \cap A\} = \pi^{-1}(B \cap A)$ is an irreducible analytic subset of A^* and $B^* = \overline{B}'$ is a branch of Γ and B^* is the graph of $f_1|B,\ldots,f_k|B$ with $B' = A^* \cap B^*$. Moreover, $B = U(B^*)$.

If $f_1,\ldots,f_k$ are φ-dependent, then rank $\psi|B^* < n + k$. Hence $f_1|B,\ldots,f_k|B$ are φ_B-dependent.

If $f_1|B\ldots,f_k|B$ are φ_B-dependent for every branch B of X, then rank $\psi|B^* < n + k$ for each branch B^* of Γ, because π maps the set of branches of Γ bijectively onto the set of branches of X. By Lemma 1.17, rank $\psi < n + k$. Hence $f_1,\ldots,f_k$ are φ-dependent;

q.e.d.

Lemma 4.5. Let X and Y be complex spaces of pure dimension m and n respectively with $q = n - n \geqq 0$. Let $\varphi\colon X \to Y$ be a holomorphic map. Let $f_1,\ldots,f_k$ be φ-dependent meromorphic functions on X which satisfy the Jacobian test $J = J(U,\beta,\omega)$ on U. Then $\omega \equiv 0$ on U.

Proof. It suffices to show, that $\omega \equiv 0$ on each component of U. Hence, it can be assumed that U is connected. Then U is contained in one and only branch B of X. The functions $f_1|B,\ldots,f_k|B$ are φ_B-dependent where $\varphi_B = \varphi: B \to Y$. If $\omega \not\equiv 0$ on U, then $f_1|B,\ldots,f_k|B$ are φ_B-independent by Lemma 4.3, which is wrong. Hence $\omega \equiv 0$ on U; q.e.d.

Lemma 4.6. Let X and Y be complex spaces of pure dimension m and n respectively with $q = m - n \geqq 0$. Let $\varphi: X \to Y$ be a holomorphic map. Let $f_1,\ldots,f_k$ be meromorphic functions on X. Suppose that they satisfy an Jacobian test $J(U_B,\beta_B,\omega_B)$ on an open subset U_B of B with $\omega_B \equiv 0$ on U for each branch B of X. Then $f_1,\ldots,f_k$ are φ-dependent.

Proof. Define $A = X - (P_{f_1} \cup \ldots \cup P_{f_k})$. Define $f: A \to \mathbb{P}^k$ by (4.1). Define $A^* = \{(x,f(x)) | x \in A\}$ and $\Gamma = \overline{A^*}$. The projection $\pi: \Gamma \to X$ is proper, surjective and holomorphic, and maps the set $\mathfrak{B}^*$ of branches of the pure m-dimensional analytic set Γ onto the set $\mathfrak{B}$ of branches of X bijectively. Define $\psi: \Gamma \to Y \times \mathbb{P}^k$ by $\psi(x,z) = (\varphi(x),z)$. The set $D = \{x \in \Gamma | \operatorname{rank}_x \psi < n + k\}$ is analytic in Γ. Suppose that $D \neq \Gamma$.

A branch B^* of Γ exists such that $B^* \cap D$ is a thin analytic subset of B^* (and Γ). Then $B = \pi(B^*) \in \mathfrak{B}$. Take $J(U_B,\beta_B,\omega_B)$. Then $U \subseteq B \cap A$. The set $U^* = \pi^{-1}(U) \subseteq B^* \cap A^*$ is open in Γ and B^* and consists of simple points of Γ only. The restriction $\pi_1 = \pi: U^* \to U$ is biholomorphic. Now, $U_0^* = U^* - D$ is open and dense in U^* and $U_0 = \pi(U^*-D)$ is open and dense in U. The map $\pi_0 = \pi: U_0^* \to U_0$ is biholomorphic. If $z \in U_0^*$, then $\operatorname{rank}_z \psi | U_0^* = \operatorname{rank}_z \psi = n + k$.

The map

$$\beta_0 = \beta_B \times \text{Id}: U_{\beta_B} \times \mathbb{P}^k \to U'_{\beta_B} \times \mathbb{P}^k$$

is biholomorphic. Hence $\psi_0 = \beta_0 \circ \psi \circ \pi_0^{-1}: U_0 \to U'_{\beta_B} \times \mathbb{P}^k$ has pure rank $n + k$. Therefore ψ_0 is an open map by Theorem 1.22. If $x \in U_0$, then

$$\psi_0(x) = (g_1(x), \ldots, g_k(x), f_1(x), \ldots, f_k(x)).$$

The set $T = \{x \in U_0 | \omega_B(x) = 0\}$ is the set of non-regular points of ψ_0 by (4.2). By assumption $T = U_0$. Because ψ_0 is open, $\psi_0(T)$ is open in $U'_{\beta_B} \times \mathbb{P}^k$. By Sard's theorem, $\psi_0(T)$ has measure zero in $U'_{\beta_B} \times \mathbb{P}^k$, which is impossible. Therefore $\Gamma = D$; q.e.d.

<u>Lemma 4.7.</u> Let X and Y be complex spaces of pure dimension m and n respectively with $q = m - n \geqq 0$. Let $\varphi: X \to Y$ be a holomorphic map. Let $f_1, \ldots, f_k$ be φ-independent meromorphic functions on X which satisfy the Jacobian test $J(U, \beta, \omega)$. Then $\omega \not\equiv 0$ on each connectivity component (= branch) of U.

<u>Proof.</u> Suppose that a component U_0 of U exists such that $\omega | U_0 \equiv 0$. Then $f_1, \ldots, f_k$ satisfy $J(U_0, \beta, \omega | U_0)$ and U_0 is contained in one and only one branch B of X. Define $\varphi_B = \varphi: B \to Y$. By Lemma 4.6, $f_1 | B, \ldots, f_k | B$ are φ-dependent. By Lemma 4.4, $f_1 | B, \ldots, f_k | B$ are φ-independent. This is a contraction. Hence $\omega | U_0 \not\equiv 0$ for each branch U_0 of U; q.e.d.

Let X and Y be complex spaces of pure dimension m and n respectively with $m - n = q \geqq 0$. Let $\varphi: X \to Y$ be a holomorphic map. Then $J^{\emptyset} = J^{\emptyset}(U,\beta,\omega)$ is <u>a Jacobian test on U for φ</u> if and only if the following conditions are satisfied.

$J_1^{\emptyset}$: The set $U \neq \emptyset$ is open in X and consists of simple points of X only.

$J_2^{\emptyset}$: A schlicht chart $\beta: U_\beta \to U'_\beta \subseteq \mathbb{C}^n$ of Y is given with $\varphi(U) \subseteq U_\beta$ and $\beta \circ \varphi = (g_1,\ldots,g_n)$.

$J_3^{\emptyset}$: Define $\omega = dg_1 \wedge \ldots \wedge dg_n$.

<u>Lemma 4.8.</u> Let X and Y be complex spaces of pure dimension m and n respectively with $m - n = q \geqq 0$. Let $\varphi: X \to Y$ be a holomorphic map. Let S be a thin analytic subset of X. Let B be a branch of X. Let S_Y be the set of non-simple points of Y. Suppose that $\varphi(B) \nsubseteq S_Y$. Then φ satisfies a Jacobian test $J^{\emptyset}(U,\beta,\omega)$ on an open subset U of B - S.

<u>Proof.</u> The set $\varphi^{-1}(S_Y) \cap B = S_Y^*$ is thin and analytic in B. Let S_X be the set of non-simple points of X. Then $B_0 = B - (S \cup S_Y^* \cup S_X)$ is open and dense in B and open in X with $\varphi(B_0) \subseteq Y - S_Y$. Take $a \in B_0$ and define $b = \varphi(a)$. A schlicht chart $\beta: U_\beta \to U'_\beta$ of Y at b exists. An open subset U of B_0 exists such that $\beta(U) \subseteq U_\beta$ and $a \in U$. Then U is open in X and $U \cap S_X = \emptyset$. Define $\beta \circ \varphi = (g_1,\ldots,g_u)$. Define $\omega = dg_1 \wedge \ldots \wedge dg_n$. Then $J^{\emptyset}(U,\beta,\omega)$ is satisfied with $U \subseteq B - S$; q.e.d.

Lemma 4.9. Let X and Y be complex spaces of pure dimension m and n respectively with $m - n \geqq q$. Let $\varphi: X \to Y$ be a holomorphic map. Then φ has strict rank n, if and only if each branch B of X contains an open subset $U \neq \emptyset$, such that φ satisfies a Jacobian test $J^{\emptyset}(U,\beta,\omega)$ for some schlicht chart β of Y with $\omega(x) \neq 0$ for all $x \in U$.

Proof. The proof proceeds as the proof of Lemma 4.2 with $k = 0$ and

4.2	A	A*	Γ	π	ψ	B*	$J(U,\beta,\omega)$	U*	π_1	γ
Here	X	X	X	Id	φ	B–B*	$J^{\emptyset}(U,\beta,\omega)$	U	Id	$\gamma=\beta\circ\varphi$

4.2	ψ_1	B	S^*_Y	S^*_X	π_2	ψ_2
Here	$\varphi_1=\varphi:U \to U_\beta$	B=B*	$S^*_Y=\varphi^{-1}(S_Y)$	S_X	Id	$\varphi_2=\varphi:A' \to Y-S_Y$

4.2	ψ_1
Here	$\varphi_1=\varphi:U \to U_\beta$

q.e.d.

Remark 4.10. The condition $\omega(x) \neq 0$ for all $x \in U$ in Lemma 4.9 can be replaced by the condition $\omega \not\equiv 0$ on U.

Lemma 4.11. Let X and Y be complex spaces of pure dimension m and n respectively with $m - n = q \geqq 0$. Suppose that $\varphi: X \to Y$ is a holomorphic map with $\text{rank}_\varphi < n$ which satisfies $J^{\emptyset}(U,\beta,\omega)$. Then $\omega \equiv 0$ on U.

Proof. It suffices to show that $\omega \equiv 0$ on each component of U. Hence, it can be assumed that U is connected. Then U is contained in a branch B of X. By Lemma 1.17, rank $\varphi|B < n$. If $\omega \not\equiv 0$ on U, then rank $\varphi|B = n$ by Remark 4.10. This is a contradiction. Hence $\omega \equiv 0$ on U; q.e.d.

Lemma 4.12. Let X and Y be complex spaces of pure dimension m and n respectively with $m - n = q \geqq 0$. Let $\varphi: X \to Y$ be a holomorphic map which satisfies a Jacobian test $J^{\emptyset}(U_B, \beta_B, \omega_B)$ on an open subset U_B of B with $\omega_B \equiv 0$ for each branch B of X. Then rank $\varphi < n$.

The proof proceeds as the proof of Lemma 4.6 with

4.6	A	A*	Γ	π	$\mathscr{L}^*$	ψ	k	B*	$J(U_B, \beta_B, \omega_B)$	π_1
Here	X	X	X	Id	$\mathscr{L}$	φ	0	B*=B	$J^{\emptyset}(U_B, \beta_B, \omega_B)$	Id

4.6	U*	U*	π_0	β_0	ψ_0
Here	$U^*=U$	$U_0^*=U_0$	Id	$\beta_0=\beta_B$	$\psi_0=\beta\circ\varphi\|U_0$

q.e.d.

Lemma 4.13. Let X and Y be complex spaces of pure dimension m and n respectively with $q = m - n \geqq 0$. Let φ be a holomorphic map of strict rank n which satisfies a Jacobian test $J^{\emptyset}(U, \beta, \varphi)$. Then $\omega \not\equiv 0$ on each connectivity component of U.

Proof. Suppose that a component U_0 of U exists such that $\omega|U_0 \equiv 0$. Then $U_0 \subseteq B$ for one and only one branch B of X and $\varphi|B$ satisfies $J^{\emptyset}(U_0, \beta, \omega|U_0)$. By Lemma 4.12, rank $\varphi|B < n$. By Lemma 1.16 rank $\varphi|B = n$. This is a contradiction. Hence $\omega|U_0 \not\equiv 0$;

q.e.d.

Lemma 4.14. Let X and Y be complex spaces of pure dimension m and n respectively with $m - n = q \geqq 0$. Let $\varphi: X \to Y$ be a holomorphic map of strict rank n. Let $U \neq \emptyset$ and V be open subsets of X and Y respectively with $\varphi(U) \subseteq V$. Then $\varphi_0 = \varphi|U: U \to V$ has strict rank n. Moreover, if $f_1,\ldots,f_k$ are meromorphic functions on X, then

a) If $f_1,\ldots,f_k$ are φ-independent, then $f_1|U,\ldots,f_k|U$ are φ_0-independent.

b) If $f_1,\ldots,f_k$ are φ-dependent, then $f_1|U,\ldots,f_k|U$ are φ_0-dependent.

c) If $U \cap B \neq \emptyset$ for each branch B of X and if $f_1|U,\ldots,f_k|U$ are φ_0-independent, then $f_1,\ldots,f_k$ are φ-independent.

d) If $U \cap B \neq \emptyset$ for each branch B of X and if $f_1|U,\ldots,f_k|U$ are φ_0-dependent, then $f_1,\ldots,f_k$ are φ-dependent.

Proof. The set $D = \{x \in X | \operatorname{rank}_x \varphi < n\}$ is thin and analytic in X. Then $D \cap U = \{x \in U | \operatorname{rank}_x \varphi_0 < n\}$ is thin and analytic. Hence φ_0 has strict rank n. (Lemma 1.16). Let $\mathfrak{L}$ be the set of branches of U. Define $S = P_{f_1} \cup \ldots \cup P_{f_k}$. By Lemma 4.8 φ_0 satisfies a Jacobian test $J^{\emptyset}(U_C,\beta_C,\omega_C)$ on an open connected subset U_C of $C - S$. Define $\tilde{\omega}_C = \omega_C \wedge df_1 \wedge \cdots \wedge df_k$ on U_C for each $C \in \mathfrak{L}$.

a) Suppose that $f_1,\ldots,f_k$ are φ-independent. By Lemma 4.7, $\tilde{\omega}_C \not\equiv 0$ on U_C. By Lemma 4.3, $f_1|U,\ldots,f_k|U$ are φ_0-independent.

b) Suppose that $f_1,\ldots,f_k$ are φ-dependent. By Lemma 4.5 $\tilde{\omega}_C \equiv 0$ on U_C. By Lemma 4.6, $f_1|U,\ldots,f_k|U$ are φ_0-dependent.

c) Suppose that $U \cap B \neq \emptyset$ for each branch of B. Suppose that $f_1|U,\ldots,f_k|U$ are φ_0-independent. Then $f_1|U \cap B,\ldots,f_k|U \cap B$

are $(\varphi_0|B \cap U)$-independent, by a). By b) $f_1|B,\ldots,f_k|B$ are not $(\varphi|B)$-dependent. Hence $f_1|B,\ldots,f_k|B$ are $(\varphi|B)$-independent, because B is irreducible. This holds for each branch B. Hence $f_1,\ldots,f_k$ are φ-independent by Lemma 4.4.

d) Suppose that $U \cap B \neq \emptyset$ for each branch B of X. Suppose that $f_1|U,\ldots,f_k|U$ are φ_0-dependent. Then $f_1|U \cap B,\ldots,f_k|U \cap B$ are $(\varphi_0|B \cap U)$-dependent. By b), $f_1|B,\ldots,f_k|B$ are not $(\varphi|B)$-independent. Hence, they are $(\varphi|B)$-dependent, because B is irreducible. This holds for each branch B of X. Hence $f_1,\ldots,f_k$ are φ-dependent by Lemma 4.4; q.e.d.

Let X and Y be complex spaces of pure dimension m and n respectively with $m - n = q \geqq 0$. Let $\varphi\colon X \to Y$ be a holomorphic map of strict rank n. Let F be a set of meromorphic functions on X. A meromorphic function $f \in \mathfrak{K}(X)$ is said to be <u>φ-dependent on F</u> if and only if either f is φ-dependent or if finitely many φ-independent functions $f_1,\ldots,f_k$ in F exist such that $f_1,\ldots,f_k,f$ are φ-dependent. Define

$$\mathfrak{K}_\varphi(X;F) = \{f \in \mathfrak{K}(X) \mid f \ \varphi\text{-dependent on } F\}$$
$$\mathfrak{K}_\varphi(X) = \mathfrak{K}_\varphi(X;\emptyset) = \{f \in \mathfrak{K}(X) \mid f \ \varphi\text{-dependent}\}.$$

Obviously $\mathfrak{K}_\varphi(X) \subseteq \mathfrak{K}_\varphi(X,F)$. If $F \subseteq G$, then

$$\mathfrak{K}_\varphi(X;F) \subseteq \mathfrak{K}_\varphi(X;G).$$

By Lemma 4.8 and Lemma 4.9, f, f are φ-dependent if $f \in \mathfrak{K}(X)$, hence $F \subseteq \mathfrak{K}_\varphi(X;F)$.

Lemma 4.15. Let X and Y complex spaces of pure dimension m and n respectively with $m - n = q \geqq 0$. Let $\varphi: X \to Y$ be a holomorphic map of strict rank n. Let $f_1,\dots,f_k$ be φ-independent meromorphic functions on X. Fefine $F = \{f_1,\dots,f_k\}$, then

$$\mathfrak{K}_\varphi(X,F) = \{f \in \mathfrak{K}(X) \mid f_1,\dots,f_k, f \text{ are } \varphi\text{-dependent}\}.$$

Proof. If $f \in \mathfrak{K}(X)$ and if $f_1,\dots,f_k$, f are φ-dependent, then $f \in \mathfrak{K}_\varphi(X,F)$ by definition. Take $f \in \mathfrak{K}_\varphi(X,F)$. Then $f_{\mu_1},\dots,f_{\mu_k}$ with $\mu_i \neq \mu_j$ exists such that $f_{\mu_1},\dots,f_{\mu_k}$ are φ-independent and such that $f_{\mu_1},\dots,f_{\mu_k},f$ are φ-dependent ($k = 0$ allowed!). Let B be a branch of X. By Lemma 4.8, $J^\emptyset(U,\beta,\omega)$ exists with $U \subseteq B - (P_{f_1} \cup \dots \cup P_{f_k} \cup P_f)$. Define

$$\omega_1 = \omega \wedge df_{\mu_1} \wedge \dots \wedge df_{\mu_k}$$

$$\omega_2 = \omega \wedge df_{\mu_1} \wedge \dots \wedge df_{\mu_k} \wedge df = \omega_1 \wedge df$$

$$\omega_3 = \omega \wedge df_1 \wedge \dots \wedge df_k$$

$$\omega_4 = \omega \wedge df_1 \wedge \dots \wedge df_k \wedge df = \omega_3 \wedge df \ .$$

Then $J(U,\beta,\omega_\mu)$ is satisfied for $\mu = 1,\dots,4$.

By Lemma 4.7, $\omega_1 \not\equiv 0$ and $\omega_3 \not\equiv 0$. By Lemma 4.5, $\omega_2 \equiv 0$. Hence, $\omega_4 \equiv 0$ for each B. By Lemma 4.6, $f_1,\dots,f_k,f$ are φ-dependent; q.e.d.

Lemma 4.16. Let X be a connected complex manifold. Let

$g_1,\ldots,g_n$, $h_1,\ldots,h_p$, $f_1,\ldots,f_k$ and f be holomorphic functions on X. Suppose that the differential forms

$$\omega = dg_1 \wedge \ldots \wedge dg_n \wedge dh_1 \wedge \ldots \wedge dh_p$$

$$\chi = dg_1 \wedge \ldots \wedge dg_n \wedge df_1 \wedge \ldots \wedge df_k$$

are not identically zero on X. Suppose that $\omega \wedge df \equiv 0$ and $\chi \wedge dh_\mu \equiv 0$ for $\mu = 1,\ldots,p$. Then $\chi \wedge df \equiv 0$ on X.

Proof. Let T be the holomorphic cotangent bundle. Define $T^q = T \wedge \ldots \wedge T$ (q-times). Then ω is a section in T^{n+p} over X. The sets $S_1 = \{x \in X | \omega(x) = 0\}$ and $S_2 = \{x \in X | \chi(x) = 0\}$ and $S_1 \cup S_2 = S$ are thin and analytic. The set $A = X - S$ is open and dense in X. Then $\gamma_\mu = dg_\mu$ for $\mu = 1,\ldots,p$ and $\varphi_\mu = df_\mu$ for $\mu = 1,\ldots,k$ and $\nu_\mu = dh_\mu$ for $\mu = 1,\ldots,p$ and $\varphi = df$ are sections in T over X. Take $x \in A$. Let T_x be the fiber X over x. Now, $\gamma_1(x),\ldots,\gamma_n(x)$, $\nu_1(x),\ldots,\nu_p(x)$ is a base of a vector space $D_x \subseteq T_x$ over $\mathbb{C}$, because $\omega(x) \neq 0$. Now, $\omega(x) \wedge \varphi(x) = 0$ implies $\varphi(x) \in D_x$. The vectors $\gamma_1(x),\ldots,\gamma_n(x)$, $\varphi_1(x),\ldots,\varphi_k(x)$ are a base of a vector space $E_x \subseteq T_x$, because $\chi(x) \neq 0$. Now, $\chi(x) \wedge \gamma_\mu(x) = 0$ for $\mu = 1,\ldots,n$ and $\chi(x) \wedge \nu_\mu(x) = 0$ for $\mu = 1,\ldots,p$ implies $D_x \subseteq E_x$. Hence, $\varphi(x) \in E_x$ which means $\chi(x) \wedge \varphi(x) = 0$. Therefore, $\chi \wedge df = 0$ on A. Hence, $\chi \wedge df = 0$ on $X = \overline{A}$;

q.e.d.

Lemma 4.17. Let X and Y be complex spaces of pure dimension m and n respectively with $q = n - n \geqq 0$. Let $\varphi: X \to Y$ be a holomorphic map of strict rank n. Let $f_1,\ldots,f_k$ be φ-independent meromorphic functions on X. Let $h_1,\ldots,h_p$ be φ-independent meromorphic functions on X such that $f_1,\ldots,f_k,h_\mu$ are φ-dependent for $\mu = 1,\ldots,p$. Let f be a meromorphic function on X such that $h_1,\ldots,h_p,f$ are φ-dependent. Then $f_1,\ldots,f_k,f$ are φ-dependent.

Proof. Let $\{B_\lambda\}_{\lambda\in\Lambda}$ be the family of branches of X. Define

$$S = \left(\bigcup_{\mu=1}^{k} P_{f_\mu}\right) \cup \left(\bigcup_{\mu=1}^{p} P_{h_\mu}\right) \cup P_f$$

Then S is a thin analytic subset of X. By Lemma 4.8 select $J^\emptyset(U_\lambda,\beta_\lambda,\omega_\lambda)$ with $U_\lambda \subseteq B_\lambda - S$ for each λ. Here U_λ can be taken as a connected open subset of $B_\lambda - S$. Define

$$\tilde{\omega}_\lambda = \omega_\lambda \wedge dh_1 \wedge \cdots \wedge dh_p$$

$$\chi_\lambda = \omega_\lambda \wedge df_1 \wedge \cdots \wedge df_k$$

on U_λ. Then $J(U_\lambda,\beta_\lambda,\tilde{\omega}_\lambda)$ and $J(U_\lambda,\beta_\lambda,\chi_\lambda)$ are satisfied for each $\lambda\in\Lambda$. By Lemma 4.7, $\tilde{\omega}_\lambda \not\equiv 0$ and $\chi_\lambda \not\equiv 0$. By Lemma 4.5, $\tilde{\omega}_\lambda \wedge df \equiv 0$ and $\chi_\lambda \wedge dh_\mu \equiv 0$ for $\mu = 1,\ldots,p$ on U_λ. By Lemma 4.16, $\chi_\lambda \wedge df \equiv 0$ for each $\lambda\in\Lambda$. By Lemma 4.6, $f_1,\ldots,f_k,f$ are φ-dependent; q.e.d.

Lemma 4.18. Let X be an irreducible complex space of dimension m. Let Y be a complex space of pure dimension on n with $q = m - n \geqq 0$. Let $\varphi: X \to Y$ be a holomorphic map of rank n. Take $F \subseteq \mathfrak{K}(X)$. Suppose that $\mathfrak{K}_\varphi(X,F) \neq \mathfrak{K}_\varphi(X)$. Then φ-independent meromorphic functions $f_1,\ldots,f_k$ exist in F with $\mathfrak{K}_\varphi(X;F) = \mathfrak{K}_\varphi(X;F')$

where $F' = \{f_1,\ldots,f_k\}$.

<u>Proof.</u> Define

$$M = \{r \in \mathbb{N} \mid f_1,\ldots,f_r \ \varphi\text{-independent and in } F\}.$$

Take $f^0 \in \mathfrak{K}_\varphi(X,F) - \tilde{\mathfrak{K}}_\varphi(X)$, then f^0 is not φ-dependent. Hence $f_1^0,\ldots,f_r^0$ in F exist such that $f_1^0,\ldots,f_r^0$ are φ-independent and $f_1^0,\ldots,f_r^0,f^0$ are φ-dependent. Hence $r \in M$ and $M \neq \emptyset$. By Lemma 4.1, q is an upper bound of M. Hence, $k = \operatorname{Max} M$ exists. There are φ-independent meromorphic functions $f_1,\ldots,f_k$ in F. If $g \in F$, then $f_1,\ldots,f_k,g$ are not φ-independent, hence φ-dependent, because X is irreducible.

Define $F' = \{f_1,\ldots,f_k\}$. By Lemma 4.15, $\tilde{\mathfrak{K}}_\varphi(X,F') \subseteq \tilde{\mathfrak{K}}_\varphi(X,F)$. Take $f \in \tilde{\mathfrak{K}}_\varphi(X,F)$. If f is φ-dependent, then $f \in \tilde{\mathfrak{K}}_\varphi(X,F')$. Assume that f is not φ-dependent. Then f is φ-independent because X is irreducible. Now, φ-independent functions $h_1,\ldots,h_p$ in F exists such that $h_1,\ldots,h_p,f$ are φ-dependent on X. Moreover, $f_1,\ldots,f_k,h_\mu$ are φ-dependent on X for $\mu = 1,\ldots,p$. By Lemma 4.17, $f_1,\ldots,f_k,f$ are φ-dependent. Hence $f \in \tilde{\mathfrak{K}}_\varphi(X,F')$. Therefore, $\tilde{\mathfrak{K}}_\varphi(X,F) \subseteq \tilde{\mathfrak{K}}_\varphi(X,F')$ which implies $\tilde{\mathfrak{K}}_\varphi(X,F) = \tilde{\mathfrak{K}}_\varphi(X,F')$; q.e.d.

Of course, it is possible to take $F = \tilde{\mathfrak{K}}(X)$. Hence, either $\tilde{\mathfrak{K}}(X) = \tilde{\mathfrak{K}}_\varphi(X)$ or

$$\tilde{\mathfrak{K}}(X) = \tilde{\mathfrak{K}}_\varphi(X,F')$$

where $F' = \{f_1,\ldots,f_k\}$ and where $f_1,\ldots,f_k$ are φ-independent. Here $k \leqq q$ by Lemma 4.1. Recall, that X irreducible was assumed.

Let M be a set. Let $\mathfrak{P}(M)$ be the set of subsets of M. A map $s: \mathfrak{P}(M) \to \mathfrak{P}(M)$ is said to be a <u>dependence relation</u> (Zariski-Samuel [27], p. 97) if and only if

1. If $F \subseteq M$, then $F \subseteq s(F)$.
2. If $F \subseteq G \subseteq M$, then $s(F) \subseteq s(G)$.
3. If $F \subseteq M$ and $a \in s(F)$, then $a \in s(F')$ for some finite subset F' of F.
4. If $F \subseteq M$, then $s(s(F)) = s(F)$.
5. If $F \subseteq M$, if $a \in s(F \cup \{b\}) - s(F)$, then $b \in s(F \cup \{a\})$.

Say $a \in M$ depends on $F \subseteq M$, if and only if $a \in s(F)$. Then these axioms can be expressed in the language of van der Waerden [26], p. 204:

1. If $a \in M$, then a depends on $\{a\}$.
2. If $a \in M$, if a depends on F, if $F \subseteq G$, then a depends on G.
3. If $a \in M$, if a depends on F, then a depends on a finite subset F' of F.
4. If $a \in M$, if a depends on F, if each element of F depends on G, then a depends on G.
5. If $a \in M$, if a depends on $F \cup \{b\}$, but not on F, then b depends on $F \cup \{a\}$.

Take $M = \mathfrak{K}(X)$ and define s by $s(F) = \mathfrak{K}_\varphi(X,F)$ if $F \subseteq \mathfrak{K}(X)$. Then it will be shown that φ-dependence is a dependence in the sense of these 5 axioms.

Theorem 4.19. Let X be an irreducible complex space of dimension m. Let Y be a pure n-dimensional complex space with $q = m - n \geqq 0$. Let $\varphi: X \to Y$ be a holomorphic map of rank n. Define $s: \mathfrak{P}(\mathfrak{K}(X)) \to \mathfrak{P}(\mathfrak{K}(X))$ by $s(F) = \mathfrak{K}_\varphi(X,F)$ if $F \subseteq \mathfrak{K}(X)$. Then s defines a dependence relation on $\mathfrak{K}(X)$.

Proof. Consider the Zariski-Samuel version of the axioms.

Now, $F \subseteq \mathfrak{K}_\varphi(X,F)$ is already proved and $\mathfrak{K}_\varphi(X,F) \subseteq \mathfrak{K}_\varphi(X,G)$ if $F \subseteq G$ is trivial. Hence axiom 1 and 2 holds.

3. If $f \in \mathfrak{K}_\varphi(X,F)$, then either $f \in \mathfrak{K}_\varphi(X) = \mathfrak{K}_\varphi(X,\emptyset)$ or $f_1,\ldots,f_k$ in F exist such that $f_1,\ldots,f_k,f$ are φ-dependent. Define $F' = \{f_1,\ldots,f_k\}$ then $f \in \mathfrak{K}_\varphi(X,F')$.

4. Take $F \subseteq M$. Define $G = \mathfrak{K}_\varphi(X,F)$. Then $\mathfrak{K}_\varphi(X,F) \subseteq \mathfrak{K}_\varphi(X,G)$. If $\mathfrak{K}_\varphi(X,F) = \mathfrak{K}_\varphi(X)$, then all elements of $\mathfrak{K}_\varphi(X,F)$ are φ-dependent. Hence, all elements of $\mathfrak{K}_\varphi(X,G)$ are φ-dependent which implies $\mathfrak{K}_\varphi(X;G) = \mathfrak{K}_\varphi(X) - \mathfrak{K}_\varphi(X;F)$.

Suppose that $\mathfrak{K}_\varphi(X;F) \neq \mathfrak{K}_\varphi(X)$. By Lemma 4.18, φ-independent function $f_1,\ldots,f_k$ exist such that $\mathfrak{K}_\varphi(X;F) = \mathfrak{K}_\varphi(X;F')$ with $F' = \{f_1,\ldots,f_k\}$. Take $f \in \mathfrak{K}_\varphi(X;G)$. If f is φ-dependent, then $f \in \mathfrak{K}_\varphi(X;F)$. Assume that f is not φ-dependent on X. Then φ-independent function $h_1,\ldots,h_p$ in G exist such that $h_1,\ldots,h_p,f$ are φ-dependent. Now $h_\mu \in \mathfrak{K}_\varphi(X;F) = \mathfrak{K}_\varphi(X,F')$ implies that $f_1,\ldots,f_k,h_\mu$ are φ-dependent for $\mu = 1,\ldots,p$. By Lemma 4.17, $f_1,\ldots,f_p,f$ are φ-dependent. Hence, $f \in \mathfrak{K}_\varphi(X;F)$. Therefore, $\mathfrak{K}_\varphi(X,F) = \mathfrak{K}_\varphi(X;G)$.

5. Take $F \subseteq M$ and $h \in \mathfrak{K}(X)$. Suppose that

$$f \in \mathfrak{K}_\varphi(X;F \cup \{h\}) - \mathfrak{K}_\varphi(X;F)$$

Then f is not φ-dependent. Finitely many φ-independent functions $f_1,\dots,f_k$ in $F \cup \{h\}$ exist such that $f_1,\dots,f_k,f$ are φ-dependent. Because $f \notin \mathfrak{K}_\varphi(X;F)$, one of these functions $f_1,\dots,f_k$ is h. Without loss of generality $f_k = h$ can be assumed. Then $f_\mu \in F$ for $\mu = 1,\dots,k-1$ and $f_1,\dots,f_{k-1}$ are φ-independent (Lemma 4.2). Also $f_1,\dots,f_k,f$ are φ-independent because $f \notin \mathfrak{K}_\varphi(X,F)$. But $f_1,\dots,f_{k-1},f,h$ are φ-dependent. Hence $h \in \mathfrak{K}_\varphi(X,F \cup \{f\})$; q.e.d.

The results of Zariski-Samuel [27], p. 50-52 hold. The subset F of $\mathfrak{K}(X)$ is said to be <u>free</u>, if $f \in F$ implies $f \notin \mathfrak{K}_\varphi(X,F-\{f\})$. If $F = \{f_1,\dots,f_k\}$ is finite with $k = \#F$, then F is free if and only if $f_1,\dots,f_k$ are φ-independent. Observe, that a dependence relation $s: \mathfrak{P}(M) \to \mathfrak{P}(M)$ induces a dependence relation $s: \mathfrak{P}(s(F)) \to \mathfrak{P}(s(F))$ for every $F \subseteq M$. Hence, Zariski-Samuel [26], p. 50-52 implies also: Call a subset G of $\mathfrak{K}_\varphi(X,F)$ to be a set of generators if $\mathfrak{K}_\varphi(X,G) \supseteq \mathfrak{K}_\varphi(X,F)$. Obviously

$$\mathfrak{K}_\varphi(X,G) \subseteq \mathfrak{K}_\varphi(X,\mathfrak{K}_\varphi(X,F)) = \mathfrak{K}_\varphi(X,F).$$

Hence, a subset G of $\mathfrak{K}_\varphi(X,F)$ is a set of generators of $\mathfrak{K}_\varphi(X,F)$ if and only if $\mathfrak{K}_\varphi(X,G) = \mathfrak{K}_\varphi(X,F)$. A free subset G of generators of $\mathfrak{K}_\varphi(X,F)$ is said to be a <u>base</u> of $\mathfrak{K}_\varphi(X,F)$. A subset G of $\mathfrak{K}_\varphi(X,F)$ is a base of $\mathfrak{K}_\varphi(X;F)$ if and only if G is a minimal set of generators of $\mathfrak{K}_\varphi(X;F)$, which is the case, if and only if G is a maximal free subset of $\mathfrak{K}_\varphi(X;F)$. A base G of $\mathfrak{K}_\varphi(X,F)$ with $G \subseteq F$ exists. Each base G of $\mathfrak{K}_\varphi(X,F)$ is finite and has the same number of elements k. By Lemma 4.1, $0 < k \leqq q$. Define

$$k = \varphi\text{-}\dim \mathfrak{K}_\varphi(X,F) = \varphi\text{-}\dim F$$

as the <u>φ-dimension of</u> $\mathfrak{K}_\varphi(X,F)$ or F. If L is a free subset of $\mathfrak{K}_\varphi(X,F)$, if S is a finite set of generators of $\mathfrak{K}_\varphi(X;F)$, a subset S' of S exists such that $L \cap S' = \emptyset$ and $L \cup S'$ is a base of $\mathfrak{K}_\varphi(X;F)$.

Obviously, $\mathfrak{K}_\varphi(X, \mathfrak{K}(X)) = \mathfrak{K}(X)$. Hence, the previous statements about bases and generators hold for $\mathfrak{K}(X)$. Define

$$\dim \varphi = \varphi\text{-}\dim \mathfrak{K}(X).$$

as the <u>analytic dimension</u> of φ. Obviously, $0 \leqq \dim \varphi \leqq m - n$. A base $F = \{f_1,\ldots,f_k\}$ of $\mathfrak{K}(X)$ exists with $k = \dim \varphi$.

<u>Proposition 4.20.</u> Let X and Y be complex spaces of pure dimension m and n respectively with $m - n = q \geqq 0$. Let $\varphi\colon X \to Y$ be a holomorphic map of strict rank n. Then $\mathfrak{K}_\varphi(X)$ is a subring of $\mathfrak{K}(X)$. If $f_1,\ldots,f_k$ are φ-independent meromorphic functions and if $F = \{f_1,\ldots,f_k\}$, then $\mathfrak{K}_\varphi(X,F)$ is a subring of $\mathfrak{K}(X)$. If X is irreducible and $F \subseteq \mathfrak{K}(X)$, then $\mathfrak{K}_\varphi(X,F)$ is a subfield of $\mathfrak{K}(X)$.

<u>Proof.</u> Let $\mathfrak{L}$ be the set of branches of X. Let S be a thin analytic subset of S. For each branch $B \in \mathfrak{L}$, and open connected subset U_B of $B - S$ and a schlicht chart β_B of Y exists such that $J^\emptyset(U_B,\beta_B,\omega_B)$ holds. In the three cases, S will be chosen accordingly:

<u>1. Case:</u> $\mathfrak{K}_\varphi(X)$. Take $g \in \mathfrak{K}_\varphi(X)$ and $h \in \mathfrak{K}_\varphi(X)$. Take $S = P_g \cup P_h$ and constructed $J^\emptyset(U_\beta,\beta_B,\omega_B)$ as above. By Lemma 4.5, $\omega_B \wedge dg = 0$ and $\omega_B \wedge dh = 0$ on U_B. Hence $\omega_B \wedge d(g-h) = 0$ and

$\omega_B \wedge d(g\cdot h) = 0$ on U_B. By Lemma 4.6 $g - h \in \mathfrak{K}_\varphi(X)$ and $g\cdot h \in \mathfrak{K}_\varphi(X)$. Hence $\mathfrak{K}_\varphi(X)$ is a subring.

2. Case: $F = \{f_1,\ldots,f_k)$ where $f_1,\ldots,f_k$ are φ-independent. Take $g \in \mathfrak{K}_\varphi(X,F)$ and $h \in \mathfrak{K}_\varphi(X,F)$. Define

$$S = P_g \cup P_h \cup P_{f_1} \cup \ldots \cup P_{f_p}.$$

Take $J^\emptyset(U_\beta,\beta_B,\omega_B)$ as above. Define

$$\tilde{\omega}_B = \omega_B \wedge df_1 \wedge \ldots \wedge df_k$$

on U_B. By Lemma 4.7, $\tilde{\omega}_B \not\equiv 0$. By Lemma 4.15 and by Lemma 4.5, $\tilde{\omega}_B \wedge dg \equiv 0$ and $\tilde{\omega}_B \wedge dh \equiv 0$ on U_B. Hence $\tilde{\omega}_B \wedge d(g-h) \equiv 0$ and $\tilde{\omega}_B \wedge d(g\cdot h) \equiv 0$ on U_B. By Lemma 4.6 and by Lemma 4.15, $g - h \in \mathfrak{K}_\varphi(X,F)$ and $g\cdot h \in \mathfrak{K}_\varphi(X,F)$. Hence, $\mathfrak{K}_\varphi(X,F)$ is a subring.

3. Case: X is irreducible and $F \subseteq \mathfrak{K}(X)$. By Lemma 4.18, φ-independent functions $f_1,\ldots,f_k$ exist such that $\mathfrak{K}_\varphi(X,F) = \mathfrak{K}_\varphi(X,F')$ if $F' = \{f_1,\ldots,f_k\}$ provided $\mathfrak{K}_\varphi(X,F) \not\equiv \mathfrak{K}_\varphi(X)$. In any case $\mathfrak{K}_\varphi(X,F)$ is a subring by 2 or 1. Take $g \in \mathfrak{K}_\varphi(X,F)$. Define $S = P_g \cup P_{f_1} \cup \ldots \cup P_{f_k}$. Take $U \subseteq X - S$ such that $J^\emptyset(U,\beta,\omega)$ exists. Define $\tilde{\omega} = \omega \wedge df_1 \wedge \ldots \wedge df_k$. By Lemma 4.7, $\tilde{\omega} \not\equiv \emptyset$. By Lemma 4.15 and Lemma 4.5, $\tilde{\omega} \wedge dg \equiv 0$. Assume that $g \not\equiv 0$. Then $\tilde{\omega} \wedge d\frac{1}{g} \equiv 0$. By Lemma 4.6 and by Lemma 4.15, $\frac{1}{g} \in \mathfrak{K}_\varphi(X,F)$. Hence, $\mathfrak{K}_\varphi(X,F)$ is a field. Of course, also $\mathfrak{K}_\varphi(X)$ is a field (k=0);

q.e.d.

Lemma 4.21. Let X and Y be complex spaces of pure dimension m and n respectively with $m - n = q \geqq 0$. Let $\varphi: X \to Y$ be a holomorphic map of strict rank n. Let $\{B_\lambda\}_{\lambda\in\Lambda}$ be the family of branches of X with $B_\lambda \neq B_\mu$ for $\lambda \neq \mu$. The inclusion map $j_\lambda: B_\lambda \to X$ induces a surjective homomorphism $j_\lambda^*: \mathfrak{K}(X) \to \mathfrak{K}(B_\lambda)$ with $j_\lambda^*(f) = B_\lambda | f$ if $f \in \mathfrak{K}(X)$. A map $\rho_\lambda: \mathfrak{K}(B_\lambda) \to \mathfrak{K}(X)$ is defined by $\rho_\lambda(f)(x) = 0$ if $x \in X - B_\lambda$ and $\rho_\lambda(f)|B_\lambda = f$, where $f \in \mathfrak{K}(X)$. Then

$$(4.4) \qquad \mathfrak{K}(X) = \prod_{\lambda\in\Lambda} \mathfrak{K}(B_\lambda)$$

with the maps j_λ^* as the projections and the maps ρ_λ as the injections. Define $\varphi_\lambda = \varphi \circ j_\lambda = \varphi | B_\lambda: B_\lambda \to Y$. Then $j_\lambda^*(\mathfrak{K}_\varphi(X)) = \mathfrak{K}_{\varphi_\lambda}(B_\lambda)$ and

$$(4.5) \qquad \mathfrak{K}_\varphi(X) = \prod_{\lambda\in\Lambda} \mathfrak{K}_{\varphi_\lambda}(B_\lambda)$$

is a subproduct. If $F \subseteq \mathfrak{K}(X)$, define $F_\lambda = j_\lambda^*(F)$. Then $j_\lambda^*(\mathfrak{K}_\varphi(X;F)) \subseteq \mathfrak{K}_{\varphi_\lambda}(B_\lambda, F_\lambda)$. If $F = \{f_1, \ldots, f_k\}$ is finite, and if $f_1, \ldots, f_k$ are φ-independent, then

$$(4.6) \qquad \mathfrak{K}_\varphi(X,F) = \prod_{\lambda\in\Lambda} \mathfrak{K}_{\varphi_\lambda}(B_\lambda, F_\lambda)$$

is a subproduct of (4.4).

Proof. By Lemma 3.4, $j_\lambda^*: \mathfrak{K}(X) \to \mathfrak{K}(B_\lambda)$ is well defined and a homomorphism. By Lemma 3.6, $\rho_\lambda: \mathfrak{K}(B_\lambda) \to \mathfrak{K}(X)$ is well-defined. Obviously, $j_\lambda^* \circ \rho_\lambda$ is the identity on $\mathfrak{K}(B_\lambda)$. Hence,

ρ_λ is injective and j_λ^* is surjective. If $f_\lambda \in \mathfrak{K}(B_\lambda)$ for each $\lambda \in \Lambda$ is given, one and only one $f \in \mathfrak{K}(X)$ exists with $j_\lambda^*(f) = f_\lambda$ for all $\lambda \in \Lambda$ (Lemma 3.6). Hence (4.4) holds. By Lemma 4.4, $j_\lambda^*(f) \in \mathfrak{K}_{\varphi_\lambda}(B_\lambda)$ if $f \in \mathfrak{K}_\varphi(X)$ and $\rho_\lambda(f) \in \mathfrak{K}_\varphi(X)$ if $f \in \mathfrak{K}_{\varphi_\lambda}(B_\lambda)$. If $f_\lambda \in \mathfrak{K}_\varphi(B_\lambda)$, then $j_\lambda^*(f) = f_\lambda$ for all $\lambda \in \Lambda$ for one and only one $f \in \mathfrak{K}(X)$; by Lemma 4.4, $f \in \mathfrak{K}_\varphi(X)$. Hence, (4.5) is a subproduct of (4.4).

If $F \subseteq \mathfrak{K}(X)$, then $j_\lambda^*(\mathfrak{K}_\varphi(X,F)) \subseteq \mathfrak{K}_{\varphi_\lambda}(B_\lambda,F_\lambda)$ by Lemma 4.4. Assume that $F = \{f_1,\ldots,f_k\}$ where $f_1,\ldots,f_k$ are φ-independent. Then $\mathfrak{K}_\varphi(X,F)$ is a subring of $\mathfrak{K}(X)$ and $\mathfrak{K}_{\varphi_\lambda}(B_\lambda,F_\lambda)$ is a subfield of $\mathfrak{K}(B_\lambda)$. If $f \in \mathfrak{K}_{\varphi_\lambda}(B_\lambda,F_\lambda)$ then $\rho_\lambda(f) \in \mathfrak{K}_\varphi(X;F)$ by Lemma 4.4. Also, if $g_\lambda \in \mathfrak{K}_{\varphi_\lambda}(B_\lambda,F_\lambda)$ for each $\lambda \in \Lambda$, then $j_\lambda^*(f_1),\ldots,j_\lambda^*(f_k),g_\lambda$ are φ_λ-dependent for each $\lambda \in \Lambda$. Now, $g \in \mathfrak{K}(X)$ exists with $j_\lambda^*(y) = g_\lambda$ for all λ. By Lemma 4.4, $f_1,\ldots,f_k g$ are φ-dependent. Hence, $g \in \mathfrak{K}_\varphi(X,F)$. Hence, (4.6) is a subproduct of (4.5); q.e.d.

These results explain the concept of (analytic) φ-dependence. Also a concept of algebraic dependence of elements of $\mathfrak{K}(X)$ over $\varphi^*\mathfrak{K}(Y)$ exists. Since this concept is well-known, only a few remarks shall be made.

Let R be a ring. Then $R[\xi_1,\ldots,\xi_k]$ denotes the polynomial ring in the indeterminants $\xi_1,\ldots,\xi_k$ over R. If $P \in R[\xi_1,\ldots,\xi_k]$ has the form

$$P = \sum_{\mu_1+\ldots+\mu_k=r} a_{\mu,\ldots,\mu_k}\xi_1^{\mu_1},\ldots,\xi_k^{\mu_k}$$

then P is a homogeneous polynomial of degree r. This representation is unique, and $P = 0$ if and only if $a_{\mu_1,\dots,\mu_k} = 0$ for all indices $\mu_1,\dots,\mu_k$ with $\mu_1+\dots+\mu_k = r$. The homogeneous polynomials of degree r form an R-module. If $0 \neq P \in R[\xi_1,\dots,\xi_k]$, then $P = P_0 + \dots + P_r$ where P_ρ is a homogeneous polynomial of degree ρ for $\rho = 0,1,\dots,r$ and where, $P_r \neq 0$. This representation is unique and $r = \deg P$ is called the degree of P over R. Observe that the 0 polynomial is a homogeneous polynomial of any degree, but that deg 0 is not defined.

Let $R(\xi_1,\dots,\xi_k)$ be the total quotient ring of $R[\xi_1,\dots,\xi_k]$. If R is a field, also $R(\xi_1,\dots,\xi_k)$ is a field. Moreover, $\deg PQ = \deg P + \deg Q$, if $P \neq 0$ and $Q \neq 0$ belong to the integral domain $R[\xi_1,\dots,\xi_k]$.

If R is a subring of a ring S, and if $f_1,\dots,f_k$ belong to S, then $R[f_1,\dots,f_k]$ denotes the ring of polynomials in $f_1,\dots,f_k$ with coefficients in R. A substitution homomorphism $\sigma: R \to S$ is is uniquely defined by the requirements that $\sigma|R: R \to R$ is the identity and that $\sigma(\xi_\mu) = f_\mu$ for $\mu = 1,\dots,k$. The ring $R[f_1\dots,f_k]$ is the image of σ. If $P \in R[\xi_1,\dots,\xi_k]$ write $P(f_1,\dots,f_k) = \sigma(P)$. If and only if σ has a non-trivial kernel, $f_1,\dots,f_k$ are algebraically dependent over R. If so, the degree of $(f_1,\dots,f_k)$ over R is uniquely defined by

$$[(f_1,\dots,f_k): R] = \mathrm{Min}\{\deg P \mid 0 \neq P \in \ker \sigma\}.$$

This degree definition may be unusual, but will be helpful later. Observe, that the degree depends on $f_1,\dots,f_k$ and R and is not defined for $R[f_1,\ldots,f_k]$ over R.

The ring of quotients of $R[f_1,\ldots,f_k]$ is denoted by $R(f_1,\ldots,f_k)$ and is contained in the ring of quotients of S. If $f_1,\ldots,f_k$ are not algebraically dependent over R, then $f_1,\ldots,f_k$ are <u>said to be algebraically independent over R</u>, this is the case if and only if σ is a monomorphism. If R is a field, then $R[f_1,\ldots,f_k]$ is an integral domain and $R(f_1,\ldots,f_k)$ is a field, which is a pure transcendental extension of transcendence degree k over R, if $f_1,\ldots,f_k$ are algebraically independent over R.

If $k = 1$, and if $f = f_1$ is algebraically dependent over the ring R, then f is said to be <u>algebraic over R of degree</u> $[f:R] = [(f_1):R]$. Moreover, f is said to be integral over R, if $0 = P \in R[\xi]$ with $P = \sum_{\mu=0}^{n-1} a_\mu \xi^\mu + \xi^n$ exists such that $\sigma(P) = P(f) = 0$. Obviously, $[f:R] \leqq n$. If R is a field, equality can be obtained. If S is also a field, then P is unique if $n = [f:R]$. If S is a ring, and if every element of S is algebraic over R, then S is said to be <u>algebraic over R</u> (or an algebraic extension of R).

Let L be a subfield of the field K. Then K is a vector space over L whose dimension over L is denoted by $[K:L]$. If $[K:L] < \infty$, then K is algebraic over L, and K is called a finite algebraic extension of L. If $L \subseteq M \subseteq K$ and if M is a subfield of K, then

$$[K:M][M:L] = [K:L].$$

If $f \in K$, then $L(f)$ is a subfield of K. If f is algebraic over L, then $[L(f):L] = [f:L] < \infty$ and $[f:L]$ depends really only on $L(f)$ and L in this particular case. If f is not algebraic over L, then $[L(f):L] = \infty$.

Most textbooks omit the following well-known Lemma, which is easily proved using the existence of a primitive element:

Lemma 4.22. Let L be a subfield of the field K. Let K be algebraic over L. Suppose that L (and K) has characteristic zero. Suppose a positive integer $q \in \mathbb{N}$ exists such that $[f:L] \leq q$ for each $f \in K$. Then K is a finite algebraic extension of L with $[K:L] \leq q$.

Let L be a subfield of the field K. If K is not algebraic over L, then K is called transcendental over L. If $F \subseteq K$, then L(F) is the field generated by F, i.e., the intersection of all subfields of K containing $L \cup F$. If $f \in K$ is algebraic over L(F), then f is said to be algebraically dependent on L(F). This concept defines a dependence relation on K. A free set F in this relation is said to be a transcendence set over L. A set G of generators in this relation is a set G such that K is algebraic over L(G). A base F of K is a transcendental set of generators over L, i.e., F is a transcencence set and K is algebraic over L(F). Then $\#F = tr(K:L)$ is independent of the base F of K and is called the transcendence degree of K over L. If $tr(K:L) < \infty$, then K is a finite transcendental extension of L and K is algebraic over L(F). If F is a transcendental base of K over L with $K = L(F)$, then K is called a pure transcendental extension of L. If F is a finite transcendental base of K over L and if $[K:L(F)] < \infty$, then K is called an algebraic function field over L.

Now, it shall be shown, that a subfield K' of K which contains L is again an algebraic function field over L. Some preparations have to be made. Let R be an integral domain with unique prime factorization. Then the polynomial ring $R[x_1,\ldots,x_n]$ in n indeterminants is again a unique factorization ring. If

$0 \neq A \in R[x_1,\ldots,x_n]$, then the degree of A is well defined with $\deg(A \cdot B) = \deg A + \deg B$ if $A \in R[x_1,\ldots,x_n]$ and $B \in R[x_2,\ldots,x_n]$. If $x_1,\ldots,x_n$, $y_1,\ldots,y_q$ are indeterminants over R, then $A \in R[x_1,\ldots,x_n,y_1,\ldots,y_q]$ can be considered as a polynomial in $y_1,\ldots,y_q$ over $R[x_1,\ldots,x_n] = S$, hence, $A \in S[y_1,\ldots,y_q]$ and as such has a degree denoted by $\deg_S A$. Especially, if A and B belong to $R[x_1,\ldots,x_n,y_1,\ldots,y_q]$ and if $C \in R[x_1,\ldots,x_n]$ such that $C = A \cdot B$ then A and B belong to $R[x_1,\ldots,x_n]$, because

$$0 = \deg_S C = \deg_S A + \deg_S B.$$

Hence, $\deg_S A = \deg_S B = 0$, which implies $A \in S$ and $B \in S$.

If $0 \neq P(x) = \sum_{\mu=0}^{m} a_\mu x^\mu$ is a polynomial in R[x], then P is called primitive, if and only if the coefficients $a_0,\ldots,a_m$ are coprime, i.e., they have 1 as greatest common divisor. The product of primitive polynomials in R[s] is primitive. Let Q be the field of quotients of R. If $0 \neq A \in Q[x]$ is a polynomial, a primitive polynomial $B \in R[x]$ and an element $C \in Q$ exists such that $A = CB$. Then C and B are uniquely defined up to a factor which is a unit in R. Especially, if $A \in R[x]$ and $B \in R[x]$ are primitive and if $C \in Q$ such that $A = CB$ then C is a unit in R.

Let $x_1,\ldots,x_n,y_1,\ldots,y_q,x$ be indeterminants over R. Define $S = R[x_1,\ldots,x_n]$ and $T = R[x_1,\ldots,x_n,y_1,\ldots,y_p] = S[y_1,\ldots,y_p]$. Let $A \in S[x]$ be primitive. Then $A \in T[x]$ is also primitive. For, consider $A = \sum_{\mu=0}^{m} A_\mu x^\mu$ with $A_\mu \in S$. Suppose that $B \in T$ is a greatest common divisor of $A_0,\ldots,A_m$ in T. Then $A_\mu = B \cdot C_\mu$ with $C_\mu \in T$ for $\mu = 0,\ldots,m$. Hence $B \in S$ with $C_\mu \in S$. Therefore B is a common divisor of $A_0,\ldots,A_m$ in S. Hence B is a unit in S. Therefore B is a unit in R, and certainly in T. The polynomial A is primitive in $T[x]$.

Proposition 4.23. Let L be a subfield of the field K. Let $x_1,\ldots,x_n\ y_1,\ldots,y_q$ be elements of K which are algebraically independent over L. Define $S = L[x_1,\ldots,x_n]$ and $\hat{S} = L(x_1,\ldots,x_n)$. Define $T = L[x_1,\ldots,x_u,y_1,\ldots,y_q] = S[y_1,\ldots,y_q]$ and $\hat{T} = L(x_1,\ldots,x_n,y_1,\ldots,y_q) = \hat{S}(y_1,\ldots,y_q)$. Suppose that K is a finite algebraic extension of $\hat{T}$ with $[K:\hat{T}] = r$. Let $f \in K$ be an element which is algebraic over $\hat{S}$. Then $[f:\hat{S}] \leqq r$.

Proof. Observe that $\hat{S}$ and $\hat{T}$ are pure transcendental extensions of degree n and $n + q$ of L respectively. Observe that S and T can be considered as polynomial rings of n respectively $n + q$ variables over L and that $\hat{S}$ and $\hat{T}$ are the fields of quotients of S and T respectively. Moreover, S and T are integral domains with unique prime factorization because L is a field.

Let x be an indeterminant over $\hat{T}$. A polynomial $0 \neq \hat{A} \in \hat{S}[x]$ of minimal degree $m > 0$ exists such that $\hat{A}(f) = 0$, because f is algebraic over S. Then $\hat{A} = a\,A$ where A is primitive in $S[x]$ and where $a \in \hat{S}$. Then $a \neq 0$ and $a\,A(f) = 0$ in K. Hence $A(f) = 0$. The degree of A is m. Because $f \in K$ is algebraic over $\hat{T}$ with $[f:\hat{T}] \leqq r$, a polynomial $0 \neq \hat{B} \in \hat{T}[x]$ exists such that $\hat{B}(f) = 0$ and $p = \deg \hat{B} \leqq r$

is minimal. Obviously $m \geqq p$. Again $\hat{B} = bB$ with $b \in \hat{T}$ such that B is primitive in $T[x]$. Then $b \neq 0$ and $bB(f) = 0$ in K. Hence $B(f) = 0$. The degree of B is p. A polynomial $\hat{Q} \in \hat{T}[x]$ and a polynomial $\hat{R} \in \hat{T}[x]$ with $\deg \hat{R} < p$ exists such that $A = \hat{Q}B + \hat{R}$ in $\hat{T}[x]$. Then $\hat{R}(f) = A(f) - \hat{Q}(f)B(f) = 0$. Therefore $R = 0$ since p was minimal. A primitive polynomial $Q \in T[x]$ and an element $g \in \hat{T}$ exist such that $\hat{Q} = gQ$. Then A and $Q \cdot B$ are primitive in $T[x]$ and $A = gQ \cdot B$. By Gauss' Lemma, g is a unit in T; hence $g \in L$. Now, Q and B belong to $L[x_1,\ldots,x_n,y_1,\ldots,y_q,x]$ and A belongs to $L[x_1,\ldots,x_n,x]$. Therefore, Q and B belong to $L[x_1,\ldots,x_n,x] = S[x]$ with $B(f) = 0$. Hence $p = \deg_T B = \deg_S B \geqq m$ because m was minimal. Therefore

$$[f:\hat{S}] = m = p \leqq r = [K:\hat{T}] \qquad \text{q.e.d.}$$

<u>Proposition 4.24.</u> Let K be an algebraic function field over the subfield L. Suppose that K has characteristic 0. Let K' be a subfield of K with $K' \supseteq L$. Then K' is an algebraic function field over L.

<u>Proof.</u> Let $x_1,\ldots,x_n$ be a transcendence base of K' over L. Then K' is algebraic over $L(x_1,\ldots,x_n) = \hat{S}$ and $\hat{S}$ is the field of quotients of $S = L[x_1,\ldots,x_n]$. The base $x_1,\ldots,x_n$ of K' extends to a transcendence base $x_1,\ldots,x_n$, $y_1,\ldots,y_q$ of K. Then $\hat{T} = L(x_1,\ldots,x_n,y_1,\ldots,y_q)$ is a pure transcendental extension of L and $\hat{S}$. The field K is finite algebraic over $\hat{T}$. Define $r = [K:\hat{T}]$. The field $\hat{T}$ is the field of quotients of $T = L[x_1,\ldots,x_n,y_1,\ldots,y_p]$. The field K' is algebraic over $\hat{S}$. Take $f \in K'$. Then f is algebraic

over $\hat{S}$. By Proposition 4.23, $[f:\hat{S}] \leqq r$. Lemma 4.22 implies that K' is a finite algebraic extension of $\hat{S}$ with $[K':S'] \leqq r$. Hence K' is an algebraic function field over L; q.e.d.

Let X and Y be complex spaces of pure dimension m and n respectively with $m - n = q \geqq 0$. Let $\varphi: X \to Y$ be a holomorphic map of strict rank n. Let $U \neq \emptyset$ be an open subset of X. Then

$$r_U^X: \mathfrak{K}(X) \to \mathfrak{K}(U)$$

is the restriction map. Write also $\mathfrak{K}(X)|U = r_U^X(\mathfrak{K}(X))$. If U intersects each branch of X, then r_U^X is injective and an isomorphism onto $\mathfrak{K}(X)|U$. If X is irreducible, $\mathfrak{K}(X)|U$ is a field.

Let V be an open subset of Y with $\varphi(U) \subseteq V$. Define $\varphi_0 = \varphi: U \to V$. Then φ_0 has strict rank n (Lemma 4.14), and $\varphi_0^*: \mathfrak{K}(V) \to \mathfrak{K}(U)$ is an injective homomorphism. Here $\varphi_0^*(\mathfrak{K}(V))$ is a subring of $\mathfrak{K}(U)$. If V is irreducible $\varphi_0^*(\mathfrak{K}(V)) = \varphi_0^*\mathfrak{K}(V)$ is a field.

If $P \in \mathfrak{K}(V)[\xi_1,\ldots,\xi_k]$, then $\varphi_0^*P \in \varphi_0^*\mathfrak{K}(V)[\xi_1,\ldots,\xi_k]$ is defined by

$$P = \sum_{\mu_1+\ldots+\mu_k \leqq p} a_{\mu_1,\ldots,\mu_k} \xi_1^{\mu_1},\ldots,\xi_k^{\mu_k}$$

$$\varphi^*P = \sum_{\mu_1+\ldots+\mu_k \leqq p} \varphi^*(a_{\mu_1,\ldots,\mu_k}) \xi_1^{\mu_1},\ldots,\xi_k^{\mu_k}.$$

Then $\varphi_0^*: \mathfrak{K}(V)[\xi_1,\ldots,\xi_k] \to \varphi_0^*\mathfrak{K}(V)[\xi_1,\ldots,\xi_k]$ is an isomorphism.

<u>Proposition 4.25.</u> Let X be a complex space of pure dimension m. Let Y be an irreducible complex space of dimension n with

$m - n = q \geqq 0$. Let $\varphi: X \to Y$ be a holomorphic map of strict rank n. Let $f_1, \ldots, f_k$ be meromorphic functions on X which are algebraically dependent over $\varphi^* \mathfrak{K}(Y)$. Then $f_1, \ldots, f_k$ are φ-dependent.

Proof. A polynomial $P \in \mathfrak{K}(Y)[\xi_1, \ldots, \xi_k]$ with $P \not\equiv 0$ and with positive minimal degree exists such that $\varphi^* P(f_1, \ldots, f_k) = 0$ on X.

Let $\zeta_1, \ldots, \zeta_k$ be the meromorphic functions on $\mathbb{P}^k$, which are holomorphic on $\mathbb{C}^k$ such that $\zeta_\mu(z_1, \ldots, z_k) = z_\mu$ if $(z_1, \ldots, z_k) \in \mathbb{C}^k$. Let $\chi: Y \times \mathbb{P}^k \to Y$ and $\eta: Y \times \mathbb{P}^k \to \mathbb{P}^k$ be the projections. Then $\chi^*: \mathfrak{K}(Y) \to \mathfrak{K}(Y \times \mathbb{P}^k)$ and $\eta^*: \mathfrak{K}(\mathbb{P}^k) \to \mathfrak{K}(Y \times \mathbb{P}^k)$ are injective homomorphisms. Define $u_\mu = \eta^*(\zeta_\mu)$. Then $P^* = \chi^* P(u_1, \ldots, u_k)$ is a meromorphic function on $Y \times \mathbb{P}^k$. Let Y_0 be the largest open subset of Y, such that all coefficients of P are holomorphic on Y_0. Then $Y - Y_0 = S$ is a thin analytic subset of Y. Moreover, P can be regarded as a holomorphic function $P_0: Y_0 \times \mathbb{C}^k \to \mathbb{C}$. Because $P \not\equiv 0$, $P_0 \not\equiv 0$ on $Y_0 \times \mathbb{C}^k$. Moreover, $P_0 = P^* | Y \times \mathbb{C}^k$. Hence $P^* \not\equiv 0$ on the irreducible complex space $Y \times \mathbb{P}^k$. The zero set $N = N_{P*}(0)$ of $P*$ is a thin analytic subset of $Y \times \mathbb{P}^k$ (of pure dimension n+k-1) with $N \cap (Y_0 \times \mathbb{C}^k) = P_0^{-1}(0)$.

The analytic set $T = P_{f_1} \cup \ldots \cup P_{f_k}$ is thin and $f_1, \ldots, f_k$ are holomorphic on $A = X - T$. Define

$$f(x) = (f_1(x), \ldots, f_k(x))$$

if $x \in A$ and $A^* = \{(x, f(x)) \mid x \in A\}$. Then $\Gamma = \overline{A}^*$ is a pure m-dimensional analytic subset of $X \times \mathbb{P}^k$. The projection $\pi: \Gamma \to X$ is proper, surjective and holomorphic. Define $\psi: \Gamma \to Y \times \mathbb{P}^k$ by

$\psi(x,z) = (\varphi(x),z)$ if $(x,z) \in \Gamma$. Because φ has strict rank n, the set $\varphi^{-1}(Y_0)$ is open and dense in X. (Lemma 1.25). Hence $A_0 = \varphi^{-1}(Y_0) \cap A$ is open and dense in A and X. Therefore $A_0^* = \pi^{-1}(A_0)$ is open and dense in A^* and Γ. Pick $(x,z) \in A_0^*$. Then $x \in A$ and $z = f(x) = (f_1(x),\ldots,f_k(x)) \in \mathbb{C}^k$ with $y = \varphi(x) \in Y_0$. Hence

$$P_0(y,z) = (\varphi^* P(f_1,\ldots,f_k))(x) = 0.$$

Therefore, $\psi(x,z) = (y,z) \in N$ which implies $\psi(A_0^*) \subseteq N$. By continuity $\psi(\Gamma) \subseteq N$. Define $\psi_0 = \psi\colon \Gamma \to N$. By Lemma 1.5 and Lemma 1.8

$$\operatorname{rank}_z \psi = \operatorname{rank}_z \psi_0 \leqq \dim N = n + k - 1.$$

Therefore, $f_1,\ldots,f_k$ are φ-dependent; q.e.d.

Lemma 4.14 and Proposition 4.25 imply

<u>Corollary 4.26.</u> Let X and Y be complex spaces of pure dimension m and n respectively. Let $\varphi\colon X \to Y$ be a holomorphic map of strict rank n. Let U and V be open subsets of X and Y respectively such that $\varphi(U) \subseteq V$. Define $\varphi_0 = \varphi\colon U \to V$. Suppose that V is irreducible and that $U \cap B \neq \emptyset$ for each branch B of X. Let $f_1,\ldots,f_k$ be meromorphic functions on X. Suppose that $f_1|U,\ldots,f_k|U$ are algebraically dependent over $\varphi_0^* \mathfrak{K}(V)$. Then $f_1,\ldots,f_k$ are φ-dependent on X.

<u>Proposition 4.27.</u> Let X be a complex space of pure dimension m. Let Y be an irreducible complex space of dimension n with

$m - n = q \geqq 0$. Let $\varphi: X \to Y$ be a holomorphic map of strict rank n. Then $\varphi^* \mathfrak{K}(Y) \subseteq \mathfrak{K}_\varphi(X)$.

<u>Proof.</u> Take $f \in \mathfrak{K}(Y)$. Then $\varphi^*(f) \in \mathfrak{K}(X)$ in algebraic over $\varphi^* \mathfrak{K}(X)$. By Proposition 4.25, $\varphi^*(f)$ is φ-dependent. Hence $\varphi^*(f) \in \mathfrak{K}_\varphi(X)$; q.e.d.

§5. Proper, light, holomorphic maps

Let X be a complex space of pure dimension n. Let Y be an irreducible complex space of rank n. Let S_Y be the set of non-simple points of Y. Let $\varphi: X \to Y$ be a proper, holomorphic map of strict rank n. The set

$$D = \{x \in X \mid \operatorname{rank}_x \varphi \leqq n-1\}$$

is thin and analytic. The set $D' = \varphi(D)$ is analytic with dim D $\leqq$ n - 2 (Theorem 1.1 and Proposition 1.24). The set $S' = S_Y \cup D'$ is analytic and thin in Y. By Lemma 1.25, $S = \varphi^{-1}(S')$ is a thin analytic subset of X. Then $X_0 = X - S$ and $Y_0 = Y - S'$ are open and dense in X respectively Y. Moreover, Y_φ is a connected complex manifold of dimension n. The map $\varphi_0: X_0 \to Y_0$ is proper, light, surjective, open and holomorphic by Lemma 2.2. Therefore, φ_0 has a finite sheet number s. Define s to be the <u>sheet number</u> of $\varphi: X \to Y$.

<u>Proposition 5.1.</u> Let Y be an irreducible complex space of dimension n. Let X be a pure n-dimensional analytic subset of $Y \times \mathbb{P}$. Let $\varphi: X \to Y$ and $f: X \to \mathbb{P}$ be the projections. Suppose that $P_f = f^{-1}(\infty)$ is a thin analytic subset of X. Suppose that φ has strict rank n and sheet number p.

Let S_Y be the set of non-simple points of Y. Define $A = X - P_f$. Then

1. The map $f: A \to \mathbb{P}$ defines a meromorphic function f on X.
2. The set $E' = S_Y \cup \varphi(P_f)$ is thin and analytic in Y and

$Y_0 = Y - E'$ is an open, connected, dense subset of Y.

3. The set $E = \varphi^{-1}(E')$ is thin and analytic in X and $X_0 = X - E$ is open and dense in X with $\varphi(X_0) \subseteq Y_0$.

4. The map $\varphi_0 = \varphi\colon X_0 \to Y_0$ is proper, light, open, surjective and holomorphic with sheet number p. Moreover, $f|X_0 = f_0$ is holomorphic.

5. Let $\nu_{\varphi_0}(x,y)$ be the y-multiplicity of φ_0 at x. Define

$$(5.1)\qquad P_0(y,z) = \prod_{x\in X_0} (z-f_0(x))^{\nu_{\varphi_0}(x;y)}$$

for $(y,z) \in Y_0 \times \mathbb{C}$. Then

$$(5.2)\qquad P_0(y,z) = \sum_{\mu=0}^{p-1} a_\mu(y)z^\mu + z^p$$

where $a_\mu \in \mathfrak{H}(Y_0)$ are holomorphic on Y_0 for $\mu = 0,1,\ldots,p-1$.

6. The map $P_0\colon Y_0 \times \mathbb{C} \to \mathbb{C}$ is holomorphic and

$$(5.3)\qquad P_0^{-1}(0) = \{(\varphi(x),f(x))\,|\,x \in X_0\} \subseteq Y_0 \times \mathbb{C}.$$

7. The functions $a_\mu \in \mathfrak{H}(Y_0)$ continue to meromorphic functions $a_\mu \in \mathfrak{K}(Y)$. Then P_0 continues to a polynomial

$$(5.4)\qquad P = P(\xi) = \sum_{\mu=0}^{p-1} a_\mu \xi^\mu + \xi^p$$

in $\mathfrak{K}(Y)[\xi]$, where ξ is an indeterminant over $\mathfrak{K}(Y)$.

8. The function $f \in \mathfrak{K}(X)$ is algebraic over $\varphi^*\mathfrak{K}(Y)$ with $\varphi^*P(f) = 0$ and $[f\colon \varphi^*\mathfrak{K}(Y)] = p$.

<u>Proof.</u> 1. Clearly $f \in \mathcal{E}(X)$ because f or $\frac{1}{f}$ is holomorphic at every point of X.

2. Obviously φ is proper. Hence $\varphi(P_f)$ is analytic in Y. If $Y = \varphi(P_f)$, then $\varphi: P_f \to Y$ surjective with $\dim P_f \leqq n - 1$ which contradicts Lemma 1.2. Hence $Y \neq \varphi(P_f)$. Because Y is irreducible, $\varphi(P_f)$ is thin. Therefore, $E' = S_Y \cup \varphi(P_f)$ is a thin analytic subset of Y and $Y_0 = y - E'$ is an open, connected, dense subset of Y and a complex manifold.

3. By Lemma 1.25, $E = \varphi^{-1}(E')$ is a thin analytic subset of X. Hence $X_0 = X - E$ is open and dense in X with $\varphi(X_0) \subseteq Y_0$.

4. The map $\varphi_0 = \varphi: X_0 \to Y_0$ is proper and holomorphic, where X_0 and Y_0 have pure dimension n. If $y \in Y_0$, then $\varphi_0^{-1}(y) = \varphi^{-1}(y)$ is analytic in $\{y\} \times \mathbb{P}$ and contained in $\{y\} \times \mathbb{C}$. Hence $\varphi_0^{-1}(y)$ is finite. The map φ_0 is light. Because Y_0 is a connected complex manifold, φ_0 is open and surjective by Lemma 2.2 and has a sheet number s. Define

$$D = \{x \in X \mid \operatorname{rank}_x \varphi \leqq n-1\}.$$

Then D is thin and analytic on X and $\varphi(D) = D'$ is thin and analytic on Y. If $y \in D'$, then $x \in D$ with $\varphi(x) = y$ exists.

$$n - 1 \geqq \operatorname{rank}_x \varphi = \dim_x X - \dim_x \varphi^{-1}(\varphi(x)) = n - \dim_x \varphi^{-1}\varphi(x)$$
$$\dim_x \varphi^{-1}\varphi(x) \geqq 1.$$

where $\varphi^{-1}(\varphi(x)) \subseteq \{y\} \times \mathbb{P}$. Hence $\varphi^{-1}(\varphi(x)) = \{y\} \times \mathbb{P}$. Hence, $(y, \infty) \in P_f$ and $y \in E'$. Therefore $E' \subseteq D'$. If $y \in Y_0$, then

$$s = \sum_{x \in X_0} \nu_{\varphi_0}(x;y) = \sum_{x \in X_0} \nu_{\varphi}(x;y) = p$$

because $y \in Y - (D' \cup S_Y)$. Because $X_0 \cap P_f = \emptyset$, the function $f|X_0 = f_0$ is holomorphic.

5. Define $P_0 \colon Y_0 \times \mathbb{C} \to \mathbb{C}$ by (5.1). Let $\rho \colon \hat{X}_0 \to X_0$ be the normalization of X_0. Then $\psi = \varphi_0 \circ \rho \colon \hat{X}_0 \to Y_0$ is proper light and surjective. By Lemma 2.2, ψ is open. By definition (2.2)

$$\nu_{\varphi_0}(x;y) = \sum_{t \in \hat{X}_0} \nu_{\psi}(t;y).$$

Hence,

$$P_0(y,z) = \prod_{t \in \hat{X}_0} (z - f_0(\rho(t)))^{\nu_{\psi}(t;y)}.$$

By [21] Proposition 3.6, $P_0(\cdot,z)$ is holomorphic on Y_0 for each fixed z. Because φ_0 has a constant sheet number over Y_0, the representation (5.2) holds. The coefficients a_μ are holomorphic on Y_0 because $P_0(\cdot,z)$ is holomorphic for each fixed z. Hence, $P_0 \colon Y_0 \times \mathbb{C} \to \mathbb{C}$ is a holomorphic map.

6. If $x \in X_0$, then $\nu_{\varphi_0}(x,\varphi(x)) = \nu_{\varphi_0}(x) \geqq 1$. Hence, $P_0(\varphi(x),f(x)) = 0$. If $(y,z) \in P_0^{-1}(0)$, then $P_0(y,z) = 0$. Hence $z = f(x)$ and $\nu_{\varphi_0}(x,y) \geqq 1$ for some $x \in X_0$. Now, $\nu_{\varphi_0}(x,y) \geqq 1$ implies $y = \varphi_0(x)$. Therefore, (5.3) is proved.

7. By Lemma 1.24, $\dim D' \leqq n - 2$ because D' is analytic. Define $Y_1 = Y - D'$. Then it will be shown that a_μ continuous to

a meromorphic function onto Y_1. Now, $Y_1 = Y_0 = E' \cap Y_1$ is analytic and thin in Y_1. Take $c \in Y_1 - Y_0$. Then $\varphi^{-1}(c)$ is finite. Take $b \in \mathbb{C}$ such that $f(x) \neq b$ for all $x \in \varphi^{-1}(c)$. For each $\delta > 0$ define $B_\delta = \{z \in \mathbb{C} \mid |z-b| \leqq \delta\}$. Because φ is proper, and because $f\colon X \to \mathbb{P}$ is holomorphic, an open neighborhood N_c of c and a number $\delta > 0$ exists such that $\overline{N}_c$ is compact with $\overline{N}_c \subseteq Y_1$ and such that $f(\varphi^{-1}(\overline{N}_c)) \cap B_{2\delta} = \emptyset$. Pick $e \in B_\delta$ and define $Q_e\colon Y_0 \cap N_c \to \mathbb{C}$ by $Q_e(y) = P_0(y,e)$. If $y \cup Y_0 \cap N_c$ and $x \in \varphi^{-1}(y)$, then

$$|f(x)-e| \geqq |f(x)-b| - |b-c| \geqq 2\delta - \delta = \delta > 0.$$

Therefore, $|Q_e(y)| \geqq \delta^p > 0$. Hence $\frac{1}{Q_e}$ is holomorphic and bounded on $Y_0 \cap N_c$, where $N_c - Y_0 = N_c \cap E'$ is thin and analytic in N_c. Hence, $1/Q_e$ continues to a meromorphic function onto N_c. Therefore, Q_e continues to a meromorphic function onto N_c for each fixed $e \in B_\delta$.

Pick complex numbers $e_0,\ldots,e_p$ in B_δ with $e_\mu \neq e_\nu$ for $\mu \neq \nu$. Then

$$Q_{e_\lambda} = \sum_{\mu=0}^{p-1} a_\mu e_\lambda^\mu + e_\lambda^p$$

on $Y_0 \cap N_c$. Because $\det(e_\lambda^\mu) = \prod_{0 \leqq \sigma < \delta \leqq p} (e_\delta - e_\sigma) \neq 0$ is a complex number. The functions $a_0,\ldots,a_{p-1}$ on $N_c \cap Y_0$ are polynomials over $\mathbb{C}$ in the functions $Q_{e_0},\ldots,Q_{e_p}$ on $N_c \cap Y_0$. Hence, $a_0,\ldots,a_{p-1}$ extend to meromorphic functions on N_c.

By Lemma 3.10, $a_0,\ldots,a_{p-1}$ extend to meromorphic functions on Y_1. By Lemma 3.7, the meromorphic functions $a_0,\ldots,a_{p-1}$ on Y_1 extend to meromorphic functions on $a_0,\ldots,a_{p-1}$ on Y. Clearly, $P_0(\xi) = \sum_{\mu=0}^{p-1} a_\mu \xi^\mu + \xi^p$ in $\mathfrak{K}(Y_0)[\xi]$ extends to $P(\xi) = \sum_{\mu=0}^{p-1} a_\mu \xi^\mu + \xi^p$ in $\mathfrak{K}(Y)[\xi]$.

8. By 6, $\varphi^* P_0(f) = 0$. Because $X - X_0$ is thin and analytic in X, also $\varphi^* P(f) = 0$ holds. Therefore f is integral algebraic over $\varphi^* \mathfrak{K}(Y)$ with $[f\colon \varphi^* \mathfrak{K}(Y)] \leqq p$. Let $Q \in \mathfrak{K}(Y)[\xi]$ with $\varphi^* Q(f) = 0$ and

$$Q(\xi) = \sum_{\mu=0}^{r} b_\mu \xi^\mu$$

and $b_r \not\equiv 0$ on Y. Then $\{y \in Y \mid b_r(y) = 0\}$ is thin analytic in Y. Hence $y_0 \in Y_0$ exists such that $\#\varphi^{-1}(y) = p$ and $b_r(y) \neq 0$. Then $\varphi^{-1}(y) = \{x_1,\ldots,x_p\}$ with $x_\mu = (y, f(x_\mu))$ for $\mu = 1,\ldots,p$. Therefore $f(x_\mu) \neq f(x_\nu)$ for $\mu \neq \nu$ and $Q(y, f(x_\mu)) = 0$ for $\mu = 1,\ldots,p$. Therefore $r \geqq p$. Consequently, $[f\colon \varphi^* \mathfrak{K}(Y)] = p$; q.e.d.

Proposition 5.2. Let X be a complex space of pure dimension n. Let Y be an irreducible complex space of dimension n. Let $\varphi\colon X \to Y$ be a proper, holomorphic map of strict rank n and with sheet number s. Let f be a meromorphic function on X. Then f is algebraic over $\varphi^* \mathfrak{K}(Y)$ with $[f\colon \varphi^* \mathfrak{K}(Y)] \leqq s$.

Proof. Define

$$D = \{x \in X \mid \operatorname{rank}_x \varphi \leqq n-1\}.$$

Then D is a thin analytic subset of X and $D' = \epsilon(D)$ is an analytic subset of Y with $\dim D' \leq n - 2$. (Theorem 1.1 and Proposition 1.24). Define $Y_0 = Y - D'$. The analytic set $\tilde{D} = \varphi^{-1}(D')$ is thin by Lemma 1.25. The open set Y_0 is irreducible and dense. The open set $X_0 = X - \tilde{D}$ is dense. The holomorphic map $\varphi_0 = \varphi: X_0 \to Y_0$ is proper and light. By Theorem 1.1 and Lemma 1.9, $\varphi_0(X_0)$ is a pure n-dimensional analytic subset of Y_0. Hence $\varphi_0(X_0) = Y_0$. The map φ_0 is surjective.

The open set $A = X_0 - P_f$ is dense in X_0 and $f|A$ is holomorphic. Define $A^* = \{(x,f(x)) | x \in A\}$. The closure Γ of A^* in $X_0 \times \mathbb{P}$ is a pure n-dimensional analytic subset of $X_0 \times \mathbb{P}$ and A^* is open and dense in Γ. The projections $\pi: \Gamma \to X_0$ and $\nu: \Gamma \to \mathbb{P}$ are holomorphic and π is proper. The restriction $\pi_0 = \pi: A^* \to A$ is biholomorphic. Define $\psi: \Gamma \to Y_0 \times \mathbb{P}$ by $\psi(x,z) = (\varphi(x),z)$ for $(x,z) \in \Gamma$.

Let K be a compact subset of $Y_0 \times \mathbb{P}$. A compact subset K_1 of Y_0 exists such that $K \subseteq K_1 \times \mathbb{P}$. Then $K_2 = \varphi_0^{-1}(K_1)$ and $K_3 = \pi^{-1}(K_2)$ are compact. If $(x,z) \in \psi^{-1}(K)$, then $(\varphi_0(x),z) \in K$ and $\varphi_0(x) \in K_1$, which implies $x \in K_2$ and $(x,z) \in K_3$. Therefore, $\psi^{-1}(K) \subseteq K_3$. Hence, $\psi^{-1}(K)$ is compact. The map $\psi: \Gamma \to Y_0 \times \mathbb{P}$ is proper. By Theorem 1.1, $\Gamma' = \psi(\Gamma)$ is an analytic subset of $Y_0 \times \mathbb{P}$. The projections $\chi: \Gamma' \to Y_0$ and $\eta: \Gamma' \to \mathbb{P}$ are holomorphic and χ is proper. Define $\psi_0 = \psi: \Gamma \to \Gamma'$. Then $\chi \circ \psi_0 = \varphi \circ \pi$ and $\eta \circ \psi_0 = \nu$. If $(y,z) \in \Gamma'$, then $\psi_0^{-1}(y,z) = (\varphi_0^{-1}(y) \times \{z\}) \cap \Gamma$ which is finite. Therefore $\psi_0: \Gamma \to \Gamma'$ is proper, light, surjective and holomorphic. By Lemma 1.9, Γ' has pure dimension n.

Suppose that a branch B of Γ' exists such that $\operatorname{rank} \chi|B \leqq n-1$. Then $\chi(B) = B'$ is a thin analytic subset of Y_0 (Theorem 1.1 and Proposition 1.23). The set $\psi_0^{-1}(B)$ is analytic in Γ and contains a non-empty open subset of Γ. A branch C of Γ with $C \subseteq \psi_0^{-1}(B)$ exists. The map $\psi_0 \colon C \to B$ is proper and light. Hence, $\psi_0(C)$ is an analytic subset of B. By Lemma 1.27, $\psi_0(C)$ is irreducible and has dimension n. Hence, $\psi_0(C) = B$. Also $\pi(C)$ is a branch of X_0. Because $\varphi_0 \colon \pi(C) \to Y_0$ is proper and light, $\varphi_0(\pi(C))$ is analytic, irreducible and has dimension on n (Theorem 1.1 and Lemma 1.27). Because Y_0 is irreducible, $\varphi_0(\pi(C)) = Y_0$, but

$$Y_0 = \varphi_0(\pi(C)) = \chi(\psi_0(C)) = \chi(B) = B' \neq Y_0.$$

This is a contradiction. Hence, rank $\chi|B = n$ for each branch B of Γ'. By Lemma 1.16, χ has strict rank n.

Suppose a branch B of Γ' is contained in $Y_0 \times \{\infty\}$. As before, a branch C of Γ exists with $\psi_0(C) = B$. Then $C \subseteq X_0 \times \{\infty\}$. Now, $C \cap A^*$ is open and dense on C with $C \cap A^* \subseteq X_0 \times C$. This is a contradiction. Hence, $P_\eta = \eta^{-1}(\infty) = \Gamma' \cap (Y_0 \times \{\infty\})$ is a thin analytic subset of Γ'. Let p be the sheet number of χ. Proposition 5.1 applies to $\eta \in \mathfrak{K}(\Gamma)$. A polynomial $P \in \mathfrak{K}(Y_0)[\xi]$ exists such that $\chi^*P(\eta) = 0$ where

$$P(\xi) = \sum_{\mu=0}^{p-1} a_\mu \xi^\mu + \xi^p$$

and where $a_\mu \in \mathfrak{K}(Y_0)$ for $\mu = 0,1,\ldots,p-1$.

Let S_Y be the set of non-simple points of Y. Then $E' = (Y_0 \cap S_Y) \cup \chi(P_\eta)$ is a thin analytic subset of Y_0 and $Y_1 = Y_0 - E'$ is a connected complex manifold which is open and dense in Y_0. The functions a_μ are holomorphic on Y_1. The inverse image $E = \chi^{-1}(E')$ is a thin analytic subset of Γ' and $\Gamma_1' = \Gamma' - E$ is open and dense in Γ'. The restriction $\chi_1 = \chi: \Gamma_1' \to Y_1$ is proper, light, surjective, open, holomorphic and has sheet number p. By Lemma 2.2 a thin analytic subset S of Y_1 exists such that $p = \#\chi_1^{-1}(y) = \#\chi^{-1}(y)$ if $y \in Y_1 - S$. Moreover, $p \geqq \#\chi_1^{-1}(y) = \#\chi^{-1}(y)$ if $y \in Y_1$.

By Lemma 4.14, φ_0 has strict rank n. By Lemma 1.25, $\tilde{E} = \varphi_0^{-1}(E)$ is thin and analytic in X_0. Then $X_1 = X_0 - \tilde{E}$ is open and dense in X_0 and $\varphi_1 = \varphi_0: X_1 \to Y_1$ is proper, light surjective and holomorphic. Because $Y_1 \cap (S_Y \cup D') = \emptyset$, the map φ_1 is open and has sheet number s. A thin analytic subset S_1 of Y_1 exists such that $s = \#\varphi_1^{-1}(y) = \#\varphi^{-1}(y)$ if $y \in Y_1 - S_1$. Moreover, $s \geqq \#\varphi_1^{-1}(y) = \#\varphi^{-1}(y)$ if $y \in Y_1$. By Theorem 1.1 and by Lemma 1.9, $\varphi_1(X_1 \cap P_f)$ is a thin analytic subset of Y_1. Take $y \in Y_1 - (\varphi(X_1 \cap P_f) \cup S \cup S_1)$. Then $\varphi^{-1}(y) \subseteq A$ and $\pi^{-1}(\varphi^{-1}(y)) \subseteq A^*$. Because $\pi: A^* \to A$ is biholomorphic $\#\pi^{-1}(\varphi^{-1}(y)) = \#\varphi^{-1}(y) = s$. Now

$$\chi^{-1}(y) = \psi_0(\pi^{-1}(\varphi^{-1}(y))).$$

Therefore, $p = \#\chi^{-1}(y) \leqq \#\pi^{-1}(\bar{\varphi}^1(y)) = s$.

Take $x \in X_1 \cap A$. Then $(x, f(x)) \in \Gamma$ and $\psi_0(x, f(x)) = (\varphi(x), f(x))$ in Γ' with $\chi(\psi(x, f(x)) = \varphi(x) \in Y_1$ and $\eta(\psi(x, f(x))) = f(x)$. Therefore,

$$P(\varphi(x), f(x)) = P(\chi(\psi(x, f(x))), \eta(\psi(x, f(x)))) = 0.$$

Hence, $\varphi^*P(f) = 0$ on $A \cap X_1$ which is dense in X_0. Hence, $\varphi^*P(f) = 0$ on X_0.

Because $D' = Y - Y_0$ is analytic with $\dim D' \leqq n - 2$, the meromorphic functions $a_\mu \in \mathfrak{K}(Y_0)$ extend to meromorphic functions $a_\mu \in \mathfrak{K}(Y)$ (Lemma 3.11) and $P \in \mathfrak{K}(Y_0)[\xi]$ becomes a polynomial $P \in \mathfrak{K}(Y)[\xi]$. Since $E = X - X_0$ is thin, $\varphi^*P(f) = 0$ on X. Therefore, f is algebraic over $\varphi^*\mathfrak{K}(Y)$ with $[f: \varphi^*\mathfrak{K}(Y)] \leqq p \leqq s$;

q.e.d.

Lemma 4.22 and Proposition 5.2 imply

<u>Theorem 5.3.</u> Let X and Y be irreducible complex spaces of dimension n. Let $\varphi: X \to Y$ be a proper, holomorphic map of strict rank n and with sheet number s. Then $\mathfrak{K}(X)$ is a finite algebraic extension of $\varphi^*\mathfrak{K}(Y)$ with $[\mathfrak{K}(X): \varphi^*\mathfrak{K}(Y)] \leqq s$.

Proposition 4.25 and Proposition 5.2 imply

<u>Proposition 5.4.</u> Let X be a complex space of pure dimension n. Let Y be an irreducible complex space of dimension n. Let $\varphi: X \to Y$ be a proper, holomorphic map of strict rank n. Then $\mathfrak{K}_\varphi(X) = \mathfrak{K}(X)$.

§6. The field $\widetilde{\mathfrak{K}}(Y \times \mathbb{P}^k)$

Let V be a complex vector space of dimension $k + 1$ with $k > 0$. Let $\rho: V - \{0\} \to \mathbb{P}(V)$ be the residual map onto the associated complex projective space. If $e_0, \ldots, e_k$ is a base of V over $\mathbb{C}$, define the coordinate functions $\varepsilon_\mu: V \to \mathbb{C}$ by

$$\varepsilon_\mu\Big(\sum_{\nu=0}^{k} z_\nu e_\nu\Big) = z_\mu.$$

The functions $\varepsilon_0, \ldots, \varepsilon_k$ are linear and holomorphic and form a base of the dual vector space V^* called the dual base to $e_0, \ldots, e_k$. The quotients $\varepsilon_\mu/\varepsilon_0$ are meromorphic functions on V. For each $\mu = 1, \ldots, k$, one and only one meromorphic function $\zeta_\mu \in \widetilde{\mathfrak{K}}(\mathbb{P}(V))$ exists such that $\rho^*(\zeta_\mu) = \varepsilon_\mu/\varepsilon_0$ on $V - \{0\}$. The functions $\zeta_1, \ldots, \zeta_k$ are called the <u>projective coordinates associated to</u> the base $e_0, \ldots, e_k$ of V.

If $V = \mathbb{C}^{k+1}$, a natural base is given by $e_\mu = (\delta_{a\mu}, \ldots, \delta_{k\mu})$ where $\delta_{\nu\mu}$ is the Kronecker symbol. Here $\varepsilon_\mu(z_0, \ldots, z_k) = z_\mu$. Let $\zeta_1, \ldots, \zeta_k$ be the projective coordinates associated to this natural base. Now, $\mathbb{P}^k = \mathbb{C}^k \cup \mathbb{P}^{k-1}$ is the disjoint union and each function ζ_μ is holomorphic on $\mathbb{C}^k$ with $\zeta_\mu(z_1, \ldots, z_k) = z_\mu$ for $\mu = 1, \ldots, k$.

Again, let V be a complex vector space of dimension $k + 1$ with $k > 0$. Let Y be a complex vector space of pure dimension n. Define

$$(6.1) \qquad \tilde{\rho} = \mathrm{Id} \times \rho: Y \times (V - \{0\}) \to Y \times \mathbb{P}(V).$$

Let

(6.2) $$\tilde{\eta}: Y \times V \to V$$

(6.3) $$\eta: Y \times \mathbb{P}(V) \to \mathbb{P}(V)$$

(6.4) $$\tilde{\chi}: Y \times V \to Y$$

(6.5) $$\chi: Y \times \mathbb{P}(V) \to Y$$

be the projections. Then $\eta \circ \tilde{\rho} = \rho \circ \tilde{\eta}$ and $\chi \circ \tilde{\rho} = \tilde{\chi}$ on $Y \times (V-\{0\})$.

Take $\lambda \in \mathbb{C} - \{0\}$. Define $\hat{\lambda}: V \to V$ by $\hat{\lambda}(z) = \lambda \cdot z$. Then $\rho \circ \hat{\lambda} = \rho$ on $V - \{0\}$. Define $\tilde{\lambda} = Id \times \hat{\lambda}: Y \times V \to Y \times V$. Both $\hat{\lambda}$ and $\tilde{\lambda}$ are biholomorphic. Moreover

(6.6) $$\tilde{\rho} \circ \tilde{\lambda} = \tilde{\rho} \qquad \text{on } Y \times (V-\{0\})$$

(6.7) $$\tilde{\chi} \circ \tilde{\lambda} = \chi$$

(6.8) $$\tilde{\eta} \circ \tilde{\lambda} = \hat{\lambda} \; .$$

The map $\tilde{\lambda}$ induces an isomorphism (of rings) $\tilde{\lambda}^*: \mathfrak{K}(Y \times V) \to \mathfrak{K}(Y \times V)$ A meromorphic function $f \in \mathfrak{K}(Y \times V)$ is said to be <u>homogeneous of degree</u> $p \in Z$ if and only if $\tilde{\lambda}^*(f) = \lambda^p \cdot f$ for all $\lambda \in \mathbb{C} - \{0\}$. Obviously, a holomorphic function f is homogeneous of degree p if and only if

(6.9) $$f(x,\lambda z) = \lambda^p f(x,z)$$

for all $(x,z) \in Y \times V$ and all $\lambda \in \mathbb{C} - \{0\}$. If $p < 0$, pick any $(x,z) \in Y \times V$. If $f(x,z) \not\equiv 0$, then $f(x,\lambda z) \to \infty$ for $\lambda \to 0$ which is impossible, hence, $f(x,z) = 0$ for all $(x,z) \in Y \times V$. Hence, if $f \not\equiv 0$ is holomorphic and of degree p then $p \geqq 0$. By continuity, (6.9) holds for $\lambda = 0$ (with $0^0 = 1$ if $p = 0$).

<u>Lemma 6.1.</u> Let Y be a complex space of pure dimension n. Let V be a complex vector space of dimension $k + 1$ with $k > 0$. Let $e_0,\dots,e_k$ be a base of V and let $\varepsilon_0,\dots,\varepsilon_k$ be the dual base. Define $\tilde{\eta}$ by (6.2) and define $\tilde{\varepsilon}_\mu = \varepsilon_\mu \circ \tilde{\eta}$ for $\mu = 0,\dots,k$. Define $\tilde{\chi}$ by (6.4). Let $f \in \mathfrak{H}(Y \times V)$ be a holomorphic function. Then f is homogeneous of degree $p \geqq 0$ if and only if holomorphic functions $f_{\mu_0,\dots,\mu_k} \in \mathfrak{H}(Y)$ exist such that

$$(6.10) \qquad f = \sum_{\mu_0+\dots+\mu_k=p} (f_{\mu_0\dots\mu_k} \circ \tilde{\chi})\, \tilde{\varepsilon}_0^{\mu_0} \cdots \tilde{\varepsilon}_k^{\mu_k}.$$

<u>Proof.</u> If $x \in Y$ is fixed and $z = \sum_{\mu=0}^{k} z_\mu e_\mu$ then $z_\mu = \varepsilon_\mu(z)$ and $f(x,\cdot)$ is entire. Hence,

$$f(x,z) = \sum_{\mu_0,\dots,\mu_k=0}^{\infty} f_{\mu_0\dots\mu_k}(x)\, z_0^{\mu_0} \cdots z_k^{\mu_k}$$

converges uniformly on every compact subset of $Y \times V$. Here

$$f_{\mu_0\dots\mu_k}(x) = \frac{1}{(2\pi)^k} \int_0^{2\pi} \cdots \int_0^{2\pi} f(x, \sum_{\mu=0}^{k} e^{i\varphi_\mu} e_\mu)\, e^{-i(\mu_0\varphi_0+\dots+\mu_0\varphi_0)}$$
$$d\varphi_0 \cdots d\varphi_k$$

if $x \in Y$. Hence, the functions $f_{\mu_0\dots\mu_k}$ are holomorphic on Y. Then $f(x,\lambda z) = \lambda^p f(x,z)$ for all $\lambda \in \mathbb{C}$, if and only if

$$(6.11) \qquad f(x,z) = \sum_{\mu_0+\dots+\mu_k=p} f_{\mu_0\dots\mu_k}(x)\, z_0^{\mu_0} \cdots z_k^{\mu_k}$$

as comparison of the terms of the power series in $f(x,\lambda z) = \lambda^p f(x,z)$ shows. Now, (6.11) and (6.10) mean the same. Hence f is homogeneous of degree $p \geqq 0$ if and only if (6.10) holds; q.e.d.

Lemma 6.1 states, that $f \in \mathfrak{H}(Y \times V)$ is homogeneous of degree p, if and only if f is a homogeneous polynomial of degree p over $\tilde{\chi}^* \mathfrak{H}(Y)$.

A meromorphic function $f \in \mathfrak{K}(Y \times V)$ is said to be <u>a homogeneous polynomial of degree $p \geqq 0$ over $\tilde{\chi}^* \mathfrak{K}(Y)$</u> if and only if for every point $a \in Y$ an open neighborhood U of a and a holomorphic function g exist such that g is not identically zero on any open subset of U and such that $(g \circ \tilde{\chi}) f | U \times V$ is holomorphic and homogeneous of degree p over U.

<u>Lemma 6.2.</u> Let Y be a complex space of pure dimension n. Let V be a complex vector space of dimension $k + 1$ with $k > 0$. Let $e_0, \ldots, e_k$ be a base of V and let $\varepsilon_0, \ldots, \varepsilon_k$ be the dual base. Define $\tilde{\eta}$ by (6.2) and define $\tilde{\varepsilon}_\mu = \varepsilon_\mu \circ \tilde{\eta}$ for $\mu = 0, \ldots, k$. Define $\tilde{\chi}$ by (6.4). Take $f \in \mathfrak{K}(Y \times V)$. Then f is a homogeneous polynomial of degree $p \geqq 0$ if and only if meromorphic functions $f_{\mu_0 \ldots \mu_k} \in \mathfrak{K}(Y)$ exist such that

(6.12) $$f = \sum_{\mu_0 + \ldots + \mu_k = p} \tilde{\chi}^*(f_{\mu_0 \ldots \mu_k}) \tilde{\varepsilon}_0^{\mu_0} \ldots \tilde{\varepsilon}_k^{\mu_k}$$

<u>Proof.</u> a) Suppose that f is given as in (6.12). Take $a \in Y$. Then an open neighborhood U of a and a holomorphic function $g \in \mathfrak{H}(U)$ exists such that g is not a zero divisor of $\mathfrak{H}(U)$ and such that $g \cdot f_{\mu_0 \ldots \mu_k} = g_{\mu_0 \ldots \mu_k} \in \mathfrak{H}(U)$ is a holomorphic function

on U. Then $(g\circ\tilde{\chi})f|U \times V$ is holomorphic and homogeneous of degree $p \geqq 0$ by Lemma 6.1.

b) Suppose that f is a homogeneous polynomial of degree $p \geqq 0$ over $\tilde{\chi}^*\mathfrak{K}(Y)$. Take $a \in Y$. By Lemma 6.1, an open neighborhood U of a and a holomorphic function g_a on U_a exists such that g_a is a non-zero divisor of (U) on U_a and such that

$$(6.14) \qquad (g_a\circ\tilde{\chi})f = \sum_{\mu_0+\dots+\mu_k=p} h_{a\mu_0\dots\mu_k}\circ\tilde{\chi}\tilde{\varepsilon}_0^{\mu_0}\dots\tilde{\varepsilon}_k^{\mu_k}$$

holds on $U_a \times V$ where $h_{a\mu_0\dots\mu_k}$ are holomorphic functions on V. Take $a \in Y$ and $b \in Y$ with $U_a \cap U_b \neq \emptyset$. Then

$$S = (P_f \cup g_a^{-1}(0) \cup g_b^{-1}(0)) \cap U_a \cap U_b$$

is a thin analytic subset of $U_a \cap U_b$. Take $x \in U_a \cap U_b - S$. Then

$$f(x,z) = \sum_{\mu_0+\dots+\mu_k=p} \frac{h_{a\mu_0\dots\mu_k}(x)}{g_a(x)} z_0^{\mu_0}\dots z_k^{\mu_k}$$

$$= \sum_{\mu_0+\dots+\mu_k=p} \frac{h_{b\mu_0\dots\mu_k}(x)}{g_b(x)} z_0^{\mu_0}\dots z_k^{\mu_k}$$

for every $z = \sum_{\mu=0}^{k} z_\mu e_\mu \in V$. Therefore

$$\frac{h_{a\mu_0\dots\mu_k}}{g_a} = \frac{h_{b\mu_0\dots\mu_p}}{g_b}$$

on $U_a \cap U_b - S$. Hence, this equation holds on $U_a \cap U_b$ within the ring $\mathfrak{K}(U_a \cap U_b)$. Therefore, one and only one meromorphic functic $f_{\mu_0 \ldots \mu_k} \in \mathfrak{K}(Y)$ exists such that

$$f_{\mu_0 \ldots \mu_k} | U_a = h_{a\mu_1 \ldots \mu_k} / g_a .$$

Now, (6.14) implies (6.12); q.e.d.

If Y is irreducible, then it will be shown, that every meromorphic function f of degree $r \in \mathbb{Z}$ is given as $f = g/h$ where g and h are meromorphic homogeneous polynomials of degree p and q over $\tilde{\chi}^* \mathfrak{K}(Y)$ respectively with $p - q = r$. For the proof, some preparations are needed.

A subset N of $Y \times V$ is said to be <u>homogeneous</u> if and only if $(y,z) \in N$ and $\lambda \in \mathbb{C}$ implies $(y,\lambda z) \in N$. A subset N of $Y \times V$ is said to be <u>cylyndric</u> if and only if $N = N' \times V$ with $N' \subseteq Y$. A subset N of $Y \times V$ is said to be <u>non-cylyndric</u> if and only if N is not cylyndric. An analytic subset N of $Y \times V$ is said to be <u>strictly non-cylyndric</u> if every branch of N is non-cylyndric.

<u>Lemma 6.3.</u> Let Y be a complex space of pure dimension n. Let V be a complex vector space of dimension k + 1 with $k > 0$. Let N be a homogeneous analytic subset of $Y \times V$. Define

$$(6.15) \qquad N_0 = N - Y \times \{0\}.$$

Define $\tilde{\rho}$ by (6.1), $\tilde{\chi}$ by (6.4) and χ by (6.5). Then

1. The image $\tilde{\rho}(N_0) = N_1$ is analytic in $Y \times \mathbb{P}(V)$.

2. If N_0 is pure m-dimensional, then N_1 is pure (m-1)-dimensional. If N_0 is irreducible, then N_1 is irreducible.

3. The image $N' = \tilde{\chi}(N_0) = \chi(N_1)$ is analytic.

4. If $\overline{N}_0 = N$, then $\tilde{\chi}(N) = N'$.

5. If N is pure m-dimensional with $m > n$, then $\overline{N}_0 = N$. Especially, this is true if $m = n + k$.

6. Each branch of N is homogeneous.

Proof. 1. Pick $b \in \mathbb{P}(V)$. Take $a \in V - \{0\}$ with $\rho(a) = b$. Let W be a k-dimensional linear subspace of V with $a \notin W$. Then $V = \mathbb{C}a \oplus W$. Also $W_a = a - W$ is a k-dimensional complex plane in V with $a \in W_a$ and $0 \notin W_a$. The image $W_a' = \rho(W_a)$ is an open, connected neighborhood of b in $\mathbb{P}(V)$. The maps $\rho_a = \rho\colon W_a \to W_a'$ and $\tilde{\rho}_a = \rho_a \times \mathrm{Id}\colon Y \times W_a \to Y \times W_a'$ are biholomorphic. Therefore, $N_a' = \tilde{\rho}_a((Y \times W_a) \cap N_0)$ is an analytic subset of $Y \times W_a'$ which is pure (m-1)-dimensional if N_0 is pure m-dimensional, and if $N_a' \neq \emptyset$. Take $(y,w) \in N_1 \cap (Y \times W_a')$. Then $w = \rho(x) = \rho(z)$ such that $(y,x) \in N_0$ and $z \in W_a$. Then $z = \lambda x$ with $\lambda \in \mathbb{C}$. Hence, $(y,z) \in N_0$ and $(y,w) = \tilde{\rho}_a(y,z) \in N_a'$. Therefore $N_a' = N_1 \cap (Y \times W_a')$. The set N_1 is analytic. If N_0 is pure m-dimensional, N_1 is pure (m-1)-dimensional. If N_0 is irreducible, then $N_1 = \tilde{\rho}(N_0)$ is irreducible by Lemma 1.27. Because χ is proper

$$\tilde{\chi}(N_0) = \chi(\tilde{\rho}(N_0)) = \chi(N_1) = N'$$

is analytic. This proves 1. - 3.

4. Suppose that $\overline{N}_0 = N$. Then $\tilde{\chi}(N) \supseteq \tilde{\chi}(N_0) = N'$. Take $y \in \tilde{\chi}(N)$. A point $x \in V$ exists such that $(y,x) \in N$. If $x \neq 0$, then $(y,x) \in N_0$ and $y \in \tilde{\chi}(N_0) = N'$. If $x = 0$, a sequence $(y_\nu, x_\nu) \in N_0$ converges to $(y,0)$ for $\nu \to \infty$. Let $|\ \ |$ be a norm on V. By taking a subsequence, it can be assumed that $z = {}^{x_\nu}/|x_\nu| \to z \neq 0$ for $\nu \to \infty$. Then $(y_\nu, z_\nu) \in N_0$. Hence, $(y,z) \in N_0$ and $y \in \tilde{\chi}(N_0) = N'$. Therefore, $N' = \tilde{\chi}(N)$.

5. Suppose that N has pure dimension $m > n$. Since $Y \times \{0\}$ is pure n-dimensional, $N - N_0 = N \cap (Y \times \{0\})$ is a thin analytic subset of N. Hence, $\overline{N}_0 = N$.

6. Let B be a branch of N. Let $(y,z) \in B$ be a simple point of N. An open neighborhood U of (y,z) exists such that $B \cap U = N \cap U$. Define $L = \{y\} \times \{\lambda z | \lambda \in \mathbb{C}\}$. Then L is an irreducible analytic subset of N. Hence, $L \subseteq B_1$ where B_1 is a branch of N. Now $(y,z) \in B_1 \cap U \subseteq N \cap U = B \cap U$. Hence, $B_1 = B$. Therefore, $(y,\lambda z) \in B$ for all $\lambda \in \mathbb{C}$ if $(y,z) \in B$ is a simple point of N.

Take any point $(y,z) \in B$ and any number $\lambda \in \mathbb{C}$. A sequence $(y_\nu, z_\nu) \in B$ of simple points of N converges to (y,z). Hence, $(y_\nu, \lambda z_\nu) \in B$. Because B is closed, $(y,\lambda z) \in B$. The branch B is homogeneous. q.e.d.

<u>Lemma 6.4.</u> Let Y be an irreducible complex space of dimension n. Let V be a complex vector space of dimension $k + 1$ with $k > 0$. Let N be a pure $(n+k)$-dimensional, homogeneous, analytic subset of $Y \times V$. Then

1. The statements a), b), c) are equivalent:
   a) The map $\tilde{\chi}|N$ has pure rank $n - 1$.
   b) rank $\tilde{\chi}|N = n - 1$.
   c) N is cylyndric.

2. The set N is strictly non-cylyndric if and only if $\tilde{\chi}|N$ has strict rank n.

3. If N is strictly non-cylyndric, N consists of finitely many branches $B_1,\ldots,B_r$. Moreover, $\tilde{\chi}(B_\lambda) = Y$ for $\lambda = 1,\ldots,r$.

<u>Proof.</u> 1. The image set $N' = \tilde{\chi}(N)$ is analytic by Lemma 6.3.4 and 6.3.5. a) implies b). Assume b). Take $y \in N'$; then $(\tilde{\chi}|N)^{-1}(y) = (\{y\} \times V) \cap N$. If $x \in (\tilde{\chi}|N)^{-1}(y)$, then

$$k + 1 \geqq \dim_x (\tilde{\chi}|N)^{-1}(y) = n + k - \operatorname{rank}_x \tilde{\chi}|N \geqq k + 1$$

by Lemma 1.13. Hence, $\{y\} \times V = (\tilde{\chi}|N)^{-1}(y)$. Therefore $N = N' \times V$ which proves c). Assume c). Then $N = N' \times V$ and $(\tilde{\chi}|N)^{-1}(y) = \{y\} \times V$ if $y \in N'$. Therefore, $\operatorname{rank}_x \tilde{\chi}|N = (n+k) - (k-1) = n - 1$ if $x \in N$, which proves a).

2. The set N is strictly non-cylyndric, if and only if every branch B of N is non-cylyndric, which is the case, if and only if rank $\tilde{\chi}|B = n$ for each branch B of N, which is the case, if and only if $\tilde{\chi}$ has strict rank n (Lemma 1.16).

3. Suppose that N is strictly non-cylyndric. Let B be a branch of N. Then B is homogeneous but not cylyndric. Hence, $\tilde{\chi}|B$ has rank n. By Lemma 1.25, $\tilde{\chi}(B)$ is not thin. By Lemma 6.3, $\tilde{\chi}(B)$ is analytic. Because Y is irreducible, $Y = \tilde{\chi}(B)$.

Pick $y \in Y$. Then $(\{y\} \times V) \cap B \neq \emptyset$. Let $|\ \ |$ be a norm on V. Then $K = \{y\} \times \{z \in V |\ |z| \leqq 1\}$ is compact. Because $(\{y\} \times V) \cap B \neq \emptyset$ and because B is homogeneous, $K \cap B \neq \emptyset$. Any compact subset of $Y \cap V$ intersects at most finitely many branches of N, and the compact subset K intersects all branches of N. Hence, N has only finitely many branches; q.e.d.

Lemma 6.5. Let Y be an irreducible, normal complex space of dimension n. Let S_Y be the set of non-simple points of Y. Let V be a complex vector space of dimension k + 1. Define $\tilde{S}_Y = S_Y \times V$. Let W be a k-dimensional linear subspace of V. Let $e \in V - W$ and $V = W \oplus \mathbb{C}e$. Let $\pi: V \to W$ be the projection defined by $\pi(w+ze) = w$ for $w \in W$ and $z \in \mathbb{C}$. Define

$$\tilde{\pi} = \mathrm{Id} \times \pi: Y \times V \to Y \times W. \tag{6.16}$$

Let N be a homogeneous, pure (n+k)-dimensional analytic subset of $Y \times V$. Suppose that $\pi_0 = \pi: N \to Y \times W$ is proper. Let S_N be the set of non-simple points of N. Then

1. The sets $(\tilde{S}_Y \cap N) \cup S_N = S$ is thin and analytic in N.

2. The restriction π_0 is proper, light, open and surjective. Let p be the sheet number of π_0.

3. A holomorphic function $f \in \mathfrak{H}(Y \times V)$ exists such that

   a) The function f is homogeneous of degree p

   b) $N = f^{-1}(0)$

   c) If $x \in N - S$, then[8] $\nu_f(x) = 1$

   d) Holomorphic function $a_\mu \in \mathfrak{H}(Y \times W)$ exist for $\mu = 0,\ldots,p - 1$ such that

$$f(y,w+ze) = \sum_{\mu=0}^{p-1} a_\mu(y,w)z^\mu + z^p \tag{6.17}$$

if $y \in Y$, $w \in W$ and $z \in \mathbb{C}$.

Proof. 1. Because Y is normal, $\dim \tilde{S}_Y \leqq n + k - 1$. Hence, $E = \tilde{S}_Y \cap N$ is thin on N. Then $S = E \cup S_N$ is thin. Define $N_1 = N - E$.

The vector space V can be identified with $W \oplus Ce = W \times C$ such that $\tilde{\pi}: Y \times W \times C \to Y \times W$ is the projection. Then $Y \times W \times C$ is open in $Y \times W \times P$. The projection $\tilde{\pi}$ extends to $\tilde{\pi}: Y \times W \times P \to Y \times W$ and π extends to $\pi: W \times P \to W$. Because π_0 is proper, N is an analytic subset of $Y \times W \times P$.

2. If $y \in Y \times W$, then $\pi_0^{-1}(y) = (\{y\} \times C) \cap N$ is compact and analytic, hence, finite. The map π_0 is light. By Lemma 2.2, π_0 is open and surjective and has a finite sheet number p. Especially, π_0 has pure rank $n + k$.

3. Let $g: N \to C$ be the projection. Here g is a holomorphic function. Define $Y_1 = Y - S_Y$. Then $\pi_0(N_1) \subseteq Y_1$. Define $\pi_1 = \pi_0: N_1 \to Y_1$. Apply Proposition 5.1:

Prop. 5.1	Y	n	X	φ	f	P_f	p	S_Y	A	E'
Here	$Y \times W$	$n+k$	N	π_0	g	$\emptyset$	p	$S_Y \times W$	N	$S_Y \times W$

Prop. 5.1	Y_0	E	X_0	φ_0	P_0	a_μ
Here	$Y_1 \times W$	E	N_1	π_1	f_1	a_μ

The holomorphic functions a_μ on $Y_1 \times W$ extend to holomorphic functions on $Y \times W$, because $\dim S_Y \times W \leqq n + k - 2$ and because $Y \times W$ is normal. Hence, $f \in \ (Y \times W \times C) = \ (Y \times V)$ with

$$f(y,w,z) = f(y,w+ze) = \sum_{\mu=0}^{p-1} a_\mu(y,w)z^\mu + z^p$$

if $(y,w,z) \in Y \times W \times C$.

If $(y,w,z) \in Y_1 \times W \times C$, define $\nu(y,w,z) = \nu_{\pi_0}(y,w,z)$ if $(y,w,z) \in N_1$ and $\nu(y,w,z) = 0$ if $(y,w,z) \notin N_1$. Take

$x = (y,w,z) \in N_1$ and $t = (r,s) \in Y_1 \times W$. Then $\pi_0(x) = (y,w,) \in Y_1 \times W$ and

$$\nu_{\pi_0}(x;t) = \begin{cases} \nu_{\pi_0}(x) = \nu_{\pi_0}(y,w,z) = \nu(y,w,z) \text{ if } (y,w) = (r,s) = t \\ 0 \qquad\qquad\qquad\qquad\qquad \text{if } (y,w) = \pi_0(x) \neq t \end{cases}$$

Hence, (5.1 reads

$$(6.18) \qquad f(y,w,z) = \prod_{u \in \mathbb{C}} (z-u)^{\nu(y,w,u)} \quad \text{with } p = \sum_{u \in \mathbb{C}} \nu(y,w,u)$$

if $(y,w,z) \in Y_1 \times W \times \mathbb{C}$. Take $\lambda \in \mathbb{C} - \{0\}$. Because $\tilde{\lambda}: Y \times W \to Y \times W$ is biholomorphic, $\nu_{\tilde{\lambda}\circ\pi_0} = \nu_{\pi_0}$. Because $\tilde{\lambda}: Y \times V \to Y \times V$ is biholomorphic, $\nu_{\pi_0 \circ \tilde{\lambda}} = \nu_{\pi_0} \circ \tilde{\lambda}$. Now, $\tilde{\lambda} \circ \pi_0 = \pi_0 \circ \tilde{\lambda}$ implies

$$\nu_{\pi_0} = \nu_{\tilde{\lambda}\circ\pi_0} = \nu_{\pi_0 \circ \tilde{\lambda}} = \nu_{\pi_0} \circ \tilde{\lambda}.$$

Hence, $\nu(y,\lambda w,\lambda u) = \nu(y,w,u)$ if $(y,w,u) \in Y_1 \times W \times \mathbb{C}$. Therefore,

$$f(y,\lambda w,\lambda z) = \prod_{u \in \mathbb{C}} (\lambda z-u)^{\nu(y,\lambda w,u)} = \prod_{u \in \mathbb{C}} (\lambda z- u)^{\nu(y,\lambda w,\lambda u)}$$

$$= \prod_{u \in \mathbb{C}} (\lambda z-\lambda u)^{\nu(y,w,u)} = \lambda^p \prod_{u \in \mathbb{C}} (z-u)^{\nu(y,w,u)} = \lambda^p f(y,w,z)$$

if $(y,w,z) \in Y_1 \times W \times \mathbb{C}$ and $\lambda \in \mathbb{C} - \{0\}$. By continuity,

$$f(y,\lambda w,\lambda z) = \lambda^p f(y,w,z)$$

if $(y,w,z) \in Y \times W \times \mathbb{C}$ and $\lambda \in \mathbb{C}$. Hence, f is homogeneous of degree p.

By (6.18), $f^{-1}(0) - E = N - E$. Because $f^{-1}(0)$ and N are pure $n + k$ dimensional and because $\dim E \leqq n + k - 1$, this implies $f^{-1}(0) = N$.

A thin analytic subset T' of $Y_1 \times W$ exists such that $T = \pi_1^{-1}(T') = \pi_0^{-1}(T')$ is a thin analytic subset of N_1 and such that $\pi_2 = \pi\colon N_1 - T \to Y_1 \times W - T'$ is a locally biholomorphic, proper, surjective map. Pick $(a,b) \in Y_1 \times W - T'$. An open connected neighborhood Z of (a,b) in $Y_1 \times W = T'$ and open, connected subsets $X_1,\ldots,X_p$ of N_1 exist such that $\pi_2^{-1}(z) = \pi^{-1}(z) = X_1 \cup \ldots \cup X_p$ with $X_\mu \cap X_\nu = \emptyset$ if $\mu \neq \nu$, and such that $\psi_\mu = \pi_2\colon X_\mu \to Z$ is biholomorphic. Clearly, $X_\mu \subseteq N_1 - S_N$. Define $g_\mu = \psi_\mu^{-1}\colon Z \to X_\mu$, for $\mu = 1,\ldots,p$. Then

$$f(y,w,z) = \prod_{\mu=1}^{p} (z - g_\mu(y,w))$$

if $(y,w,z) \in Z \times \mathbb{C}$. Hence, $\nu_f(y,w,z) = 1$ if $(y,w,z) \in \pi_0^{-1}(Z)$. Therefore $\nu_f|(N_1-T) = 1$. Because $N_1 - T$ is open and dense in $N_1 - S = N - S$ and because ν_f is locally constant on $N - S$, this implies $\nu_f = 1$ on $N - S$; q.e.d.

Lemma 6.6. Let Y be a normal, irreducible complex space of dimension n. Let S_Y be the set of non-simple points of Y. Let V be a complex vector space of dimension $k + 1$ with $k > 0$. Define $\tilde{S}_Y = S_Y \times V$. Let N be a pure (n+k)-dimensional, homogeneous, strictly non-cylyndric, analytic subset of N. Let S_N be the set of non-simple points of N. Then $S = S_N \cup (\tilde{S}_Y \cap N)$ is a thin analytic subset of N. A meromorphic function $f \in \mathfrak{K}(Y \times V)$ exists such that

1. The meromorphic function f is a homogeneous polynomial of degree $p > 0$ over $\tilde{\chi}^* \mathfrak{K}(Y)$ where $\tilde{\chi}$ is defined by (6.4).

2. A thin analytic subset T' of Y exists such that $T = N \cap (T' \times V)$ is thin in N, such that $f_0 = f|((Y-T') \times V)$ is holomorphic with $N - T = f_0^{-1}(0)$.

3. If $x \in N - (T \cup S)$, then $\nu_{f_0}(x) = 1$.

<u>Proof.</u> For the same reason as in Lemma 6.5.1, S is thin. By Lemma 1.24 and Lemma 6.4.2

$$D = \{x \in N | \operatorname{rank}_x \tilde{\chi}|N \leqq n-1\} \tag{6.19}$$

is a thin analytic subset of N. Define $D' = \tilde{\chi}(D)$. Now

$$D' = \{y \in Y | \tilde{\chi}^{-1}(y) \cap N = \{y\} \times V\} \tag{6.20}$$

is claimed. Observe that $\tilde{\chi}^{-1}(y) \cap N = (\{y\} \times V) \cap N$. If $y \in D'$, take any $x \in D$ with $\tilde{\chi}(x) = y$. Then

$$k + 1 \geqq \dim_x \tilde{\chi}^{-1}(\tilde{\chi}(x)) \cap N = n + k - \operatorname{rank}_x \tilde{\chi}|N \geqq k + 1.$$

Therefore, $\tilde{\chi}^{-1}(y) \cap N = \{y\} \times V$. If $\{y\} \times V = \tilde{\chi}^{-1}(y) \cap N$, take $x \in \{y\} \times V$. Then $\dim_x \tilde{\chi}^{-1}(\tilde{\chi}(x)) \cap N = k + 1$ which implies $\operatorname{rank}_x \tilde{\chi}|N = n - 1$. Hence, (6.20) is proved.

Take a k-dimensional linear subspace W of V. For each $a \in V$, an analytic subset T_a' of Y is defined by $T_a' \times \{a\} = (Y \times \{a\}) \cap N$.

Then $T'_a = Y$ if and only if $(Y \times \{a\}) \subseteq N$. Define

$$G = \{a \in V - W | T'_a \neq Y\}.$$

If $y \in D'$ and $a \in G$, then $(y,a) \in (Y \times \{a\}) \cap N$, because $\{y\} \times V \subseteq N$. Hence, $y \in T'_a$. If $y \in Y - D'$, then $a \in V - W$ exists such that $(y,a) \notin N$. Hence, $y \notin T'_a$ where $a \in G$. Therefore,

$$D' = \bigcap_{a \in G} T'_a$$

consequently, D' is analytic in Y. By Proposition 1.24, $\dim D' \leqq n - 2$.

For $a \in G$, define $A_a = Y - T'_a$. Then A_a, is open and dense in Y. Also, A_a is a normal, irreducible complex space of dimension n. Moreover

$$A = \bigcup_{a \in G} A_a = Y - D'.$$

Because $\tilde{\chi}|N$ has strict rank n, the analytic set $T_a = \tilde{\chi}^{-1}(T'_a) \cap N$ is thin on N and $N_a = N - T_a$ is open and dense on N (Lemma 6.4.2 and Lemma 1.25).

Take $a \in G$. Then $V = W \oplus \mathbb{C}a$. Define $\tilde{\pi}_a: V \to W$ by $\pi_a(w+za) = w$ if $w \in W$ and $z \in \mathbb{C}$. Define $\tilde{\pi}_a: N_a \to A_a \times W$ by $\tilde{\pi}_a(y,v) = (y,\pi_a(v))$ if $(y,v) \in N_a$. Now, it is claimed that $\tilde{\pi}_a$ is proper.

By construction $(A_a \times \{a\}) \cap N = \emptyset$. Let K be a compact subset of $A_a \times W$. Compact subsets K_1 of A_a and K_2 of W exist such that $K \subseteq K_1 \times K_2$. Let $| \cdot |$ be a norm on W. For $t > 0$ define

$B_t = \{w \in W \mid |w| < t\}$. Because $(K_1 \times \{a\}) \cap N = \emptyset$, a number $r > 0$ exists such that $(K_1 \times (a+\overline{B}_r)) \cap N = \emptyset$. A number $s > 0$ exists such that $K_2 \subset B_s$. Define $K_3 = \{z \in \mathbb{C} \mid |z| \leqq {}^s/_r\}$ and $K_4 = K_2 + K_3 a$. Then K_4 is compact in V. Take $(y, w+za) \in \tilde{\pi}_a^{-1}(K) \subseteq N$ with $w \in W$ and $z \in \mathbb{C}$. Then $(y,w) \in K \subseteq K_1 \times K_2$. Hence, $y \in K_1$ and $|w| \leqq s$. Suppose $z \notin K_3$. Then $|z| \geqq \frac{s}{r} > 0$. Hence, $(y, \frac{w}{z} + a) \in N$ with $|\frac{w}{z}| \leqq r$. Hence,

$$(y, \frac{w}{z} + a) \in N \cap (K_1 \times (\overline{B}_r + a)) = \emptyset.$$

Contradiction! Therefore, $(y, w+za) \in K_1 \times (K_2 + K_3 a) = K_1 \times K_4$. Hence, $\tilde{\pi}_a^{-1}(K) \subseteq K_1 \times K_4$. Because $(K_1 \times K_4) \cap N_a = (K_1 \times K_4) \cap N$ is compact in N_a, the map $\tilde{\pi}_a$ is proper.

By Lemma 6.5, a holomorphic function f_a on $A_a \times V$ exists such that f_a is homogeneous of degree p_a, such that $f_a^{-1}(0) = N_a$ and such that $\nu_{f_a}(x) = 1$ if $x \in N_a - S$. Let $a_0, \ldots, a_{k-1}$ be a base of W. Then $a_0, \ldots, a_{k-1}, a_k = a$ is a base of V. By Lemma 6.1 holomorphic functions $c_{a\mu_0 \ldots \mu_k}$ on A_a exist such that

$$f_a(y,z) = \sum_{\mu_0 + \ldots + \mu_k = p_a} c_{a\mu_0 \ldots \mu_k}(y) z_0^{\mu_0} \ldots z_p^{\mu_k}$$

$$= \sum_{\nu=0}^{p_a} d_\nu(y, z') z_k^\mu$$

if $y \in A_a$ and $z = \sum_{\mu=0}^{k} z_\mu a_\mu$ with $z_\mu \in \mathbb{C}$ and if $z' = \sum_{\mu=0}^{p-1} z_\mu \mathbf{a}_\mu$ and

$$d_\nu(y,z') = \sum_{\mu_0+\dots+\mu_{k-1}=p_a-\nu} c_{a\mu_0\dots\mu_k,\nu}(y) z_0^{\mu_0}\dots z_{k-1}^{\mu_{k-1}}.$$

By (6.17), $d_{p_a} = 1$ on A_a which implies $c_{a,0\dots0,p_a} \equiv 1$ on A_a.

Take $a \in G$ and $b \in G$. Then $A_a \cap A_b = Y - T_a' \cup T_b'$ is open and dense in Y. Since $T_a' \cup T_b'$ is thin and analytic, $A_a \cap A_b$ is irreducible. Moreover

$$f_a^{-1}(0) \cap ((A_a \cap A_b) \times V) = f_b^{-1}(0) \cap ((A_a \cap A_b) \times V)$$

where $\nu_{f_a}(x) = 1 = \nu_{f_b}(x) = s$ on $(N_a \cap N_b - S)$. Hence, $f_a = gf_b$ on $(A_a \cap A_b) \times V$, where g is holomorphic on $(A_a \cap A_b) \times V$ with $g(y,z) \neq 0$ if $(y,z) \in (A_a \cap A_b) \times V$. Take $(y,z) \in (A_a \cap A_b) \times V$. Take $\lambda \in \mathbb{C} - \{0\}$. Then

$$\lambda^{p_a} f_a(y,z) = f_a(y,\lambda z) = g(y,\lambda z) f_b(y,\lambda z) = g(y,\lambda z)\lambda^{p_b} f_b(y,z)$$

$$= \frac{g(y,\lambda z)}{g(y,z)} \lambda^{p_a} f_a(y,z).$$

$$\lambda^{p_a-p_b} g(y,z) = g(y,\lambda z) \to g(y,0) \neq 0 \quad \text{for} \quad \lambda \to 0.$$

Hence, $p_a = p_b = p > 0$ does not depend on $a \in G$. Moreover, $g(y,z) = g(y,0) = h(y)$ for all $(y,z) \in (A_a \cap A_b) \times V$. Here h is holomorphic on $A_a \cap A_b$ with $h(y) \neq 0$ if $y \in A_a \cap A_b$.

Now, keep $a \in G$ fixed. Take any $b \in G$ with $b \neq a$. Then $A_a - A_b$ and $A_b - A_a$ are thin analytic subsets of A_a and A_b

respectively. Let $a_0, \ldots, a_{k-1}, a_k = a$ be the base of V selected for a. Then

$$f_a(y,z) = \sum_{\mu_0+\ldots+\mu_k=p} c_{a\mu_0\ldots\mu_k} z_0^{\mu_0} \ldots z_k^{\mu_k}$$

if $(y,z) \in A_a \times V$ and $z = \sum_{\mu=0}^{k} z_\mu a_\mu$ and

$$f_b(y,z) = \sum_{\mu_0+\ldots+\mu_k=p} \tilde{c}_{b\mu_0\ldots\mu_k} z_0^{\mu_0} \ldots z_k^{\mu_k}$$

if $(y,z) \in A_b \times V$ and $z = \sum_{\mu=0}^{k} z_\mu a_\mu$. Here $c_{a\mu_0\ldots\mu_k}$ are holomorphic on A_a with $c_{a,0..0,p} = 1$. The functions $\tilde{c}_{b\mu_0\ldots\mu_k}$ are holomorphic on A_b. A holomorphic function $h \in \mathfrak{H}(A_a \cap A_b)$ exists such that $h(y) \neq 0$ if $y \in A_a \cap A_b$ and such that $f_a(y,z) = h(y) f_b(y,z)$ if $y \in A_a \cap A_b$ and $z \in V$. Hence,

$$c_{a\mu_0\ldots\mu_k}(y) = h(y)\tilde{c}_{b\mu_0\ldots\mu_k}(y)$$

if $y \in A_a \cap A_b$. Hence, $h = {}^1/\tilde{c}_{b,0\ldots0,p}$ on $A_a \cap A_b$. Therefore, h continuous to a meromorphic function $H \in \mathfrak{K}(A_b)$ such that $H\tilde{c}_{b\mu_0\ldots\mu_k} = c_{a\mu_0\ldots\mu_k}$ on $A_a \cap A_b$. Here $H\tilde{c}_{b\mu_0\ldots\mu_k}$ is meromorphic on A_b. Hence, $c_{a\mu_0\ldots\mu_k}$ continuous to meromorphic function $C_{a\mu_0\ldots\mu_k,b}$ on $A_a \cup A_b$. If $e \in G$ with $e \neq a$. Then $A_a \cap A_b \cap A_e$ is open and dense in $A_a \cup A_b \cup A_e$ and

$$C_{a\mu_0\ldots\mu_k,b}|A_a \cap A_b \cap A_c = c_{a,\mu_0\ldots\mu_k}|A_a \cap A_b \cap A_e$$

$$= C_{a,\mu_0\ldots\mu_k,e}|A_a \cap A_b \cap A_e .$$

Hence,

$$C_{a\mu_0\ldots\mu_k,b}|(A_a \cup A_b) \cap (A_a \cup A_c) = C_{a\mu_0\ldots\mu_k,e}|(A_a \cup B_b) \cap (A_a \cup A_e).$$

Hence, one and only one meromorphic function $C_{a\mu_0\ldots\mu_k}$ on $A = \bigcup_{b\in G}(A_a \cup A_b)$ exists such that $C_{a\mu_0\ldots\mu_k}|A_a \cup A_b = C_{a,\mu_0\ldots\mu_k,b}$. Therefcre, $c_{a\mu_0\ldots\mu_k}$ continues to $C_{a\mu_0\ldots\mu_k} \in \mathfrak{K}(A)$. Because $Y - A = D'$ is analytic with $\dim D' \leqq n - 2$, $C_{a\mu_0\ldots\mu_k}$ continues to a meromorphic function $C_{\mu_0\ldots\mu_k} \in \mathfrak{K}(Y)$. Moreover $f_a \in \tilde{\mathfrak{K}}(A_a \times V)$ continues to the meromorphic function

$$f = \sum_{\mu_0+\ldots+\mu_k=p} \tilde{\chi}^*(C_{\mu_0\ldots\mu_k})\tilde{\alpha}_0^{\mu_0}\ldots\tilde{\alpha}_k^{\mu_k} \in \tilde{\mathfrak{K}}(Y \times V)$$

which is a homogeneous polynomial of degree p over $\tilde{\chi}^* \tilde{\mathfrak{K}}(Y)$. (Lemma 6.2). Here $\alpha_0,\ldots,\alpha_k$ is the dual base to $a_0,\ldots,a_k$ and $\tilde{\alpha}_\mu = \alpha_\mu \circ \tilde{\eta}$.

Define $T' = T_a$. Then T' is a thin analytic subset of Y and $T = N \cap (T' \times V) = T_a = (\tilde{\chi}|N)^{-1}(T')$ is a thin analytic subset of N. Moreover,

$$f_0 = f|(Y-T') \times V = f_a$$

is holomorphic and $f_0^{-1}(0) = f_a^{-1}(0) = N_a = N - T$. If $x \in N - (T \cup S)$ $= N_a - S$, then $\nu_{f_0}(x) = \nu_{f_a}(x) = 1$; q.e.d.

Lemma 6.7. Let Y be a normal complex space of pure dimension n. Let V be a complex vector space of dimension k + 1 with $k > 0$. Let $f \in \mathfrak{K}(Y \times V)$ be a meromorphic function which is homogeneous of degree $p \in \mathbb{Z}$. Then the zero set $N_f(0)$ and the pole set $P_f = N_f(\infty)$ are homogeneous. The set P_f is either empty or has pure dimension n + k. If f is not a zero-divisor, then $N_f(0)$ is either empty or has pure dimension n + k.

Proof. Take $\lambda \in \mathbb{C} - \{0\}$. Then $\tilde{\lambda}: Y \times V \to Y \times V$ is biholomorphic and $\tilde{\lambda}^*(f) = \lambda^p f$. Hence $N_f(a) = N_{\tilde{\lambda}^* f}(a) = \tilde{\lambda}^{-1} N(a)$ if $a = 0, \infty$. Hence, $(y,z) \in N_f(a)$ implies $(y, \lambda z) \in N_f(a)$ is $\lambda \neq 0$. Because $N_f(a)$ is closed, this remains true for $\lambda = 0$. Hence, $N_f(a)$ is homogeneous; q.e.d.

Theorem 6.8. Let Y be an irreducible complex space of dimension n. Let V be a complex vector space of dimension k + 1 with $k > 0$. Let $f \in \mathfrak{K}(Y \times V)$ be a meromorphic function which is homogeneous of degree $p \in \mathbb{Z}$. Then meromorphic functions g and $h \not\equiv 0$ on $Y \times V$ exist, which are homogeneous polynomials of degree p_0 and p_∞ respectively over $\tilde{\chi}^* \mathfrak{K}(Y)$ such that $f = g/h$ and $p = p_0 - p_\infty$.

Proof. 1. At first, assume that Y is normal. Let $N_f(\infty) = P_f$ and $N_f(0)$ be the pole and zero sets of f respectively. If

$f \equiv 0$, define $g \equiv 0$ and $h \equiv 1$. If $N_f(0) = \emptyset$, then $1/f$ is holomorphic. Hence, $1/f$ is a homogeneous polynomial of degree-$p \geqq 0$ $\tilde{\chi}^*\mathfrak{K}(Y)$. Define $g \equiv 1$ and $h \equiv 1/f$. If $N_f(\infty) = \emptyset$, then f is holomorphic and a homogeneous polynomial of degree $p \geqq 0$ over $\tilde{\chi}^*\mathfrak{K}(Y)$. Take $g = f$ and $h = 1$. Therefore, it can be assumed that $N_f(a)$ has pure dimension $n + k$ for $a = 0, \infty$.

Let $\mathfrak{B}_a$ be the set of branches of $N_f(a)$ is $a = 0, \infty$. By Lemma 6.3 each branch $B \in \mathfrak{B}_a$ is homogeneous. Define $\tilde{N} = N_f(a) \cup N_f(\infty)$. $\mathfrak{B} = \mathfrak{B}_0 \cup \mathfrak{B}_\infty$ is the set of branches of N with $\mathfrak{B}_0 \cap \mathfrak{B}_\infty = \emptyset$. Let $\mathfrak{A}_a$ be the set of non-cylyndric branches of $N_f(a)$. Then

$$N = \Big(\bigcup_{B\in\mathfrak{A}_0} B\Big) \cup \bigcup_{B\in\mathfrak{A}_\infty} B$$

is a homogeneous, pure $(n+k)$-dimensional analytic subset of $Y \times V$ and $\mathfrak{A}_0 \cup \mathfrak{A}_\infty$ is the set of branches of N. Hence, N is strictly non-cylyndric. By Lemma 6.4, $\mathfrak{A}_0$ and $\mathfrak{A}_\infty$ are finite. Let $\mathfrak{L}_a$ be the set of cylyndric branches of $N_f(a)$ for $a = 0, \infty$. If $B \in \mathfrak{L}_a$ then $B = B' \times V$ where $B' = \tilde{\chi}(B)$ is irreducible and analytic with $\dim B' = n - 1$. Let K be a compact subset of Y. Define $K_0 = K \times \{0\}$. Then $B' \cap K \neq \emptyset$ if and only if $B \cap K_0 \neq \emptyset$. But, $\{B \in \mathfrak{L}_a \mid B \cap K_0 \neq \emptyset\}$ is a finite set because K_0 is compact. Hence, $B' \cap K \neq \emptyset$ for at most finitely many $B \in \mathfrak{L}_a$. Hence

$$M' = \bigcup_{B\in\mathfrak{L}_0} B' \cup \bigcup_{B\in\mathfrak{L}_\infty} B'$$

is a pure $(n-1)$-dimensional analytic subset of Y or empty. Moreover, $M = M' \times V$ is analytic with

$$M = \bigcup_{B \in \mathfrak{L}_0} B \cup \bigcup_{B \in \mathfrak{L}_\infty} B.$$

Hence, M is a pure (n+k)-dimensional, cylyndric analytic subset of $Y \times V$ with $M \cup N = \tilde{N}$.

Let S_Y, $S_{\tilde{N}}$, S_N, S_M, S_B be the set of non-simple points of Y, $\tilde{N}$, N, M and $B \in \mathfrak{L}_0 \cup \mathfrak{L}_\infty$ respectively. Then $\tilde{S}_Y = S_Y \times V$ is the set of non-simple points of $Y \times V$. Let $\nu_f(x,a)$ be the a-multiplicity of f at $x \in Y \times V - \tilde{S}_Y$. Then $\nu_f(x) = \nu_f(x,0) - \nu_f(x,\infty)$ is the multiplicity of f at $x \in Y \times V - \tilde{S}_Y$. If $B \in \mathfrak{L}_0 \cup \mathfrak{L}_\infty$, then B is a branch of $\tilde{N}$ and $\tilde{S}_Y \cap B$ is thin in B, because $\dim \tilde{S}_Y \leqq n + k - 1$. An integer $t(B) \in \mathbb{Z}$ exists such that $\nu_f(x) = t(B)$ if $x \in B - (\tilde{S}_Y \cup S_{\tilde{N}})$. If $B \in \mathfrak{L}_0$, then $t(B) > 0$, if $B \in \mathfrak{L}_\infty$ then $t(B) < 0$.

Take $B \in \mathfrak{A}_0 \cup \mathfrak{A}_\infty$. By Lemma 6.6, a homogeneous polynomial $F_B \in \mathfrak{K}(Y \times V)$ of degree $p(B)$ over $\tilde{\chi}^* \mathfrak{K}(Y)$ and a thin analytic subset $T'(B)$ of Y exist such that $F_{B0} = F_B|(Y-T') \times V$ is holomorphic, such that $T(B) = (T'(B) \times V) \cap B$ is thin on B and such that $F_{B0}^{-1}(0) = B - T(B)$ with $\nu_{F_{B0}}(x) = 1$ if $x \in B - (T(B) \cup S_B \cup \tilde{S}_Y)$. Especially, $F_B \not\equiv 0$. Because $\mathfrak{A} = \mathfrak{A}_0 \cup \mathfrak{A}_\infty$ is finite,

$$F = \prod_{B \in \mathfrak{A}} F_B^{t(B)} \in \mathfrak{K}(Y \times V).$$

is defined and meromorphic on $\mathfrak{K}(Y \times V)$. Moreover,

$$F_0 = \prod_{B\in\mathfrak{A}_0} F_B^{t(B)} \in \mathfrak{K}(Y \times V)$$

$$F_\infty = \prod_{B\in\mathfrak{A}_\infty} F_B^{-t(B)} \in \mathfrak{K}(Y \times V)$$

are homogeneous polynomials of degree $p_a = \sum_{B\in\mathfrak{A}_a} p(B)|t(B)| > 0$ over $\tilde{\chi}^*\mathfrak{K}(Y)$ with $F = F_0/F_\infty$.

The set $T' = \bigcup_{B\in\mathfrak{A}} T'(B)$ is thin and analytic in Y. Define $T = T' \times V$. If $x \in N - (S_{\tilde{N}} \cup T \cup \tilde{S}_Y)$ then $x \in B - (S \cup T(B) \cup \tilde{S}_Y)$ for one and only one $B \in \mathcal{L}_0 \cup \mathcal{L}_\infty$ and $B \in \mathfrak{A}$. Moreover,

$$\nu_F(x) = t(B)\nu_{F_B}(x) = t(B) = \nu_f(x).$$

If $x \in Y \times V - (\tilde{N} \cup T)$ then f and F are holomorphic at x with $F(x) \neq 0 \neq f(x)$. Hence, $G = f/_F \in \mathfrak{K}(Y \times V)$ is meromorphic on $Y \times V$ and holomorphic on $Y \times V - (T \cup M)$ with $G(x) \neq 0$ if $x \in Y \times V - (T \cup M)$. Moreover, G is homogeneous of degree $p + p_\infty - p_0$. Because $G(x) \neq 0$ if $x \in Y \times V - (T \cup M)$, Lemma 6.1 implies, that $p + p_\infty - p_0 = 0$ and that $G(y,z) = G(y,0) = G_0(y)$ for all $y \in Y - (T' \cup M')$, where G_0 is holomorphic on $Y - (T' \cup M')$. Define $j: Y \to Y \times V$ by $j(y,0) = y$. Then $j^{-1}(P_G) \subseteq T' \cup M'$. Therefore $j^*(G) = G_1$ exists in $\mathfrak{K}(Y)$ with $G_1|(Y - T' \cup M') = G_0$. Now, $G = \tilde{\chi}^*(G_1)$ on $Y \times V - T \cup M$. Therefore, $G = \tilde{\chi}^*(G_1)$ on $Y \times V$. Then $g =$ $\tilde{\chi}^*(G_1)F_0 = GF_0 \in \mathfrak{K}(Y \times V)$ is a homogeneous polynomial of degree p_0 over $\tilde{\chi}^*\mathfrak{K}(Y)$. Now, $h = F_\infty \neq 0$ is a homogeneous polynomial of degree p_∞ over $\tilde{\chi}^*\mathfrak{K}(Y)$ with $p = p_0 - p_\infty$. Moreover, $f = G\cdot h =$

$GF_0/F_\infty = g/h$. Hence, the theorem is proved if Y is normal.

2. Now, consider the general case. Let $\pi: \hat{Y} \to Y$ be the normalization of Y. Define $\tilde{\eta}$ by (6.2), $\tilde{\chi}$ by (6.4) and let $\hat{\eta}: \hat{Y} \times V \to V$ and $\hat{\chi}: \hat{Y} \times V \to \hat{Y}$ be the projections. Now, $\tilde{\pi} = \pi \times \mathrm{Id}: \hat{Y} \times V \to Y \times V$ is the normalization of $Y \times V$ with $\tilde{\eta}\circ\tilde{\pi} = \hat{\eta}$, $\tilde{\chi}\circ\tilde{\pi} = \pi\circ\hat{\chi}$. Take $\lambda \in \mathbb{C} - \{0\}$. Then $\tilde{\lambda}: Y \times V \to Y \times V$ is defined by $\tilde{\lambda}(y,z) = (y,\lambda z)$ if $(y;z) \in Y \times V$ and $\check{\lambda}: \hat{Y} \times V \to \hat{Y} \times V$ is defined by $\check{\lambda}(y,z) = (y,\lambda z)$. Then $\tilde{\pi}\circ\check{\lambda} = \hat{\lambda}\circ\tilde{\pi}$. Moreover, $\tilde{\lambda}^*(f) = \lambda^p\cdot f$. Because $\tilde{\lambda}^*: \mathfrak{K}(Y \times V) \to \mathfrak{K}(Y \times V)$, and $\check{\lambda}^*: \mathfrak{K}(\hat{Y} \times V) \to \mathfrak{K}(\hat{Y} \times V)$ and $\tilde{\pi}^*: \mathfrak{K}(Y \times V) \to \mathfrak{K}(\hat{Y} \times V)$ are isomorphisms, $\tilde{\pi}^*\circ\hat{\lambda}^* = \check{\lambda}^*\circ\tilde{\pi}^*$. Hence, $\check{\lambda}^*(\tilde{\pi}^*(f)) = \tilde{\pi}^*(\hat{\lambda}^*(f)) = \tilde{\pi}^*(\lambda^p f) = \lambda^p\cdot\tilde{\pi}^*(f)$. Therefore, $\tilde{\pi}^*(f) \in \mathfrak{K}(\hat{Y} \times V)$ is homogeneous of degree p. Homogeneous polynomials $\tilde{g}$ and $\tilde{h} \not\equiv 0$ over $\hat{\chi}^*\mathfrak{K}(\hat{Y})$ exist in $\mathfrak{K}(\hat{Y} \times V)$ such that $\tilde{\pi}^*(f) = \tilde{g}/\tilde{h}$ and such that $p = p_0 - p_\infty$, where $p_0 = \tilde{g} \geqq 0$ and $p_\infty =$ degree $\tilde{h} \geqq 0$.

Let $e_0,\ldots,e_k$ be a base of V. Let $\varepsilon_0,\ldots,\varepsilon_k$ be the dual base. Define $\tilde{\varepsilon}_\mu = \varepsilon_\mu\circ\tilde{\eta}$ and $\hat{\varepsilon}_\mu = \varepsilon_\mu\circ\hat{\eta} = \tilde{\varepsilon}_\mu\circ\tilde{\pi} = \tilde{\pi}^*(\tilde{\varepsilon}_\mu)$. Meromorphic functions $\tilde{c}_{a\mu_0\ldots\mu_k} \in \mathfrak{K}(\hat{Y})$ such that

$$\tilde{g} = \sum_{\mu_0+\ldots+\mu_k=p_0} \hat{\chi}^*(\tilde{c}_{0\mu_0\ldots\mu_k})\hat{\varepsilon}_0^{\mu_0}\ldots\hat{\varepsilon}_k^{\mu_k}$$

$$\tilde{h} = \sum_{\mu_0+\ldots+\mu_k=p_\infty} \hat{\chi}^*(\tilde{c}_{\infty\mu_0\ldots\mu_k})\hat{\varepsilon}_0^{\mu_0}\ldots\hat{\varepsilon}_k^{\mu_k}$$

on $\hat{Y} \times V$. Because $\pi^*: \mathfrak{K}(Y) \to \mathfrak{K}(\hat{Y})$ is an isomorphism, meromorphic functions $c_{a\mu_0\ldots\mu_k} \in \mathfrak{K}(Y)$ exist such that $\pi^*(c_{a\mu_0\ldots\mu_k}) =$

$\tilde{c}_{a\mu_0\ldots\mu_k}$. Define

$$g = \sum_{\mu_0+\ldots+\mu_k=p_0} \tilde{\chi}^*(c_{0\mu_0\ldots\mu_k})\tilde{\varepsilon}_0^{\mu_0} \ldots \tilde{\varepsilon}_k^{\mu_k} \in \mathfrak{K}(Y \times V)$$

$$h = \sum_{\mu_0+\ldots+\mu_k=p_\infty} \tilde{\chi}^*(c_{\infty\mu_0\ldots\mu_k})\tilde{\varepsilon}_0^{\mu_0} \ldots \tilde{\varepsilon}_k^{\mu_k} \in \mathfrak{K}(Y \times V).$$

Then g and h are homogeneous polynomials of degree p_0 and p_∞ respectively over $\tilde{\chi}^*(\mathfrak{K}(Y))$. Because

$$\hat{\chi}^*(\tilde{c}_{a\mu_0\ldots\mu_k}) = \hat{\chi}^*(\pi^*(c_{a\mu_0\ldots\mu_k})) = \tilde{\pi}^*(\tilde{\chi}^*(c_{a\mu_0\ldots\mu_k}))$$

and because $\hat{\varepsilon}_\nu = \tilde{\pi}^*(\tilde{\varepsilon}_\nu)$, this implies $\tilde{\pi}^*(g) = \tilde{g}$ and $\tilde{\pi}^*(h) = \tilde{h} \not\equiv 0$. Hence, $\tilde{\pi}^*(f) = \tilde{\pi}^*(g)/\tilde{\pi}^*(h) = \tilde{\pi}^*(g/h)$. Therefore, $f = g/h$; q.e.d.

<u>Theorem 6.9.</u> Let Y be an irreducible complex space of dimension n. Let V be a complex vector space of dimension k + 1 with $k > 0$. Let $e_0,\ldots,e_k$ be any base of V. Let $\zeta_1,\ldots,\zeta_k$ be the associated projective coordinates on $\mathbb{P}(V)$. Let $\chi: Y \times \mathbb{P}(V) \to Y$ and $\eta: Y \times \mathbb{P}(V) \to \mathbb{P}(V)$ be the projections. Define $\tilde{\zeta}_\mu = \eta^*(\zeta_\mu)$. Then

$$(6.21) \qquad \mathfrak{K}(Y \times \mathbb{P}(V)) = \chi^*\mathfrak{K}(Y)(\tilde{\zeta}_1,\ldots,\tilde{\zeta}_k)$$

is a pure transcendental extension of degree k of $\chi^*\mathfrak{K}(Y)$.

<u>Proof.</u> Let $\varepsilon_0,\ldots,\varepsilon_k$ be the dual base to $e_0,\ldots,e_k$. Let $\rho: V - \{0\} \to \mathbb{P}(V)$ be the residual map. Then $\rho^*(\zeta_\mu) = \varepsilon_\mu/\varepsilon_0$.

Define $\tilde{\rho}$ by (6.1), $\tilde{\eta}$ by (6.2) and $\tilde{\chi}$ by (6.4). Then $\eta\circ\tilde{\zeta} = \rho\circ\tilde{\eta}$ and $\chi\circ\tilde{\rho} = \tilde{\chi}$. Define $\tilde{\varepsilon}_\mu = \varepsilon_\mu\circ\tilde{\eta} = \tilde{\eta}^*(\varepsilon_\mu)$. Then

$$(6.22)\qquad \tilde{\rho}^*(\tilde{\zeta}_\mu) = \tilde{\rho}^*(\eta^*(\zeta_\mu)) = \tilde{\eta}^*(\rho^*(\zeta_\mu)) = \tilde{\eta}^*({}^{\varepsilon_\mu}/_{\varepsilon_0}) = {}^{\tilde{\varepsilon}_\mu}/_{\tilde{\varepsilon}_0}.$$

Take $f \in \mathfrak{K}(Y \times \mathbb{P}(V))$. Then $F = \tilde{\rho}^*(f) \in \mathfrak{K}(Y \times V)$. If $\lambda \in \mathbb{C} - \{0\}$, then $\tilde{\lambda}: Y \times V \to Y \times V$ is defined with $\tilde{\rho}\circ\tilde{\lambda} = \tilde{\rho}$. Hence, $\tilde{\lambda}^*(F) = \tilde{\lambda}^*\circ\tilde{\rho}^*(f) = \tilde{\rho}^*(f) = F$. Therefore, F is homogeneous of degree 0. By Theorem 6.8, homogeneous polynomials $G \in \mathfrak{K}(Y \times V)$ of degree p and $H \in \mathfrak{K}(Y \times V) - \{0\}$ of degree p over $\chi^*\mathfrak{K}(Y)$ exist such that $F = {}^G/_H$. By Lemma 6.2

$$G = \sum_{\mu_0+\ldots+\mu_k=p} \tilde{\chi}^*(a_{\mu_0\ldots\mu_k})\tilde{\varepsilon}_0^{\mu_0} \ldots \tilde{\varepsilon}_k^{\mu_k}$$

$$H = \sum_{\mu_0+\ldots+\mu_k=p} \tilde{\chi}^*(b_{\mu_0\ldots\mu_k})\tilde{\varepsilon}_0^{\mu_0} \ldots \tilde{\varepsilon}_k^{\mu_k}$$

where $a_{\mu_0\ldots\mu_k}$ and $b_{\mu_0\ldots\mu_k}$ are meromorphic functions on Y. Define

$$g = \sum_{\mu_0+\ldots+\mu_k=p} \chi^*(a_{\mu_0\ldots\mu_k})\tilde{\zeta}_1^{\mu_1} \ldots \tilde{\zeta}_k^{\mu_k}$$

$$h = \sum_{\mu_0+\ldots+\mu_k=p} \chi^*(b_{\mu_0\ldots\mu_k})\tilde{\zeta}_1^{\mu_1} \ldots \tilde{\zeta}_k^{\mu_k}.$$

Now, $\chi\circ\tilde{\rho} = \tilde{\chi}$ and (6.22) imply

$$\tilde{\rho}^*(g) = G\cdot\tilde{\varepsilon}_0^{-p} \qquad \text{and} \qquad \tilde{\rho}^*(h) = G\cdot\tilde{\varepsilon}_0^{-p} \not\equiv 0.$$

Hence, $\tilde{\rho}^*(f) = F = {}^G/_H = \tilde{\rho}^*(g)/_{\tilde{\rho}^*}(g) = \tilde{\rho}^*({}^g/_h)$. Because $\tilde{\rho}^*$ is injective $f = g/h \in \chi^* \mathfrak{S}(Y)(\zeta_1,\ldots,\zeta_k)$. Hence, (6.21) is proved.

It remains to be shown that $\tilde{\zeta}_1,\ldots,\tilde{\zeta}_k$ are algebraically independent over $\chi^* \mathfrak{S}(Y)$. Suppose that $P \in \mathfrak{S}(Y)(\xi_1,\ldots,\xi_k)$ is given with $\chi^*(\tilde{\zeta}_1,\ldots,\tilde{\zeta}_k) = 0$. Then

$$P = P(\xi_1,\ldots,\xi_k) = \sum_{\mu_0+\ldots+\mu_k=p} c_{\mu_0\ldots\mu_k} \xi_1^{\mu_1} \ldots \xi_k^{\mu_k}$$

$$0 = \chi^* P(\tilde{\zeta}_1,\ldots,\tilde{\zeta}_k) = \sum_{\mu_0+\ldots+\mu_k=p} \chi^*(c_{\mu_0\ldots\mu_k})\tilde{\zeta}_1^{\mu_1}\ldots\tilde{\zeta}_k^{\mu_k}$$

$$0 = \varepsilon_0^p \tilde{\rho}^*(\chi^* P(\tilde{\zeta}_1,\ldots,\tilde{\zeta}_k)) = \sum_{\mu_0+\ldots+\mu_k=p} \tilde{\chi}^*(c_{\mu_0\ldots\mu_k})\tilde{\varepsilon}_0^{\mu_0} \ldots \tilde{\varepsilon}_k^{\mu_k}.$$

Let A be the largest open subset such that each $c_{\mu_0\ldots\mu_k}$ is holomorphic on A. Then A is open and dense in Y. If $y \in A$ and $z = z_0 e_0 + \ldots + z_p e_k \in V$, then

$$\sum_{\mu_0+\ldots+\mu_k=p} c_{\mu_0\ldots\mu_k}(y) z_0^{\mu_0} \ldots z_k^{\mu_0} = 0.$$

Hence, $c_{\mu_0\ldots\mu_k}(y) = 0$ for each $y \in A$. Therefore, $c_{\mu_0\ldots\mu_p} \equiv 0$ on Y. Hence, P and $\chi^*(\xi_1,\ldots,\xi_k)$ are the zero polynomials. Therefore, $\tilde{\zeta}_1,\ldots,\tilde{\zeta}_k$ are algebraically independent over $\chi^* \mathfrak{S}(Y)$; q.e.d.

§7. Semi-proper maps

Let X and Y be complex spaces. A map $\varphi: X \to Y$ is said to be semi-proper if and only if for every compact subset K of Y a compact subset K' of Y exists such that $\varphi(K') = K \cap \varphi(X)$. This concept was introduced by Kuhlmann in [6], [7] and [8]. A second proof is outlined in [6] and [7]. A third complicated proof is given in [8] which avoids the Remmert-Stein continuation theorem. For the readers assurance and convenience a proof shall be given here along the lines of Kuhlmann's second proof. An extension theorem of Shiffman [14] is used:

Theorem 7.1. (Bishop-Shiffman). Let U be an open subset of $\mathbb{C}^n$. Let E be a subset of U which is closed in the space U. Suppose that the (2k-1)-dimensional Hausdorff measure of E is zero. Let A be a pure k-dimensional analytic subset of U - E. Then the closure $\overline{A} \cap U$ of A in U is a pure k-dimensional analytic subset of U.

If $M \subseteq \mathbb{C}^n$, let $\mu_p(M)$ be its p-dimensional Hausdorff measure. If M is a smooth manifold of real dimension $q < p$ and if M is of class $\mathbb{C}^\infty$, then $\mu_p(M) = 0$. Because each analytic set A of an open subset U of $\mathbb{C}^n$ with $\dim A \leqq k$ is the at most countable union of smooth manifolds of dimension $\leqq 2k$, $\mu_p(A) = 0$ if $p > 2k$. If E is almost thin of dimension k, then $E \subseteq \bigcup_{\mu=0}^{\infty} A_\mu$ where each A_μ is analytic in an open subset U_μ of $\mathbb{C}^n$ with $\dim A_\mu \leqq k$. Hence, $\mu_p(E) = 0$ if $p > 2k$. Therefore, Theorem 7.1 is true, if E is closed in U and almost thin of real dimension 2k-2.

Theorem 7.2. Let X be a complex space. Let S be an analytic subset of an open A of S such that $\dim_x S \geqq n$ for every point $x \in S$.

Suppose that $E = X - A$ is almost thin of dimension $n - 1$. Then $\overline{S}$ is analytic in X. If S is pure k-dimensional, then $\overline{S}$ is pure k-dimensional.

<u>Proof.</u> At first assume, that S is pure k-dimensional. If $S = \emptyset$, the theorem is trivial. If $S \neq \emptyset$, then $k \geqq n$. Take $a \in \overline{S}$. An open neighborhood U of a and a biholomorphic map $\beta: U \to U'$ of U onto an analytic subset U' of an open subset V of $\mathbb{C}^s$ exists. The set $E' = \beta(U \cap E)$ is closed in U' and in V and is almost thin of dimension $n - 1$. Hence, $\mu_{2k-1}(E') = 0$, since $k \geqq n$. The set $S' = \beta(S \cap U)$ is not empty and is analytic and pure k-dimensional in $U' - E'$ and in $V - E'$. By Theorem 7.1, $\overline{S}' \cap V$ is a pure k-dimensional analytic subset of V. Obviously, $\overline{S}' \cap V \subseteq U'$ is the closure of S' in U'. Therefore, $\beta^{-1}(\overline{S}' \cap V)$ is the closure of $S \cap U$ in U. Hence, $\overline{S} \cap U = \beta^{-1}(\overline{S}' \cap V)$ and $\overline{S} \cap U$ is analytic and pure k-dimensional. Therefore, $\overline{S}$ is analytic and pure k-dimensional.

Now, consider the general case. Here $S = \bigcup_{k=n}^{\infty} S_k$, where S_k is empty or a pure k-dimensional analytic subset of A. Then $\overline{S}_k$ is empty or a pure k-dimensional analytic subset of X. If $a \in X$, an open neighborhood U of a exists such that $\dim U = m < \infty$. Hence, $\overline{S}_k \cap U = \emptyset$ if $k > m$. Therefore,

$$\overline{S} \cap U = \bigcup_{k=n}^{m} \overline{S}_k \cap U$$

is analytic in U. Hence, $\overline{S}$ is analytic; q.e.d.

<u>Lemma 7.3.</u> Let S be an analytic subset of the complex space X. Define $A = X - S$. Let $\mathfrak{L}$ be the set of branches of X. Define

$\mathfrak{B}_1 = \{B \in \mathfrak{B} \mid B \cap S \neq B\}$. Then $X_1 = \bigcup_{B \in \mathfrak{B}_1} B$ is an analytic subset of X and $\mathfrak{B}_1$ is the set of branches of X_1. Moreover, $\overline{A} = X_1$.

<u>Proof.</u> Obviously, X_1 is analytic and $\mathfrak{B}_1$ is the set of branches of X_1. Now

$$A = X - S = \bigcup_{B \in \mathfrak{B}} (B-S) = \bigcup_{B \in \mathfrak{B}_1} (B-S) = X_1 - S \subseteq X_1.$$

Hence, $\overline{A} \subseteq X_1$. Take $a \in X_1$. Then $B \in \mathfrak{B}_1$ with $a \in B$ exists. Now, $B \cap S$ is thin in B. Hence, $a \in B = \overline{B - S} = \overline{B \cap A} \subseteq \overline{A}$. Therefore, $X_1 \subseteq \overline{A}$, which implies $\overline{A} = X_1$; q.e.d.

<u>Lemma 7.4.</u> Let X and Y be complex spaces. Let $\varphi: X \to Y$ be a holomorphic map. Then φ is semi-proper, if and only if for every compact subset K of Y a compact subset K' of X exists such that $\varphi^{-1}(y) \cap K' \neq \emptyset$ if $y \in K \cap \varphi(X)$.

<u>Proof.</u> a) Suppose that the condition is satisfied for every compact subset K of Y. Define $K'' = \varphi^{-1}(K) \cap K'$, then K" is compact with $\varphi(K'') \subseteq \varphi(X) \cap K$. If $y \in \varphi(X) \cap K$, then $x \in \varphi^{-1}(y) \cap K'$ exists. Then $\varphi(x) = y \in K$. Hence, $x \in K' \cap \varphi^{-1}(K) = K''$. Therefore, $y \in \varphi(K'')$ which implies $\varphi(K'') \supseteq \varphi(X) \cap K$. Hence, $\varphi(K'') = \varphi(X) \cap K$. The map φ is semi-proper.

b) Suppose that φ is semi-proper. Let K be compact in Y. A compact subset K' of X exists such that $\varphi(K') = K \cap \varphi(X)$. If $y \in K \cap \varphi(X)$, then $\varphi^{-1}(y) \cap K' \neq \emptyset$; q.e.d.

<u>Lemma 7.5.</u> Let X and Y be complex spaces. Let $\varphi: X \to Y$ be a

semi-proper, holomorphic map. Then, $\varphi(X)$ is closed.

Proof. Take $a \in \overline{\varphi(X)}$. Let U be an open neighborhood of a such that $\overline{U}$ is compact. A compact subset K' of X exists such that $\varphi(K') = \overline{U} \cap \varphi(X)$. Then $U \cap \varphi(X) = \varphi(K') \cap U$ is closed in U. Hence $U \cap \varphi(X) = \overline{\varphi(X)} \cap U$, because U is open. Therefore, $a \in \varphi(X)$;

q.e.d.

Lemma 7.6. Let X and Y be complex spaces. Let $\varphi: X \to Y$ be a semi-proper, holomorphic map of strict rank n. Then $\varphi(X)$ is a pure n-dimensional analytic subset of Y.

Proof. By Lemma 1.16 and Lemma 1.21, the analytic set $D = \{x \in X | \operatorname{rank}_x \varphi \leqq n-1\}$ is thin in X and $D' = \varphi(D)$ is almost thin of dimension n - 2 in Y. Let N be the set of all points $y \in \varphi(X)$ such that $\varphi(X) \cap U$ is a pure n-dimensional analytic subset of an open neighborhood U of y. Obviously N is open in $\varphi(X)$. Take $a \in \varphi(X) - D'$. Let K be a compact neighborhoood of a. A compact subset K' of X exists such that $\varphi(X) \cap K = \varphi(K')$. Since $\varphi^{-1}(a) \cap K' \subseteq X - D$, every point $b \in \varphi^{-1}(a) \cap K'$ has an open neighborhood U(b) such that $\varphi(U(b)) = W(b)$ is a pure n-dimensional analytic subset of an open neighborhood V(b) of a with $V(b) \subseteq K$. (Proposition 1.21). Finitely many points $b_1, \ldots, b_s$ exist in $\varphi^{-1}(a) \cap K'$ such that

$$\varphi^{-1}(a) \cap K \subseteq U(b_1) \cup \ldots \cup U(b_s).$$

Define $V_0 = V(b_1) \cap \ldots \cap V(b_s)$. Then $W_0 = V_0 \cap (W(b_1) \cup \ldots \cup W(b_s))$ is a pure n-dimensional analytic subset of the open neighborhood V_0 of a with $V_0 \subseteq K$.

It is claimed, that an open neighborhood V of a in V_0 exists such that $\varphi(X) \cap V = W_0 \cap V$. Clearly, $\varphi(X) \cap V \supseteq W_0 \cap V$ for any neighborhood V. Suppose that $\varphi(X) \cap V \supset W_0 \cap V$ for every open neighborhood V of a. Then a sequence $\{y_\nu\}_{\nu \in \mathbb{N}}$ converges to a with $y_\nu \in (\varphi(X) - W_0) \cap V_0 \subseteq K$. A point $x_\nu \in K'$ with $\varphi(x_\nu) = y_\nu$ exists. A subsequence $\{x_{\nu_\lambda}\}_{\lambda \in \mathbb{N}}$ converges to a point $b \in K'$ with $\varphi(b) = a$. Hence, $b \in \varphi^{-1}(a) \cap K$ and $b \in U(b_\mu)$ for some index μ. A number λ_0 exists such that $x_{\nu_\lambda} \in U(b_\mu)$ if $\lambda \geqq \lambda_0$. Hence, $y_{\nu_x} = \varphi(x_{\nu_\lambda}) \in W(b_\mu) \cap V_0 \subseteq W_0$, which is wrong. Therefore, an open neighborhood V of a in V_0 exists such that $\varphi(X) \cap V = W_0 \cap V$. Especially, $\varphi(X) \cap V$ is a pure n-dimensional analytic subset of the open neighborhood $a \in V$. Therefore, $a \in N$. Consequently, $\varphi(X) - D' \subseteq N$ and $\varphi(X) - N \subseteq D'$.

The closed subset $E = \varphi(X) - N$ of Y is almost thin of dimension $n - 2$ and N is analytic and pure n-dimensional in $Y - E$. (Lemma 7.5). By Theorem 7.2, $\overline{N}$ is a pure n-dimensional analytic subset of X. Now, $\overline{N} \subseteq \varphi(X)$, because $N \subseteq \varphi(X)$ and because $\varphi(X)$ is closed.

Take $y \in \varphi(X)$. Let Z be any open neighborhood of y. Take $x \in \varphi^{-1}(y)$ and let U be an open neighborhood of X with $\varphi(U) \subseteq Z$. Take $v \in U - D$. By Proposition 1.21 an open neighborhood G of v in $U - D$ exists such that $\varphi(G)$ is a pure n-dimensional analytic subset of an open neighborhood H of $\varphi(v)$. Then $\varphi(G) - D' \neq \emptyset$, because D' is almost thin of dimension $n - 2$. Take $u \in \varphi(G) - D'$, then $u \in (\varphi(X) - D') \cap Z \subseteq N \cap Z$. Hence, $N \cap Z \neq \emptyset$ for every open neighborhood of y. Therefore, $y \in \overline{N}$. Hence, $\overline{N} \supseteq \varphi(X)$. Consequently, $\overline{N} = \varphi(X)$. The set $\varphi(X)$ is analytic and pure n-dimensional;

q.e.d.

Lemma 7.7. Let X and Y be complex spaces. Let $\varphi: X \to Y$ be a holomorphic map of finite rank n. Define $D = \{x \in X \mid \mathrm{rank}_x \varphi < n\}$. Define $A = X - D$. Then $X_0 = \overline{A}$ is an analytic subset of X. The map $\varphi_0 = \varphi: X_0 \to Y$ has strict rank n. If φ is semi-proper, then φ_0 is semi-proper and $\varphi(X_0)$ is a pure n-dimensional analytic subset of Y.

Proof. By Theorem 1.14, D is analytic. Let $\mathfrak{B}$ be the set of branches of X. Define $\mathfrak{B}_0 = \{B \in \mathfrak{B} \mid B \cap D \neq B\}$. By Lemma 7.3, $X_0 = \bigcup_{B \in \mathfrak{B}_0} B$ is analytic in X with $X_0 = \overline{A}$ and $\mathfrak{B}_0$ is the set of branches of X_0. By Lemma 1.17, rank $\varphi|B \leqq n$ if $B \in \mathfrak{B}$. If $B \in \mathfrak{B}_0$, a simple point x of X exists with $x \in B - D$. Then $\mathrm{rank}_x \varphi|B = \mathrm{rank}_x \varphi = n$. Hence, rank $\varphi|B = n$ if $B \in \mathfrak{B}_0$. Because $\varphi_0|B = \varphi|B$ if $B \in \mathfrak{B}_0$, this implies rank $\varphi_0|B = n$ for all branches B of X_0. By Lemma 1.16, φ_0 has strict rank n.

Now, assume, in addition, that φ is semi-proper. Take any compact subset K of Y. An open neighborhood Z of K exists such that $\overline{Z}$ is compact. A compact subset K' of X exists such that $\varphi(X) \cap \overline{Z} = \varphi(K')$. Then $K'' = K' \cap X_0$ is a compact subset of X. Take $y \in \varphi_0(X_0) \cap K$. Let V be an open neighborhood of y with a metric d on V. A point $x \in X_0$ with $\varphi(x) = \varphi_0(x) = y$ exists. Then $x \in B$ for some $B \in \mathfrak{B}_0$. A sequence of points $\{x_\nu\}_{\nu \in \mathbb{N}}$ converges to x with $x_\nu \in B - D$ and with $y_\nu = \varphi(x_\nu) \in V$. By Proposition 1.21, an open neighborhood U_ν of x_ν exists in X such that $\varphi(U_\nu)$ is a pure n-dimensional analytic subset in some open subset V_ν of V with diameter $V_\nu < \frac{1}{\nu}$. By Lemma 1.30, $D' = \varphi(D)$ is almost thin of dimension n - 1. Hence, $z_\nu \in \varphi(U_\nu) - D'$ exists.

Because $x_\nu \to x$ for $\nu \to \infty$, $y_\nu = \varphi(x_\nu) \to y$ for $\nu \to \infty$. Now,

$z_\nu \in V_\nu$ implies $d(z_\nu, y_\nu) < 1/\nu$. Hence, $z_\nu \to y$ for $\nu \to \infty$. Hence, an index ν_0 exists such that $z_\nu \in Z$ if $\nu \geqq \nu_0$. Therefore, $w_\nu \in \varphi^{-1}(z_\nu) \cap K'$ exists. Because $z_\nu \in Y - D'$, then $\varphi^{-1}(z_\nu) \subseteq A$. Therefore, $w_\nu \in A \cap K' \subseteq X_0 \cap K' = K''$. A subsequence $\{w_{\nu_\lambda}\}_{\lambda \in \mathbb{N}}$ converges to a point $w \in K''$ with $\varphi(w) = \lim_{\lambda \to \infty} z_{\nu_\lambda} = y$. Hence, $\varphi_0^{-1}(y) \cap K'' \neq \emptyset$. By Lemma 7.4, φ_0 is semi-proper. By Lemma 7.6, $\varphi_0(X_0) = \varphi(X_0)$ is a pure n-dimensional analytic subset of Y; q.e.d.

<u>Lemma 7.8.</u> Let X and Y be complex spaces. Let $\varphi: X \to Y$ be a semi-proper holomorphic map. Let S be an analytic subset of Y. Suppose that $T = \varphi^{-1}(S) \neq X$. Define $A = X - T$. Then $X_1 = \overline{A}$ is an analytic subset of X and $\varphi_1 = \varphi: X_1 \to Y$ is semi-proper.

<u>Proof.</u> Let K be a compact subset of Y. An open neighborhood Z of K exists such that $\overline{Z}$ is compact. A compact subset K' of X exists such that $\varphi(K') = \varphi(X) \cap \overline{Z}$. Then $K'' = X_1 \cap K'$ is a compact subset of X_1. Take $y \in \varphi_1(X_1) \cap K = \varphi(X_1) \cap K$. A point $x \in X_1$ exists such that $y = \varphi(x)$. A sequence $\{x_\nu\}_{\nu \in \mathbb{N}}$ of points $x_\nu \in A$ converges to x. Then $y_\nu = \varphi(x_\nu) \to \varphi(x) = y$ for $\nu \to \infty$. An index ν_0 exists such that $y_\nu \in Z_\nu$ for $\nu \geqq \nu_0$. A point $w_\nu \in K'$ with $\varphi(w_\nu) = y_\nu$ exists. Because $x_\nu \in X - \varphi^{-1}(S)$, the point $y_\nu = \varphi(x_\nu)$ does not belong to S. Hence, $w_\nu \in X - \varphi^{-1}(S) = A$. Therefore, $w_\nu \in X_1 \cap K' = K''$. A subsequence of $\{w_\nu\}_{\nu \in \mathbb{N}}$ converges to a point $w \in K''$ with $\varphi_1(w) = \varphi(w) = y$. Hence, $\varphi_1^{-1}(y) \cap K'' = \emptyset$. The map φ_1 is semi-proper. By Lemma 7.3, X_1 is analytic; q.e.d.

Theorem 7.9. (Kuhlmann). Let X and Y be complex spaces. Let $\varphi\colon X \to Y$ be a semi-proper, holomorphic map. Then $\varphi(X)$ is analytic.

Proof. At first, the theorem will be proved by induction for holomorphic maps of finite rank. Suppose that rank $\varphi = 0$. Let B be a branch of X. Take a simple point x of X with $x \in B$. Then $\dim_x \varphi^{-1}(\varphi(x)) = \dim B - \operatorname{rank}_x \varphi = \dim B$. Hence, $B \subseteq \varphi^{-1}(\varphi(x))$. Therefore, φ is constant on each branch of X. Therefore, $\varphi(X)$ is countable. If $a \in Y$, take a compact neighborhood K. A compact set K' in X exists with $\varphi(K') = \varphi(X) \cap K$. Because $K' \cap B \neq \emptyset$ for only finitely many branches of X, $\varphi(X) \cap \varphi(K')$ is finite. Therefore, $\varphi(X)$ is a pure 0-dimensional analytic subset of Y.

Now, let φ be a semi-proper holomorphic map of rank n and suppose that the theorem is true for all semi-proper, holomorphic maps of rank less than n. Then,

$$D = \{x \in X \mid \operatorname{rank}_x \varphi < n\}$$

is an analytic subset of X with $D \neq X$. Define $A = X - D$. Then $X_0 = \overline{A}$ is analytic and $S = \varphi(X_0)$ is a pure n-dimensional analytic subset of Y by Lemma 7.7. If $\varphi(X) = S$, the image $\varphi(X)$ is analytic. Suppose that $\varphi(X) \neq S$. Then $T = \varphi^{-1}(S) \neq X$. Define $A_1 = X - T$. Then $X_1 = \overline{A}_1$ is an analytic subset of X and $\varphi_1 = \varphi\colon X_1 \to Y$ is semi-proper. Let $\mathfrak{B}$ be the set of branches of X. Define $\mathfrak{B}_0 = \{B \in \mathfrak{B} \mid B \cap D \neq B\}$ and $\mathfrak{B}_1 = \{B \in \mathfrak{B} \mid B \cap T \neq B\}$. By Lemma 7.3, $X_0 = \bigcup_{B \in \mathfrak{B}_0} B$ and $X_1 = \bigcup_{B \in \mathfrak{B}_1} B$. Take $B \in \mathfrak{B}_1$. If $B \cap D \neq B$, then $B \subseteq X_0$ and $\varphi(B) \subseteq S$ and $B \subseteq T$ which is wrong. Hence, $B \subseteq D$.

Because B is irreducible, B is contained in a branch C of D. Because B is a maximal irreducible analytic subset of X, B = C. Hence, B is a branch of D. By Lemma 1.30, rank $\varphi|D \leqq n - 1$. By Lemma 1.17 rank $\varphi|B \leqq n - 1$ for every $B \in \mathcal{B}_1$. Now, $\mathcal{B}_1$ is the set of branches of X_1, hence, Lemma 1.17 implies rank $\varphi_1 \leqq n - 1$. By induction, $\varphi_1(X_1) = \varphi(X_1)$ is an analytic subset of Y. If $y \in \varphi(X) - S$, then $y = \varphi(x)$ with $x \in X - T \subseteq X_1$. Hence, $y \in \varphi(X_1)$. Therefore, $\varphi(X) = S \cup \varphi(X_1)$ is an analytic subset of Y.

The theorem is proved, if φ has finite rank. Suppose that φ has infinite rank. Take $a \in Y$. An open neighborhood V of a exists such that $\dim V < \infty$. Define $U = \varphi^{-1}(V)$. Then $\varphi_2 = \varphi: U \to V$ is semi-proper with $\varphi_2(U) = V \cap \varphi(X)$. By Lemma 1.8, rank $\varphi_2 \leqq \dim V < \infty$. Hence, $\varphi_2(U) = V \cap \varphi(X)$ is analytic in V. Therefore, $\varphi(X)$ is analytic; q.e.d.

The theorem of Shiffman implies an improvement of Lemma 3.11.

<u>Proposition 7.10.</u> Let Y be a complex space of pure dimension n. Let E be a closed subset which is almost thin of dimension n - 2. Define A = Y - E. Let f be a meromorphic function on A. Then one and only one meromorphic function F on Y exists such that $F|A = f$.

<u>Proof.</u> Let P_f be the set of poles of f. Then $A_0 = A - P_f$ is open and dense in A. Define $A_0^* = \{(x,f(x)) | x \in A_0\}$ and let $\Gamma = \overline{A_0^*}$ be the closure in $Y \times \mathbb{P}$. Then $\Gamma_1 = (A \times \mathbb{P}) \cap \Gamma$ is the closure of A_0^* in $A \times \mathbb{P}$. By Proposition 3.8, Γ_1 is a pure n-dimensional analytic subset of $A \times P$. Now, $E \times \mathbb{P} = Y \times \mathbb{P} - A \times \mathbb{P}$ is

almost thin of dimension n - 1. Therefore, $\overline{\Gamma}_1$ is a pure n-dimensional analytic subset of $Y \times \mathbb{P}$. Now, $A_0^* \subseteq \overline{\Gamma}_1$ and $\Gamma \supseteq \Gamma_1$ implies $\Gamma \subseteq \overline{\Gamma}_1$. Hence, $\Gamma = \overline{\Gamma}_1$. By Proposition 3.8, a meromorphic function F exists such that $F|A_0 = f|A_0$. Then $F|A = f$, because A_0 is dense in A. Obviously, F is unique; q.e.d.

Lemma 7.11. Let X and Y be complex spaces of pure dimension m and n respectively with $m - n = q \geqq 0$. Let $\varphi: X \to Y$ be a surjective, semi-proper holomorphic map. Let B be a branch of Y and take $b \in B$. Then $c \in \varphi^{-1}(b)$ exist. A branch C of X with $c \in C$ and $\varphi(C) \subseteq B$ exist. If $x \in C$, then

$$\mathrm{rank}_x \varphi|C \geqq \mathrm{rank}_x \varphi.$$

Proof. The set $Z = \varphi^{-1}(B)$ is analytic. If $K \subseteq B$ is compact, a compact subset K' of X with $\varphi(K') = K$ exists. Then $K' \subseteq Z$. Hence, $\psi = \varphi: Z \to B$ is semi-proper, surjective and holomorphic with $\psi^{-1}(y) = \varphi^{-1}(y)$ if $y \in B$. The set $D = \{x \in Z | \mathrm{rank}_x \psi \leqq n-1\}$ is analytic. If $D = Z$, then $\psi(D) = \psi(Z) = B$ is almost thin of dimension n - 1 by Proposition 1.23, which contradicts dim B = n. Hence, $D \neq Z$. Define $A = Z - D$. Because $n = \dim Y$ and $D \neq Z$, rank $\psi = n$. By Lemma 7.7, $Z_0 = \overline{A}$ is analytic and $\psi_0 = \psi: Z_0 \to B$ is semi-proper and has strict rank n. The image $\psi(Z_0)$ is analytic and pure n-dimensional. Hence, $\psi_0(Z_0) = B$. Let S_Z be the set of non-simple points of Z. Let C be any branch of Z. By Lemma 7.3 C is a branch of Z. Hence, $C \cap (S_Z \cup D)$ is a thin analytic subset of C. Take $x \in C - (S_Z \cup D)$. Then $\dim C = \dim_x Z$ and

$$m \geqq \dim C = \dim_x Z = \tilde{\text{rank}}_x \psi + \dim_x \psi^{-1}(\psi(x))$$

$$= \text{rank}_x \psi + \dim_x \varphi^{-1}(\varphi(x))$$

$$= n + m - \text{rank}_x \varphi \geqq m$$

by Lemma 1.8. Hence, $\dim C = m$. The analytic set Z_0 is pure m-dimensional. Because ψ_0 is surjective, a point $c \in Z_0$ and a branch C of Z_0 with $c \in C$ and $\varphi(c) = b$ exist. Then $\varphi(C) \subseteq B$. Because C is an irreducible, m-dimensional analytic subset of X, the set C is a branch of X. If $x \in C$, then

$$\text{rank}_x \varphi | C = m - \dim_x \varphi^{-1}(\varphi(x)) \cap C$$

$$\geqq m - \dim_x \varphi^{-1}(\varphi(x)) = \text{rank}_x \varphi \qquad \text{q.e.d.}$$

<u>Lemma 7.12.</u> Let X and Y be complex spaces. Let $\varphi\colon X \to Y$ be a holomorphic map. Suppose that every point b of $\overline{\varphi(X)}$ has an open neighborhood W such that $V = \varphi^{-1}(W)$ and such that $\tilde{\varphi} = \varphi\colon V \to W$ is semi-proper. Then φ is semi-proper.

<u>Proof.</u> Let K be a compact subset of Y. Then open subsets $W_1, \ldots, W_p$ of Y with $V_\mu = \varphi^{-1}(W_\mu) \neq \emptyset$ exist such that

$$K \cap \overline{\varphi(X)} \subseteq W_1 \cup \ldots \cup W_p$$

and such that $\varphi_\mu = \varphi\colon V_\mu \to W_\mu$ are semi-proper. Open subsets U_μ of W_μ exist such that $\overline{U}_\mu \subset W_\mu$ and $\overline{U}_\mu$ is compact and

$$K \cap \overline{\varphi(X)} \subseteq \overline{U}_1 \cup \ldots \cup \overline{U}_p.$$

Compact subsets K'_μ of $V_\mu \subseteq X$ exist such that $\varphi(K'_\mu) = \overline{U}_\mu \cap \varphi(V_\mu) = \overline{U}_\mu \cap \varphi(X)$. Then $K' = K'_1 \cup \ldots \cup K'_p$ is compact. Take $y \in K \cap \varphi(X)$. Then $y \in \varphi(K'_\mu)$ for some μ. Hence, $x \in K'_\mu$ exists such that $y = \varphi(x)$. Hence, $x \in \varphi^{-1}(y) \cap K'$. By Lemma 7.4, φ is semi-proper; q.e.d.

§8. Quasi-proper maps

Let X and Y be complex spaces. A holomorphic map $\varphi: X \to Y$ is said to be quasi-proper if and only if for every compact subset K of Y a compact subset K' of X exists such that $B \cap K' \neq \emptyset$ for every branch B of $\varphi^{-1}(y)$ if $y \in K \cap \varphi(X)$. By Lemma 7.4, a quasi-proper map is semi-proper. Hence, the following Lemma is true.

Lemma 8.1. Let X and Y be complex spaces. Let $\varphi: X \to Y$ be a quasi-proper holomorphic map. Then $\varphi(X)$ is analytic. If φ has strict rank n, then $\varphi(X)$ is pure n-dimensional.

The restriction of a quasi-proper map to an analytic subset may not be quasi-proper. However, in many important cases, information about the restriction can be obtained.

Lemma 8.2. Let X and Y be complex spaces. Let S be a closed subset of X. Let $\varphi: X \to Y$ be a quasi-proper holomorphic map. Assume that $\varphi^{-1}\varphi(x) \cap S$ contains a branch of $\varphi^{-1}(\varphi(x))$ if $x \in S$. Then $\varphi(S)$ is closed. If S is also analytic, then $\varphi_0 = \varphi: S \to Y$ and $\varphi_1 = \varphi: S \to \varphi(S)$ are semi-proper and then $\varphi(S)$ is analytic in Y.

Proof. 1. Take $b \in \overline{\varphi(S)}$. Let U be an open neighborhood of b such that $\overline{U}$ is compact. A compact subset K' of X exists such that $K' \cap B \neq \emptyset$ if B is a branch of $\varphi^{-1}(y)$ and if $y \in \varphi(X) \cap \overline{U}$. A sequence $\{y_\nu\}_{\nu \in \mathbb{N}}$ of points of $\varphi(S)$ converges to b. An index ν_0 exists such that $y_\nu \in U$ for $\nu \geqq \nu_0$. Then $x_\nu \in \varphi^{-1}(y_\nu) \cap S$ exists for each $\nu \in \mathbb{N}$. A branch B_ν of $\varphi^{-1}(\varphi(x_\nu)) = \varphi^{-1}(y_\nu)$ is contained in S. If $\nu \geqq \nu_0$, then $R_\nu \cap K' \neq \emptyset$. Hence, $z_\nu \in B_\nu \cap K'$ exists

A subsequence of $\{z_\nu\}_{\nu\in\mathbb{N}}$ converges to a point $a \in S$. Then $b = \varphi(a) \in \varphi(S)$. The set $\varphi(S)$ is closed.

2. Let K be a compact subset of Y. A compact subset K' of X exists such that $K' \cap B \neq \emptyset$ if B is a branch of $\varphi^{-1}(y)$ and if $y \in \varphi(X) \cap K$. The subset $K'' = K \cap S$ of S is compact. If $y \in K \cap \varphi(S)$, a branch B of $\varphi^{-1}(y)$ is contained in S. Then $B \cap K' \neq \emptyset$ with $B \cap K' = B \cap K''$. Now, $\varphi_0^{-1}(y) \cap K'' = \varphi^{-1}(y) \cap S \cap K' \supseteq B \cap K'' \neq \emptyset$. By Lemma 7.4, φ_0 is semi-proper. By Theorem 7.9, $\varphi(S) = \varphi_0(S)$ is analytic in Y.

3. Let K be a compact subset of $\varphi(S)$. A compact subset K' of S exists such that $\varphi_0(K') = K \cap \varphi_0(S)$. Now, $\varphi_0(S) = \varphi_1(S) = \varphi(S)$ and $\varphi_0(K') = \varphi_1(K')$. Hence, $\varphi_1(K') = K \cap \varphi_1(S) = K$. Therefore, φ_1 is semi-proper; q.e.d.

Lemma 8.3. Let X and Y be complex spaces. Let $\varphi: X \to Y$ be a quasi-proper, holomorphic map. Let S be a φ-saturated[9] analytic subset of X. Then $\varphi_0 = \varphi: S \to Y$ is quasi-proper and $\varphi(S)$ is analytic.

Proof. Let K be a compact subset of Y. A compact subset K' of X exists such that each branch of $\varphi^{-1}(y)$ intersects K' if $y \in \varphi(X) \cap K$. Then $K'' = K' \cap S$ is a compact subset of S. Take $y \in \varphi_0(S) \cap K = \varphi(S) \cap K$. Let B be a branch of $\varphi_0^{-1}(y) = \varphi^{-1}(y) \cap S$. Then B is a branch of $\varphi^{-1}(y)$. Hence, $K'' \cap B = K' \cap B \neq \emptyset$. Therefore, φ_0 is quasi-proper. By Lemma 8.1, $\varphi(S)$ is analytic; q.e.d.

Proposition 8.4. Let X and Y be complex spaces. Suppose that X has pure dimension m. Let $\varphi: X \to Y$ be a quasi-proper, holomorphic map. Let p be a non-negative integer. Define $E = \{x \in X \mid \operatorname{rank}_x \varphi \leq p\}$. Then $\varphi_0 = \varphi: E \to Y$ is quasi-proper and $\varphi(E)$ is analytic. If E is thin, then $\dim \varphi(E) \leqq p - 1$.

<u>Proof.</u> By Lemma 1.19, E is φ-saturated. Therefore, φ_0 is quasi-proper by Lemma 8.3. Therefore, $\varphi(E) = \varphi_0(E)$ is analytic by Lemma 8.1. If E is thin, then rank $\varphi_0 \leqq p - 1$ by Lemma 1.20. Hence, $\varphi(E)$ is almost thin of dimension $p - 1$ by Proposition 1.23. Therefore, $\dim \varphi(E) \leqq p - 1$; q.e.d.

<u>Lemma 8.5.</u> Let X and Y be complex spaces. Let $\varphi: X \to Y$ be a quasi-proper, holomorphic map.

a) If T' is analytic in Y and if $T = \varphi^{-1}(T') \neq \emptyset$, then $\varphi_0 = \varphi: T \to T'$ and $\varphi_1 = \varphi: T \to Y$ are quasi-proper.

b) If V' is open in Y and if $V = \varphi^{-1}(V')$, then $\varphi_1 = \varphi: V \to V'$ is quasi-proper.

<u>Proof.</u> a) T is φ-saturated. By Lemma 8.3, φ_1 is quasi-proper. Let K be a compact subset of T'. A compact subset K' of T exists such that $K' \cap B \neq \emptyset$ for each branch of $\varphi_0^{-1}(y) = \varphi^{-1}(y) \cap T$ if $y \in K \cap \varphi_1(T)$. Then $y \in K \cap \varphi_0(T)$ and $\varphi_0^{-1}(y) = \varphi_1^{-1}(y)$. Therefore, K' intersects each branch of $\varphi_0^{-1}(y)$ for each $y \in K \cap \varphi_0(T) = K \cap \varphi_1(T)$. The map φ_0 is quasi-proper.

b) Let K be a compact subset of V'. A compact subset K' of X exists such that $K' \cap B \neq \emptyset$ for each branch B of $\varphi^{-1}(y)$ if $y \in K \cap \varphi(X)$. Then $K'' = K' \cap \varphi^{-1}(K)$ is compact and contained in V. Take $y \in K \cap \varphi_2(V)$. Then $K'' \cap B = K' \cap B \neq 0$ for every branch B of $\varphi^{-1}(y) = \varphi_2^{-1}(y)$. Hence, φ_2 is quasi-proper; q.e.d.

<u>Lemma 8.6.</u> Let $X_1,\ldots,X_p,Y_1,\ldots,Y_p$ be complex spaces. Let

$\varphi_\nu: X_\nu \to Y_\nu$ be quasi-proper holomorphic maps. Define $X = X_1 \times \dots \times X_p$ and $Y = Y_1 \times \dots \times Y_p$. Then

$$\varphi = \varphi_1 \times \dots \times \varphi_p: X \to Y$$

is a quasi-proper holomorphic map.

Proof. Let K be a compact subset of Y. Compact sets $K_\nu \subseteq Y_\nu$ exist such that $K \subseteq K_1 \times \dots \times K_p$. For each K_ν a compact subset K'_ν of X_ν exists such that each branch of $\varphi_\nu^{-1}(y_\nu)$ intersects K'_ν if $y_\nu \in K_\nu \cap \varphi_\nu(X_\nu)$. Define $K' = K'_1 \times \dots \times K'_p$. Take $y = (y_1,\dots,y_p) \in K \cap \varphi(X)$. Then $y_\nu \in K_\nu \cap \varphi_\nu(X_\nu)$ and

$$\varphi^{-1}(y) = \varphi_1^{-1}(y_1) \times \dots \times \varphi_p^{-1}(y_p).$$

Let B be a branch of $\varphi^{-1}(y)$. Then $B = B_1 \times \dots \times B_p$ where B_ν is a branch of $\varphi_\nu^{-1}(y_\nu)$. Then $B_\nu \cap K'_\nu \neq \emptyset$ for $\nu = 1,\dots,p$. Hence,

$$B \cap K' = (B_1 \cap K'_1) \times \dots \times (B_p \cap K'_p) \neq \emptyset.$$

Therefore, φ is quasi-proper; q.e.d.

For the next Lemma, recall the notation in and before Lemma 1.28.

Lemma 8.7. Let X and Y be complex spaces. Let $\varphi: X \to Y$ be a quasi-proper, holomorphic map. Let p be a positive integer. Then $\delta_\varphi: X_\varphi^p \to Y$ is quasi-proper.

<u>Proof.</u> By Lemma 8.6. $\varphi_p: X^p \to Y^p$ is quasi-proper. By Lemma 8.5, the restriction $\varphi_p: X^p_\varphi \to \Delta_Y$ is quasi-proper. Because $\delta_Y: \Delta_Y \to Y$ is biholomorphic, $\delta_\varphi = \delta_Y \circ \varphi_p: X^p_\varphi \to Y$ is quasi-proper;

q.e.d.

<u>Lemma 8.8.</u> Let X and Y be complex spaces. Let $\varphi: X \to Y$ be a holomorphic map. Let C be a connectivity component of X. Then C is φ-saturated. If φ is quasi-proper, then $\varphi_0 = \varphi: C \to Y$ is quasi-proper and $\varphi(C)$ is analytic.

<u>Proof.</u> Take $y \in \varphi(C)$. Then $\varphi^{-1}(y) \cap C = \varphi_0^{-1}(y)$ is open and closed in $\varphi^{-1}(y)$. Hence, $\varphi_0^{-1}(y)$ is a union of branches of $\varphi^{-1}(y)$. Hence, the closed set C is φ-saturated. If φ is quasi-proper, then φ_0 is quasi-proper and $\varphi(C)$ is analytic by Lemma 8.2; q.e.d.

<u>Proposition 8.9.</u> Let X and Y be complex spaces. Let $\varphi: X \to Y$ be a quasi-proper, holomorphic map. Let S be an analytic subset of $X \neq S$. Define $A = X - S$ and $\overline{A} = X_0$. Then X_0 is analytic in X and $\varphi_0 = \varphi: X_0 \to Y$ is semi-proper.

<u>Proof.</u> Let K be a compact subset of Y. Let Z be an open neighborhood of K such that $\overline{Z}$ is compact. A compact subset K' of X exists such that each branch of $\varphi^{-1}(y)$ intersects K' if $y \in \overline{Z} \cap \varphi(X)$. The set $K'' = X_0 \cap K'$ is compact.

Let $\mathfrak{B}$ be the set of branches of X. Define $\mathfrak{B}_0 = \{B \in \mathfrak{B} \mid B \cap S \neq B\}$. By Lemma 7.3, $X_0 = \bigcup_{B \in \mathfrak{B}_0} B = \overline{A}$ is analytic in X and $\mathfrak{B}_0$ is the set of branches of X_0. Take $y \in \varphi(X_0) \cap K$. Take

$x \in \varphi_0^{-1}(y) = \varphi^{-1}(y) \cap X_0$. Then $B \in \mathcal{B}_0$ exists with $x \in B$. A sequence $\{x_\nu\}_{\nu \in \mathbb{N}}$ of simple points x_ν of X with $x_\nu \in B - S = B \cap A$ converges to x. For each ν, the branch B of X is the only branch of X containing x_ν. Because $y_\nu = \varphi(x_\nu) \to y$ for $\nu \to \infty$, an index ν_0 exists such that $y_\nu \in Z$ for all $\nu \geqq \nu_0$. Let C_ν be a branch of $\varphi^{-1}(y_\nu)$ which contains x_ν. Because C_ν is irreducible in X, a branch $B_\nu \in \mathcal{B}$ with $B_\nu \supseteq C_\nu$ exists. Now, $x_\nu \in C_\nu \in B_\nu$ implies $B_\nu = B$. Hence, $C_\nu \subseteq B \subseteq X_0$. For $\nu \geqq \nu_0$, a point $w_\nu \in C_\nu \cap K' = C_\nu \cap K''$ exists. A subsequence $\{w_{\nu_\lambda}\}_{\lambda \in \mathbb{N}}$ converges to a point $w \in K''$ with $\varphi_0(w) = \varphi(w) = y$. Hence, $\varphi_0^{-1}(y) \cap K'' \neq \emptyset$. The map φ_0 is semiproper; q.e.d.

Now, an example will be given where φ_0 in Proposition 8.9 is semi-proper, but not quasi-proper.

Define

$$N = \bigcup_{n=1}^{\infty} \{z \in \mathbb{C} \mid |z| = n\}$$

$X = (\mathbb{C}^2 \times \{0\}) \cup (\{0\} \times \mathbb{C}^2) - \{0\} \times N \times \{0\}$ as a subset of C^3.

$Y = \mathbb{C}$

$S = \{0\} \times \mathbb{C}^2$

$\varphi(u,v,w) = u.$

Then

$$A = \mathbb{C}^2 \times \{0\} - \{0\} \times \mathbb{C} \times \{0\}$$

$$\overline{A} = X_0 = \mathbb{C}^2 \times \{0\} - \{0\} \times N \times \{0\}.$$

Then X_0 is a branch of X. Define $\varphi_0 = \varphi: X_0 \to \mathbb{C}$. If K is a compact subset of $\mathbb{C}$, then $K \times \{1\} \times \{0\} = K'$ is compact and contained in X_0 and X. If $0 \neq u \in K$, then $\varphi^{-1}(u) = \{u\} \times \mathbb{C} \times \{0\}$ is irreducible and intersects K'. If u = 0, then $\varphi^{-1}(0) = \{0\} \times C^2 - \{0\} \times N \times \{0\}$ is irreducible and intersects K'. Hence, φ is quasi-proper. If $0 \neq u \in K$, then $\varphi_0^{-1}(u) = \varphi^{-1}(u) = \{u\} \times \mathbb{C}$ is irreducible and intersects K'. If u = 0, then $\varphi_0^{-1}(0) = \{0\} \times (\mathbb{C}-\check{N}) \times \{0\}$ intersects K'. Hence, φ_0 is semi-proper in accordance with Proposition 8.9. Observe that $\varphi_0^{-1}(0)$ has infinitely many branches. If φ_0 would be quasi-proper, a compact subset K' of X_0 would exist such that each branch of $\varphi_0^{-1}(0)$ would intersect K'. But only finitely many branches of an analytic set intersect a compact set. Therefore, $\varphi_0^{-1}(0)$ would have only finitely many branches, which is wrong. Hence, φ_0 is not quasi-proper.

Proposition 8.10. Let X and Y be complex spaces. Let $\varphi: X \to Y$ be a q-fibering, holomorphic map. Let $S \neq X$ be an analytic subset of X. Define A = X - S. Then $X_0 = \overline{A}$ is a φ-saturated analytic subset of X. The map $\varphi_0 = \varphi: X_0 \to Y$ is q-fibering. If φ is quasi-proper, then φ_0 is quasi-proper.

Proof. Let $\mathfrak{B}$ be the set of branches of X. Define $\mathfrak{B}_0 = \{B \in \mathfrak{B} \mid B \cap S \neq B\}$. By Lemma 7.3, $\overline{A} = X_0 = \bigcup_{B \in \mathfrak{B}_0} B$ is analytic in X and $\mathfrak{B}_0$ is the set of branches of X_0. Take $a \in X_0$. Define $b = \varphi(a) = \varphi_0(a)$. Then $B \in \mathfrak{B}_0$ exists with $a \in B$. Define $\varphi_1 = \varphi: B \to Y$. An open neighborhood U of a exists such that $\operatorname{rank}_a \varphi_1 \leqq \operatorname{rank}_x \varphi_1$ for all $x \in B \cap U$ (Lemma 1.6). Take a simple point x of X with $x \in (B-S) \cap U$. Then $q = \dim_x \varphi^{-1}(b) = \dim_x \varphi^{-1}(b) \cap B = \dim_x \varphi_1^{-1}(b)$.

Hence, $\operatorname{rank}_x \varphi = \dim B - q$. Therefore,

$$\dim_a \varphi_1^{-1}(b) = \dim B - \operatorname{rank}_a \varphi_1 \geqq \dim B - \operatorname{rank}_x \varphi_1 = q$$

which implies

$$q = \dim_a \varphi^{-1}(b) \geqq \dim_a \varphi^{-1}(b) \cap X_0 = \dim_a \varphi_0^{-1}(b)$$
$$\geqq \dim_a \varphi^{-1}(b) \cap B = \dim_a \varphi_1^{-1}(b) \geqq q.$$

Therefore, $\dim_a \varphi_0^{-1}(b) = q$. The map φ_0 is q-fibering.

Again, take $a \in X_0$, then $\varphi_0^{-1}(\varphi_0(a)) = \varphi^{-1}(\varphi(a)) \cap X_0$ is a pure q-dimensional analytic subset of the pure q-dimensional analytic subset $\varphi^{-1}(\varphi(a))$. Hence, $\varphi^{-1}(\varphi(a)) \cap X_0$ is a union of branches of $\varphi^{-1}(\varphi(a))$. Therefore, X_0 is φ-saturated. Hence, if φ is quasi-proper, φ_0 is quasi-proper by Lemma 8.2; q.e.d.

The analytic subset X_0 of X in Proposition 8.9 and Proposition 8.10 is a union of branches of X by Lemma 7.3. Now, let X be a complex space. Let $\mathfrak{B}$ be the set of branches of X. Let $\mathfrak{B}_0 \neq \emptyset$ be a subset of $\mathfrak{B}$. Then $X_0 = \bigcup_{B \in \mathfrak{B}_0} B$ is an analytic subset of X and $\mathfrak{B}_0$ is the set of branches of X_0. Define $\mathfrak{B}_1 = \mathfrak{B} - \mathfrak{B}_0$. Then $S = \bigcup_{B \in \mathfrak{B}_1} B$ is an analytic subset of X with $A = X - S \neq \emptyset$. Also $\mathfrak{B}_0 = \{B \in \mathfrak{B} \mid B \cap S \neq B\}$. Hence, $\overline{A} = X_0$. Proposition 8.9 and Proposition 8.10 can be reformulated to:

<u>Proposition 8.9'.</u> Let X and Y be complex spaces. Let $\varphi\colon X \to Y$ be a quasi-proper, holomorphic map. Let $\mathfrak{B}$ be the set

of branches of X. Let $\mathcal{B}_0 \neq \emptyset$ be a subset of . Then $X_0 = \bigcup_{B \in \mathcal{B}_0} B$ is analytic in X and $\varphi_0 = \varphi: X_0 \to Y$ is semi-proper.

Proposition 8.10'. Let X and Y be complex spaces. Let $\varphi: X \to Y$ be a q-fibering, holomorphic map. Let $\mathcal{B}_0 \neq \emptyset$ be a subset of the set $\mathcal{B}$ of branches of X. Then $X_0 = \bigcup_{B \in \mathcal{B}_0} B$ is φ-saturated and analytic in X. The map $\varphi_0 = \varphi: X_0 \to Y$ is q-fibering. If φ is quasi-proper, then φ_0 is quasi-proper.

Lemma 8.11. Let X be a complex space. Let Y be an irreducible complex space of dimension n. Let $\varphi: X \to Y$ be a semi-proper, holomorphic map of rank n. Then φ is surjective.

Proof. By Theorem 7.9, $\varphi(X)$ is analytic. Define $\varphi_0 = \varphi: X \to \varphi(X)$. By Lemma 1.5, rank φ_0 = rank $\varphi = n$. By Lemma 1.7, $n = \text{rank } \varphi_0 \leqq \dim \varphi(X)$. Hence, $\varphi(X)$ is not thin in Y. Because Y is irreducible, $\varphi(X) = Y$; q.e.d.

Of course, Lemma 8.11 holds if φ is quasi-proper.

Lemma 8.12. Let X and Y be complex spaces. Suppose that X consists of finitely many branches $B_1, \ldots, B_p$ only. Let $\varphi: X \to Y$ be a holomorphic map. Suppose that $\varphi_\lambda = \varphi: B_\lambda \to Y$ is quasi-proper for each $\lambda = 1, \ldots, p$. Then φ is quasi-proper.

Proof. Let K be a compact subset of Y. For each λ, a compact subset K_λ' of B_λ exists such that each branch of $\varphi^{-1}(y) \cap B_\lambda$ intersects K_λ' if $y \in K \cap \varphi(B_\lambda)$. Then $K' = K_1' \cup \ldots \cup K_p'$ is compact.

Take $y \in K \cap \varphi(X)$. Let C be a branch of $\varphi^{-1}(y)$. Then $C \subseteq B_\lambda$ for some index λ. Hence, $C \subseteq \varphi^{-1}(y) \cap B_\lambda = \varphi_\lambda^{-1}(y)$. Because C is a maximal irreducible analytic subset of $\varphi^{-1}(y) \supseteq \varphi_\lambda^{-1}(y)$, the set C is a branch of $\varphi_\lambda^{-1}(y)$. Therefore, $C \cap K' \supseteq C \cap K_\lambda' \neq \emptyset$; q.e.d.

Lemma 8.13. Let X, Y and Z be complex spaces. Let $\varphi: X \to Y$ $\psi: Y \to Z$ quasi-proper, holomorphic maps. Suppose that φ is p-fibering, that ψ is q-fibering and that $\chi = \psi \circ \varphi$ is (p+q)-fibering. Then χ is quasi-proper.

Proof. Let K be a compact subset of Z. A compact subset K' of Y exists such that each branch of $\psi^{-1}(z)$ intersects K' if $z \in K \cap \psi(Y)$. A compact subset K" of X exists such that each branch of $\varphi^{-1}(y)$ intersects K" if $y \in K' \cap \varphi(X)$.

Take $z \in K \cap \chi(X) \subseteq K \cap \psi(Y)$. Then

$$F = \chi^{-1}(z) = \varphi^{-1}(\psi^{-1}(z)) \neq \emptyset$$

has pure dimension p + q. Let B be a branch of F. Then dim B = p + q. By Lemma 8.5, $\varphi_0 = \varphi: F \to \psi^{-1}(z)$ is quasi-proper. Obviously, φ_0 is p-fibering. By Proposition 8.10', $\varphi_1 = \varphi_0: B \to \psi^{-1}(z)$ is p-fibering and quasi-proper. The map φ_1 has pure rank q. By Lemma 8.1 and Lemma 1.27, $\varphi(B)$ is an irreducible, q-dimensional analytic subset of $\psi^{-1}(z)$. Because $\psi^{-1}(z)$ is pure q-dimensional, $\varphi(B)$ is a branch of $\psi^{-1}(z)$. A point $y \in K' \cap \varphi(B)$ exists. Let C be a branch of $\varphi_1^{-1}(y) = \varphi^{-1}(y) \cap B$. Then C is an irreducible, p-dimensional analytic subset of the pure p-dimensional analytic set $\varphi^{-1}(y)$. Therefore, C is a branch of $\varphi^{-1}(y)$, which implies $C \cap K'' \neq \emptyset$. Because $C \subseteq B$, also $B \cap K'' \neq \emptyset$. The map χ is quasi-proper; q.e.d.

<u>Lemma 8.14.</u> Let X and Y be complex spaces. Let $\varphi: X \to Y$ be a q-fibering, holomorphic map. Let S be an analytic subset of X. Let T be the set of all $x \in S$ such that a branch B of $\varphi^{-1}(\varphi(x))$ with $x \in B \subseteq S$ exists. Then T is closed.

<u>Proof.</u> Take $a \in \overline{T} \subseteq S$. A sequence $\{x_\lambda\}_{\lambda \in \mathbb{N}}$ of points $x_\lambda \in T$ converges to a. If $x_\lambda = a$ for some $\lambda \in \mathbb{N}$, then $a \in T$. Hence, $x_\lambda \neq a$ for all $\lambda \in \mathbb{N}$ can be assumed. Define $y_\lambda = \varphi(x_\lambda)$ and $b = \varphi(a)$. Then $y_\lambda \to b$ for $\lambda \to \infty$. A branch B_λ of $\varphi^{-1}(y_\lambda)$ exists with $x_\lambda \in B_\lambda \subseteq S$. Define $A = \bigcup_{\lambda \in \mathbb{N}} B_\lambda$, then A is a pure q-dimensional analytic subset of $S = \varphi^{-1}(b)$, which is singular[10)] at $a \in \varphi^{-1}(b) \cap S$. Here $\varphi^{-1}(b) \cap S$ is analytic with $q \geqq \dim \varphi^{-1}(b) \cap S$. By Remmert and Stein [10], a q-dimensional branch B of $\varphi^{-1}(b) \cap S$ with $a \in B$ exists such that A is singular at every point of B. Because $\varphi^{-1}(b)$ is pure q-dimensional, B is also a branch of $\varphi^{-1}(b)$. Therefore, $a \in T$. The set T is closed; q.e.d.

<u>Proposition 8.15.</u> Let X and Y be complex spaces of pure dimension m and n respectively with $m - n = q \geqq 0$. Let $\varphi: X \to Y$ be a quasi-proper, q-fibering, holomorphic map. Let S be a thin analytic subset of X. Let R be the set of all $y \in \varphi(S)$ such that at least one branch of $\varphi^{-1}(y) \cap S$ is also a branch of $\varphi^{-1}(y)$. Then R is a thin analytic subset of Y. (Observe that R is the set of all $y \in \varphi(S)$ such that a branch of $\varphi^{-1}(y)$ is contained in S.)

Proof. A branch B of $\varphi^{-1}(y)$ is a branch of $\varphi^{-1}(y) \cap S$ if and only if $B \subseteq S$. Let T be the set of all $x \in S$ such that a branch of $\varphi^{-1}(\varphi(x))$ with $x \in B \subseteq S$ exists. Then, $\varphi(T) = R$. By Lemma 8.14, T is closed.

Now, it shall be shown, that T is analytic. Take $a \in T$. An open connected neighborhood U of a in X and a holomorphic map $\psi: U \to \mathbb{C}^k$ exists such that $\psi^{-1}(0) = U \cap S$. Define $\chi: U \to Y \times \mathbb{C}^k$ by $\chi(x) = (\varphi(x), \psi(x))$. If $(y,z) \in Y \times \mathbb{C}^k$, then

$$\chi^{-1}(y,z) = \varphi^{-1}(y) \cap \psi^{-1}(z).$$

The set $E = \{x \in U \mid \operatorname{rank}_x \chi \leqq n\}$ is analytic in U. Take $x \in U$. Then

$$\begin{aligned}\operatorname{rank}_x \chi &= m - \dim_x \varphi^{-1}(\varphi(x)) \cap \psi^{-1}(\psi(x)) \\ &\geqq m - \dim_x \varphi^{-1}(\varphi(x)) = m - q = n .\end{aligned}$$

Therefore, $E = \{x \in U \mid \operatorname{rank}_x \chi = n\}$. Take $x \in T \cap U$. A branch B of $\varphi^{-1}(\varphi(x))$ with $x \in B \subseteq S$ exists. Hence, $B \subseteq \psi^{-1}(0)$ and $\psi(x) = 0$. Then B has dimension q. Therefore,

$$\begin{aligned}\operatorname{rank}_x \chi &= m - \dim_x \varphi^{-1}(\varphi(x)) \cap \psi^{-1}(0) \\ &\leqq m - \dim_x B \cap \psi^{-1}(0) = m - \dim_x B = m - q = n\end{aligned}$$

Hence, $x \in E \cap S$. Therefore, $U \cap T \subseteq E \cap S$.

Take $x \in E \cap S$. Then $\psi(x) = 0$ and

$$n = \text{rank}_x \chi = m - \dim_x \varphi^{-1}(\varphi(x)) \cap \psi^{-1}(0)$$

$$= m - \dim_x \varphi^{-1}(\varphi(x)) \cap S$$

or

$$\dim_x \varphi^{-1}(\varphi(x)) \cap S = m - n = q.$$

A branch B of $\varphi^{-1}(\varphi(x)) \cap S$ with $x \in B$ and $\dim B = q$ exists. Then B is an irreducible, q-dimensional analytic subset of the pure q-dimensional analytic subset $\varphi^{-1}(\varphi(x))$. Hence, B is a branch of $\varphi^{-1}(\varphi(x))$ and $x \in B \subseteq S$. Therefore, $x \in T \cap U$, which implies $T \cap U \supseteq E \cap S$. Hence, $T \cap U = E \cap S$ is analytic in U. Consequently, T is analytic in X.

Take $a \in T$. A branch B of $\varphi^{-1}(\varphi(a))$ with $a \in B \subseteq S$ exists. Take $x \in B$. Then B is a branch of $\varphi^{-1}(\varphi(x)) = \varphi^{-1}(\varphi(a))$ with $x \in B \subseteq S$. Therefore, $x \in T$. Hence, $B \subseteq T$. Therefore, B is a branch of $\varphi^{-1}(\varphi(a)) \cap T$. By Lemma 8.2, $R = \varphi(T)$ is analytic, because φ is quasi-proper. By Lemma 1.26, R is almost thin. Because R is analytic, R is thin; q.e.d.

Lemma 8.16. Let X and Y be complex spaces. Let $\varphi: X \to Y$ be a holomorphic map. Let (α, β, π) be a product representation of φ. Then $\varphi_0 = \varphi: U_\alpha \to U_\beta$ is quasi-proper.

Proof. By Lemma 2.5.5 a strictly central section S of φ in U_α over U_β exists. Then S is an analytic subset of U_α, the restriction $\varphi_1 = \varphi: S \to U_\beta$ is proper and light. If $y \in \varphi(U_\alpha)$, then each branch of $\varphi^{-1}(y) \cap U_\alpha = \varphi_0^{-1}(y)$ intersects S. Let K be a compact subset of U_β. Then $K' = \varphi_1^{-1}(K) = \varphi^{-1}(K) \cap S$ is compact. If $y \in K \cap \varphi_0(U_\alpha)$ and if B is a branch of $\varphi_0^{-1}(y)$, a point $x \in S \cap B$

exists. Then $\varphi(x) \in K$ and $x \in S \cap K = K'$. Hence, $B \cap K' \neq \emptyset$. The map φ_0 is quasi-proper; q.e.d.

Lemma 8.17. Let X and Y be complex spaces. Let $\varphi: X \to Y$ be a holomorphic map. Let $\{V_\lambda\}_{\lambda \in \Lambda}$ be an open covering of Y and define $U_\lambda = \varphi^{-1}(V_\lambda)$. Suppose that each map $\varphi_\lambda = \varphi: U_\lambda \to V_\lambda$ is quasi-proper if $U_\lambda \neq \emptyset$. Then φ is quasi-proper.

Proof. Let K be a compact subset of Y. Then $K \subseteq V_{\lambda_1} \cup \ldots \cup V_{\lambda_p}$. Open sets W_{λ_μ} exist such that $\overline{W}_{\lambda_\mu}$ is compact and contained in V_{λ_μ} with

$$K \subseteq \overline{W}_{\lambda_1} \cup \ldots \cup \overline{W}_{\lambda_p}.$$

Compact subsets K'_μ of U_{λ_μ} exist, such that each branch of $\varphi^{-1}_{\lambda_\mu}(y) = \varphi^{-1}(y)$ intersects K'_μ if $y \in \varphi(U_{\lambda_\mu}) \cap \overline{W}_{\lambda_\mu}$. Then $K' = K'_1 \cup \ldots \cup K'_p$ is compact. Take $y \in \varphi(X) \cap K$ and let B be a branch of $\varphi^{-1}(y)$. Then $y \in \overline{W}_{\lambda_\mu} \cap \varphi(X)$ for some index μ. Moreover, $x \in X$ with $\varphi(x) = y \in V_{\lambda_\mu}$ exists. Hence, $x \in U_{\lambda_\mu}$. Therefore, $y \in \varphi(U_{\lambda_\mu}) \cap \overline{W}_{\lambda_\mu}$ which implies $K' \cap B \neq \emptyset$. The map φ is quasiproper; q.e.d.

Proposition 8.18. Let X and Y be complex spaces of pure dimension m and n respectively with $m - n = q \geqq 0$. Let $\varphi: X \to Y$ be an open, q-fibering, surjective holomorphic map such that $\varphi^{-1}(y)$ is irreducible for each $y \in Y$. Then φ is quasi-proper.

Proof. Take $b \in Y$. Then $a \in \varphi^{-1}(b)$ exists. By Proposition 2.4, a product representation (α,β,π) of φ centered at a and over b exists. By Lemma 2.2, $\beta(B) = U'_\beta$ for each branch B of U_β. Since $\{b\} = \beta^{-1}(\beta(b))$, each branch of U'_β contains b. Because φ is open, $\varphi(U_\alpha)$ is a pure n-dimensional, open subset of U_β which intersects each branch of U_β (Lemma 2.5.6). Hence, $\varphi(U_\alpha) = U_\beta$. The map $\varphi_0 = \varphi: U_\alpha \to U_\beta$ is surjective and quasi-proper by Lemma 8.16. Define $\tilde{U}_\beta = \varphi^{-1}(U_\beta)$. Let K be a compact subset of U_β. A compact subset K' of U_α exists such that each branch of $\varphi_0^{-1}(y) = \varphi^{-1}(y) \cap U_\alpha$ intersect K' if $y \in U_\beta$. Especially, $\varphi^{-1}(y) \cap K' \neq \emptyset$. Because $\varphi^{-1}(y)$ is irreducible, $\varphi_1 = \varphi: \tilde{U}_\beta \to U_\beta$ is quasi-proper. By Lemma 8.17, φ is quasi-proper; q.e.d.

Let X and Y be complex spaces. Let $\varphi: X \to Y$ be a holomorphic map. For $y \in Y$, let $\tau_\varphi(y)$ be the number of branches of $\varphi^{-1}(y)$. Then $0 \leqq \tau_\varphi(y) \leqq \infty$.

Lemma 8.19. Let X and Y be complex spaces. Let $\varphi: X \to Y$ be a quasi-proper holomorphic map. Then $\tau_\varphi(y) < \infty$ if $y \in Y$.

Proof. The set $K = \{y\}$ is compact. Let $\mathcal{L}$ be the set of branches of $\varphi^{-1}(y)$. If $y \in Y - \varphi(X)$, then $\tau_\varphi(y) = 0 < \infty$. If $y \in \varphi(X)$, a compact set K' exists in X such that $B \cap K' = 0$ for $B \in \mathcal{L}$. Hence, $\mathcal{L}$ is finite. i.e., $\tau_\varphi(y) < \infty$; q.e.d.

Lemma 8.20. Let X and Y be complex spaces of pure dimension m and n respectively with $q = m - n \geqq 0$. Let $\varphi: X \to Y$ be a quasi-proper, holomorphic map of strict rank n. Suppose that Y consists of finitely many branches. Then X consists of finitely many branches.

Proof. Let $\mathfrak{B}$ and $\mathfrak{L}$ be the sets of branches of X and Y respectively. Let $B \in \mathfrak{B}$ be a branch of X. By Proposition 8.9', $\varphi: B \to Y$ is semi-proper. By Lemma 1.16, $\varphi: B \to Y$ has rank n. By Lemma 7.6 and Lemma 1.27, $\varphi(B)$ is an n-dimensional irreducible analytic subset of Y. Therefore, $\varphi(B)$ is a branch of Y. A map $\tilde{\varphi}: \mathfrak{B} \to \mathfrak{L}$ is defined by $\tilde{\varphi}(B) = \varphi(B)$.

Let $C \in \mathfrak{L}$ be a branch of Y. Define $\mathfrak{B}_C = \tilde{\varphi}^{-1}(C)$. The set $D = \{x \in X | \operatorname{rank}_x \varphi \leqq n-1\}$ is thin in X. By Proposition 8.4, $\varphi(D) = D'$ is analytic with $\dim E' \leqq n - 2$. By Lemma 1.25, $D^* = \varphi^{-1}(D')$ is a thin analytic subset of X. Take $C \in \mathfrak{L}$. Take $y \in C - D'$. If $\mathfrak{B}_C = \emptyset$, then $\mathfrak{B}_C$ is finite. If $\mathfrak{B}_C \neq \emptyset$, then $y \in \varphi(B)$ for each $B \in \mathfrak{B}_C$. The set $K = \{y\} \subseteq \varphi(X)$ is compact. Hence, a compact subset K' of X exists, such that each branch of $\varphi^{-1}(y)$ intersects K'. Take $B \in \mathfrak{B}_C$. Then $x \in \varphi^{-1}(y) \cap B$ exists. The set $\varphi^{-1}(y)$ is contained X - D. Therefore, $\varphi^{-1}(y)$ is pure q-dimensional. Then

$$q \geqq \dim_x B \cap \varphi^{-1}(y) = \dim B - \operatorname{rank}_x \varphi | B \geqq m - n = q.$$

Therefore, $B \cap \varphi^{-1}(y)$ is pure q-dimensional. Each branch of $B \cap \varphi^{-1}(y)$ is a branch of $\varphi^{-1}(y)$. Let H be a branch of $B \cap \varphi^{-1}(y) \neq \emptyset$. Then $H \cap K' \neq \emptyset$ because H is a branch of $\varphi^{-1}(y)$. Especially, $B \cap K' \neq \emptyset$. This is true for each $B \in \mathfrak{B}_C$. Therefore, $\mathfrak{B}_C$ is finite. Consequently, $\mathfrak{B} = \bigcup_{C \in \mathfrak{L}} \mathfrak{B}_C$ is finite; q.e.d.

Lemma 8.21. Let X and Y be complex spaces of pure dimension m and n respectively with $m - n = q \geqq 0$. Let $\varphi: X \to Y$ be a quasi-proper, q-fibering, holomorphic map. Then τ_φ is finite and locally bounded.

Proof. By Lemma 8.19, $\tau_\varphi(y) < \infty$ if $y \in Y$. Take $b \in Y$. By Lemma 8.1, $\varphi(X)$ is analytic. If $b \in Y - \varphi(X)$, then $\tau_\varphi = 0$ on the open neighborhood $Y - \varphi(X)$. Therefore, only $b \in \varphi(X)$ has to be considered. Define $F = \varphi^{-1}(b)$. Let U be an open neighborhood of b such that $\overline{U}$ is compact. A compact subset K' of X exists such that each branch of $\varphi^{-1}(y)$ intersects K' if $y \in \overline{U} \cap \varphi(X)$. A finite number of points $a_1, \ldots, a_p$ in $F \cap K'$ and product representations $(\alpha_i, \beta_i, \pi_i)$ of φ centered at a_λ and over b exist such that

$$K' \cap F \subseteq U_{\alpha_1} \cup \ldots \cup U_{\alpha_p}.$$

An open neighborhood V of b exists such that

$$V \subseteq U_{\beta_1} \cap \ldots \cap U_{\beta_p} \cap U$$

$$K' \cap V \subseteq U_{\alpha_1} \cup \ldots \cup U_{\alpha_p}.$$

If this would be wrong, a sequence $\{y_\nu\}_{\nu \in \mathbb{N}}$ of points of $U_{\beta_1} \cap \ldots \cap U_{\beta_p} \cap U$ converges to b such that

$$x_\nu \in K' \cap \varphi^{-1}(y_\nu) - (U_{\alpha_1} \cup \ldots \cup U_{\alpha_p})$$

exists. A subsequence $\{x_{\nu_\lambda}\}_{\lambda \in \mathbb{N}}$ converges to a point

$$x \in K' \cap \varphi^{-1}(b) - (U_{\alpha_1} \cup \ldots \cup U_{\alpha_p}) = \emptyset$$

Let s_i be the sheet number of α_i. Define $s = s_1 + \ldots + s_p$. Then $s \geqq \tau_\varphi(y)$ for all $y \in V$ is claimed. If $y \in V - \varphi(X)$, then $\tau_\varphi(y) = 0 \leqq s$. Take $y \in V \cap \varphi(X)$. Let $\mathcal{L}$ be the set of branches of $\varphi^{-1}(y)$. Let $\mathcal{L}_i$ be the set of branches of $\varphi^{-1}(y) \cap U_{\alpha_i}$. By definition, $\tau_\varphi(y) = \#\mathcal{L}$. By Lemma 2.5.3, $\#\mathcal{L}_i \leqq s_i$. If $H \in \mathcal{L}_i$, one and only one $\lambda_i(H) \in \mathcal{L}$ with $\lambda_i(H) \supseteq H$ exists. A map $\lambda_i : \mathcal{L}_i \to \mathcal{L}$ is defined. If $B \in \mathcal{L}$, then $B \cap K' \neq \emptyset$, because $y \in V \cap \varphi(X) \subseteq \overline{U} \cap \varphi(X)$. Now

$$\emptyset \neq B \cap K' \subseteq \varphi^{-1}(y) \cap K' \subseteq U_{\alpha_1} \cup \ldots \cup U_{\alpha_p}.$$

Hence, $K' \cap B \cap U_{\alpha_i} \neq \emptyset$ for some index i. Since $B \cap U_{\alpha_i} \neq \emptyset$ is a pure q-dimensional analytic subset of the pure q-dimensional analytic subset $\varphi^{-1}(y) \cap U_{\alpha_i}$ of U_{α_i}, the set $B \cap U_{\alpha_i}$ is a union of branches of $\varphi^{-1}(y) \cap U_{\alpha_i}$. Hence, $H \in \mathcal{L}_i$ with $H \subseteq B \cap U_{\alpha_i}$ exists. Therefore, $\lambda_i(H) = B$. Consequently, $\mathcal{L} = \bigcup_{i=1}^{p} \lambda_i(\mathcal{L}_i)$ and

$$\tau_\varphi(y) = \#\mathcal{L} \leqq \sum_{i=1}^{p} \#\lambda_i(\mathcal{L}_i) \leqq \sum_{i=1}^{b} \#\mathcal{L}_i \leqq \sum_{i=1}^{p} s_i = s.$$

Therefore, $\tau_\varphi(y) \leqq s$ for each $y \in V$; q.e.d.

§9. $\mathfrak{K}_\varphi(X)$ as a finite algebraic extension of $\varphi^*\mathfrak{K}(Y)$

If X and Y are irreducible complex spaces with $\dim Y = n$, if $\varphi: X \to Y$ is a quasi-proper, holomorphic map of rank n, then $\mathfrak{K}_\varphi(X)$ is a finite algebraic extension of $\varphi^*\mathfrak{K}(Y)$, as shall be shown in this paragraph.

Lemma 9.1. Let X and Y be irreducible complex spaces of dimension m and n respectively with $q = m - n \geqq 0$. Let $\varphi: X \to Y$ be a quasi-proper, q-fibering, holomorphic map. Let $W \neq \emptyset$ be open in Y such that $\overline{W}$ is compact. Define $r = \sup \tau_\varphi(\overline{W})$. Then $r < \infty$. Take $f \in \mathfrak{K}_\varphi(X)$. Let $A = X - P_f$ be the largest open subset of X where f is holomorphic. Define $A^* = \{(x,f(x)) \mid x \in X\}$ and let $\Gamma = \overline{A}^*$ be the closure of A^* in $X \times \mathbb{P}$. Let $\pi: \Gamma \to X$ and $\upsilon: \Gamma \to \mathbb{P}$ be the projections. Define $\psi: \Gamma \to Y \times \mathbb{P}$ by $\psi(x,z) = (\varphi(x),z)$ if $(x,z) \in \Gamma$. Define $F = \psi(\Gamma)$ and $\psi_0 = \psi: \Gamma \to F$. Let $\chi: F \to Y$ and $\eta: F \to \mathbb{P}$ be the projections. Then

1. The set Γ is irreducible, analytic and m-dimensional in $X \times \mathbb{P}$. The maps π and υ are holomorphic. Moreover, π is proper and $\pi_0 = \pi: A^* \to A$ is biholomorphic.

2. The diagram

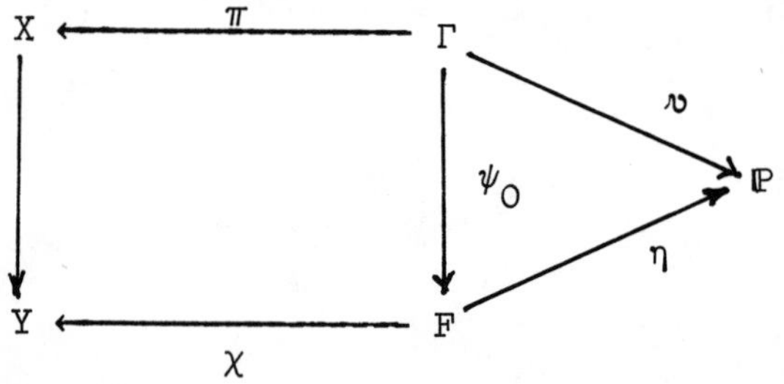

is commutative.

3. The map ψ is q-fibering and quasi-proper.

4. The set F is analytic, irreducible and has dimension n. The map $\psi_0 = \psi\colon \Gamma \to F$ is quasi-proper, q-fibering, surjective and holomorphic.

5. The projection χ is proper, surjective, holomorphic, has rank n and sheet number $s \leqq r$.

6. The pole set $P_\eta = F \cap (Y \times \{\infty\})$ is thin and analytic in F.

7. The function f is algebraic over $\varphi^*\mathfrak{K}(Y)$ with $[f\colon \varphi^*\mathfrak{K}(Y)] \leqq s \leqq r$.

<u>Proof.</u> By Lemma 8.21, τ_φ is locally bounded. Therefore, τ_φ is bounded on each compact subset of Y. Hence $r < \infty$. By Proposition 3.8, Γ is analytic. Moreover, $\pi\colon \Gamma \to X$ is proper and surjective. Because X is irreducible and m-dimensional, Γ is irreducible with $\dim \Gamma = m$. Obviously, π_0 is biholomorphic and υ is holomorphic. Trivially, the diagram commutes.

3. Because $f \in \mathfrak{K}_\varphi(X)$ is φ-dependent, rank $\psi \leqq n$. The map φ has pure rank n. By Lemma 8.11, φ is surjective. Take $(y,z) \in F = \psi(\Gamma)$. Then

$$\psi^{-1}(y,z) = (\varphi^{-1}(y) \times \{z\}) \cap \Gamma \subseteq \varphi^{-1}(y) \times \{z\}. \tag{9.1}$$

Hence, $\dim \psi^{-1}(y,z) \leqq q$. Take $(x,z) \in \psi^{-1}(y,z)$. Then

$$q \geqq \dim_{(x,z)} \psi^{-1}(y,z) = m - \operatorname{rank}_{(y,z)} \psi \geqq m - n = q.$$

Hence, ψ is q-fibering and has pure rank n.

Let K be a compact subset of $Y \times \mathbb{P}$. A compact subset K_0 of Y exists such that $K \subseteq K_0 \times \mathbb{P}$. A compact subset K' of X exists such that each branch of $\varphi^{-1}(y)$ intersects K' if $y \in K_0$. Then $K'' = \pi^{-1}(K')$ is a compact subset of Γ. Take $(y,z) \in K$ and let B be a branch of $\psi^{-1}(y,z)$. Then $B = B' \times \{z\}$, where B' is a q-dimensional, irreducible, analytic subset of $\varphi^{-1}(y)$ by (9.1). Hence, B' is a branch of $\varphi^{-1}(y)$. A point $x \in B' \cap K'$ exists. Then $(x,z) \in B \cap K''$. The map $\psi: \Gamma \to Y \times \mathbb{P}$ is quasi-proper.

4. By Lemma 8.2 and Lemma 1.27, $F = \psi(\Gamma)$ is an irreducible analytic subset of dimension n in $Y \times \mathbb{P}$, because Γ is irreducible and m-dimensional and because ψ is quasi-proper, q-fibering and has pure rank n. Obviously, $\psi_0 = \psi: \Gamma \to F$ is holomorphic, surjective, q-fibering, and has pure rank n. By Lemma 8.5 a, ψ_0 is quasi-proper.

5. Obviously, χ is holomorphic and proper. Moreover,

$$\chi(F) = \chi(\psi_0(\Gamma)) = \varphi(\pi(\Gamma)) = \varphi(X) = Y.$$

Hence, χ is surjective. Hence, rank $\chi \geqq n$ by Proposition 1.23. By Lemma 1.8, rank $\chi \leqq n$. Hence, rank $\chi = n$. As explained and defined at the beginning of §5, χ has a sheet number s, since F and Y are irreducible. The set

$$E = \{u \in F \mid \operatorname{rank}_n \chi \leqq n-1\}$$

is analytic and thin in F. The set $E' = \chi(E)$ is analytic in Y with $\dim E' \leqq n - 2$. Let S_Y be the set of non-simple points of Y.

The set $S' = S_Y \cup E'$ is thin and analytic in Y and $S = \chi^{-1}(S')$ is a thin analytic subset of F. Then $F_0 = F - S$ and $Y_0 = Y - S'$ are open, connected, and dense in F respectively Y. The map $\chi_0 = \chi\colon F_0 \to Y_0$ is proper, light, surjective, open and holomorphic. The set $N_2 = \{u \in F_0 | \nu_{\chi_0}(u) > 1\}$ is thin and analytic in F_0. The sets $N' = \chi_0(N_2)$ and $\tilde{N} = \chi_0^{-1}(N')$ are thin and analytic in Y_0 and and F_0 respectively. Moreover, for every $y \in F_0$ define $H(y) = \{z | (y,z) \in F_0\}$, then

$$\sum_{(y,z) \in H(y)} \nu_{\chi_0}(y,z) = s,$$

and $\#\chi_0^{-1}(y) \leqq s$ if $y \in Y_0$ and $\#\chi_0^{-1}(y) = s$ if $y \in Y_0 - N'$. (Lemma 2.2.) By definition s is the sheet number of χ.

If $x \in A$, then $(x,f(x)) \in A^* \subseteq X \times \mathbb{C}$ and $\psi(x,f(x)) = (\varphi(x),f(x)) \in F \cap (Y \times \mathbb{C})$. Therefore, $\eta^{-1}(\infty) = F \cap (Y \times \{\infty\}) = P_\eta$ is a thin analytic subset of F. Because χ is proper, $\chi(P_\eta) = P_\eta'$ is analytic in Y. By Lemma 1.9, $\dim_y P_\eta' \leqq n - 1$ if $y \in Y_0$. Hence, P_η' is thin in Y. (This proves 6.)

Let S_X be the set of non-simple points of X. By Proposition 8.15, the set R of all $y \in \varphi(S_X)$ such that S_X contains a branch of $\varphi^{-1}(y)$ is thin and analytic in Y.

Now, S', N', P_η' and R are nowhere dense in Y. Hence a point $y \in W - (S' \cup N' \cup P_\eta' \cup R) \subseteq W \cap Y_0$ exists. Then $s = \#\chi_0^{-1}(y) = \#\chi^{-1}(y)$ and $\chi^{-1}(y) = \{y\} \times H$, where $H \subseteq \mathbb{C}$ and $\#H = s$. Let $\mathfrak{L}$ be the set of branches of $\varphi^{-1}(y)$. Then $\#\mathfrak{L} \leqq r$, because $y \in W$. A map $\lambda: \mathcal{L} \to H$ shall be defined. Take $B \in \mathcal{L}$. Because $y \notin R$, the set $B - S_X$ is open, connected and dense in B. Take any point $x \in B - S_X$. If $x \in P_f$, a sequence $\{x_\nu\}_{\nu \in \mathbb{N}}$ of points of A converges to x such that $f(x_\nu) \to \infty$ for $\nu \to \infty$, because x is a simple point of X. Hence, $(x_\nu, f(x_\nu)) \to (x, \infty) \in \Gamma$ for $\nu \to \infty$, which implies $(y, \infty) = \psi_0(x, \infty) \in F$. Hence, $(y, \infty) \in P_\eta$ and $y \in P'_\eta$, which is wrong. Therefore, f is holomorphic on $B - S_X$; i.e., $B - S_X \subseteq A$. If $x \in B - S(X)$, then

$$\chi(\psi_0(x, f(x))) = \chi(\varphi(x), f(x)) = \chi(y, f(x)) = y.$$

Hence, $(y, f(x)) = \psi_0(x, f(x)) \in \chi^{-1}(y) = \{y\} \times H$. Therefore, $f(x) \in H$ for all $x \in B - S_X$. The holomorphic function $f|B - S_X$ assumes only finitely many values on the irreducible complex space $B - S_X$. Hence, if $f|B - S_X$ is constant. Let $\lambda(B) \in H$ be the constant function value of $f|B - S_X$. The map $\lambda: \mathcal{L} \to H$ is defined.

Take $z \in H$. Then $(y, z) \in \chi^{-1}(y)$. Let C be a branch of

$$\psi_0^{-1}(y, z) = (\varphi^{-1}(y) \times \{z\}) \cap \Gamma.$$

Because $\psi_0^{-1}(y, z)$ and $\varphi^{-1}(y)$ are pure q-dimensional, a branch B of $\varphi^{-1}(y)$ exists such that $C = B \times \{z\}$. Hence, $B \in \mathcal{L}$. If $x \in B - S_X$, then $x \in A$ and $\pi^{-1}(x) = \{(x, f(x))\}$. Also $(x, z) \in C \subseteq \Gamma$ with $(x, z) \in \pi^{-1}(x)$. Hence, $z = f(x)$. This is true for all $x \in B - S_X$. Hence, $z = \lambda(B)$. The map λ is surjective. Consequently, $r = \# \mathcal{L} \geqq \# \lambda(\mathcal{L}) = \# H = s$.

7. Because $F \cap (Y \times \{\infty\})$ is thin, η is a meromorphic function

on F. By Proposition 5.1 (or Theorem 5.3). $[\eta: \chi^*\mathfrak{K}(Y)] = s$. Meromorphic functions $a_0,\dots,a_{s-1}$ exist on Y such that

$$Q = \sum_{\mu=0}^{s-1} a_\mu \xi^\mu + \xi^p$$

with $Q \in \mathfrak{K}(Y)[\xi]$ and $\chi^*Q(\eta) = 0$, i.e.,

$$0 = \sum_{\mu=0}^{s-1} \chi^*(a_\mu)\eta^\mu + \eta^p$$

on F. By Proposition 5.1, the coefficients are holomorphic on $Y_0 = Y - S'$, where $S' = S_Y \cup E'$ was defined in 5. By Lemma 1.25, $\tilde{S} = \varphi^{-1}(S)$ is a thin analytic subset of X. Take $x \in A - \tilde{S}$. Then $\varphi(x) \in Y_0$ and $(x,f(x)) \in \Gamma$ which implies $(\varphi(x),f(x)) = \psi(x,f(x)) \in F$. Hence, $\eta(\varphi(x),f(x)) = f(x)$ and $\chi(\varphi(x),f(x)) = \varphi(x)$. Therefore,

$$\begin{aligned}\varphi^*Q(f)(x) &= \sum_{\mu=0}^{s-1} a_\mu(\varphi(x))f(x)^\mu + f(x)^p.\\ &= \sum_{\mu=0}^{s-1} a_\mu(\chi(\psi(x;f(x))))\eta(\psi(x,f(x)))^\mu\\ &\quad + \eta(\psi(x,f(x))))^p\\ &= 0.\end{aligned}$$

Therefore, $\varphi^*Q(f) = 0$ on X, because $\varphi^*Q(f)$ is meromorphic on X and $A - \tilde{S}$ is open and dense on X. Hence, $[f: \varphi^*\mathfrak{K}(Y)] \leqq s \leqq r$; q.e.d.

Let X and Y be complex spaces. Let $\varphi: X \to Y$ be holomorphic map. For $y \in Y$, let $\tau_\varphi(y)$ be the number of branches of $\varphi^{-1}(y)$. Then $0 \leqq \tau_\varphi(y) \leqq \infty$. If $\emptyset \neq M \subseteq Y$, define $\sigma_\varphi(M) = \sup \tau_\varphi(M)$. Then

$0 \leqq \sigma_\varphi(M) \leqq \infty$. Let $\mathfrak{M}$ be the set of all open sets $W \neq \emptyset$ in Y such that $\overline{W}$ is compact. Then $r_\varphi = \inf\{\sigma_\varphi(W) | W \in \mathfrak{M}\}$ is said to be the <u>branch number</u> of φ. If φ is quasi-proper, then τ_φ is bounded on each compact subset of Y. Hence, $\sigma_\varphi(W) < \infty$ for each $W \in \mathfrak{M}$, which implies $r_\varphi < \infty$.

<u>Theorem 9.2.</u> Let X and Y be irreducible complex spaces of dimension m and n respectively with $q = m - n \geqq 0$. Let $\varphi: X \to Y$ be a quasi-proper, holomorphic map of rank n. Let r_φ be the branch number of φ. Then $\mathfrak{K}_\varphi(X)$ is a finite algebraic extension field of $\varphi^*\mathfrak{K}(Y)$ with

$$[\mathfrak{K}_\varphi(X) : \varphi^*\mathfrak{K}(Y)] \leqq r_\varphi < \infty.$$

<u>Proof.</u> The set $D = \{x \in X | \operatorname{rank}_x\varphi \leqq n-1\}$ is analytic and thin in X. By Proposition 8.4, $D' = \varphi(D)$ is analytic in Y with $\dim D' \leqq n - 2$. By Lemma 1.25, $\tilde{D} = \varphi^{-1}(D')$ is analytic and thin in X. Then $X_1 = X - \tilde{D}$ and $Y_1 = Y - D'$ are open and dense in X and Y respectively. Also X_1 and Y_1 are irreducible complex spaces of dimension m and n respectively. The map $\varphi_1 = \varphi: X_1 \to Y_1$ is q-fibering, quasi-proper and holomorphic. Let $W_1 \neq \emptyset$ be open and $\overline{W}_1$ compact in Y_1. Then $\sigma(\overline{W}_1) = \sigma_{\varphi_1}(\overline{W}_1) < \infty$ by Lemma 9.1. Therefore, $r_\varphi \leqq \sigma_\varphi(\overline{W}_1) < \infty$. Because $\sigma_\varphi(W)$ either infinite or an integer $r_\varphi = \sigma_\varphi(W_0)$ for some open subset $W_0 \neq 0$ of Y with $\overline{W}_0$ compact. Now, $W_0 - D' \neq \emptyset$. Hence, take any open subset $W \neq \emptyset$ of Y, such that $\overline{W}$ is compact and contained in $W_0 - D' = W_0 \cap Y_1$. Then $r_\varphi \leqq \sigma_\varphi(W) \leqq \sigma_\varphi(\overline{W}) \leqq \sigma_\varphi(W_0) = r_\varphi$. Hence, $r_\varphi = \sigma_\varphi(W) = \sigma_{\varphi_1}(W) = \sigma_{\varphi_1}(\overline{W})$.

Take $f \in \mathfrak{K}_\varphi(X)$. By Lemma 4.14, $f_1 = f|X_1 \in \mathfrak{K}_{\varphi_1}(X_1)$. By Lemma 9.1, f_1 is algebraic over $\varphi_1^* \mathfrak{K}(Y_1)$ with $[f_1 : \varphi_1^* \mathfrak{K}(Y)] \leqq r_\varphi$ because $r_\varphi = \sup \tau_{\varphi_1}(\overline{W})$. A polynomial $Q \in \mathfrak{K}(Y_1)[\xi]$ of degree $s \leqq r_\varphi$ exists such that $\varphi_1^* Q(f_1) = 0$. Here $a_\mu \in \mathfrak{K}(Y_1)$ and

$$Q = \sum_{\mu=0}^{s-1} a_\mu \xi^\mu + \xi^s$$

with

$$0 = \sum_{\mu=0}^{s-1} \varphi_1^*(a_\mu) f_1^\mu + f_1^s$$

on Y_1. Because $Y_1 = Y - D'$, and because D' is analytic with $\dim D' \leqq n - 2$, Lemma 3.11 gives meromorphic functions $\tilde{a}_\mu \in \mathfrak{K}(Y)$ such that $\tilde{a}_\mu | Y_1 = a_\mu$. Then

$$\tilde{Q} = \sum_{\mu=0}^{s-1} \tilde{a}_\mu \xi^\mu + \xi^s \in \mathfrak{K}(Y)[\xi]$$

with $\tilde{Q}|Y_1 = Q$. By Lemma 3.4.2, $\varphi^*(\tilde{a}_\mu)|X_1 = \varphi_1^*(\tilde{a}_\mu|X_1) = \varphi_1^*(a_\mu)$. Hence,

$$0 = \sum_{\mu=0}^{s-1} \varphi^*(\tilde{a}_\mu) \cdot f^\mu + f^s$$

on X_1 and therefore, on X. Therefore, f is algebraic over $\varphi^* \mathfrak{K}(Y)$ with $[f : \varphi^* \mathfrak{K}(Y)] \leqq s \leqq r_\varphi < \infty$ for every $f \in \mathfrak{K}_\varphi(X)$ where r_φ does not depend on f. By Lemma 4.22, $\mathfrak{K}_\varphi(X)$ is a finite algebraic extension of $\varphi^* \mathfrak{K}(Y)$ with

$$[\mathfrak{K}_\varphi(X): \varphi^*\mathfrak{K}(Y)] \leqq r_\varphi \qquad \text{q.e.d.}$$

Now, some theorems about the continuation of meromorphic functions shall be proved.

Theorem 9.3. Let X be a complex space of pure dimension m. Let Y be an irreducible complex space of dimension n with $m - n = q \geqq 0$. Let $\varphi: X \to Y$ be a quasi-proper, holomorphic map of strict rank n. Let $V \neq \emptyset$ be an open subset of Y. Define $U = \varphi^{-1}(V)$. Then $U \neq \emptyset$. Define $\varphi_0 = \varphi: U \to V$. Let $f \in \mathfrak{K}(X)$ be a meromorphic function on X. Suppose that a meromorphic function $g \in \mathfrak{K}(V)$ on V exists such that $\varphi_0^*(g) = f|U$. Then g is unique. One and only one meromorphic function $h \in \mathfrak{K}(Y)$ exists such that $f = \varphi^*(h)$. Moreover, $h|V = g$.

Proof. By Lemma 8.4, the map φ is surjective. Hence, $U \neq \emptyset$ and φ_0 is defined. By Lemma 4.14, φ_0 has strict rank n. By Remark 3.5.2, $\varphi^*: \mathfrak{K}(Y) \to \mathfrak{K}(X)$ and $\varphi_0^*: \mathfrak{K}(V) \to \mathfrak{K}(U)$ are defined and injective homomorphisms. Hence, g is unique. Also h is unique if it exists. If h exists, Lemma 3.4.2 implies $\varphi_0^*(h|V) = \varphi^*(h)|U = f|U = \varphi_0^*(g)$. Hence, $g = h|V$. Only the existence of h remains to be proved.

1. Case: The space X is irreducible and φ is q-fibering: Define $A = X - P_f$ and $A^* = \{(x,f(x)) | x \in A\}$. Then $\Gamma = \overline{A^*}$ is an irreducible, m-dimensional analytic subset of $X \times \mathbb{P}$. The projection $\pi: \Gamma \to X$ is proper, surjective and holomorphic. The restriction $\pi_0 = \pi: A^* \to A$ is biholomorphic. Define $\psi: \Gamma \to Y \times \mathbb{P}$ by $\psi(x,z) = (\varphi(x),z)$ if $(x,z) \in \Gamma$. Define $F = \psi(\Gamma)$ and $\psi_0 = \psi: \Gamma \to F$. Let $\chi: F \to Y$ be the projection. By Lemma 9.1, F is an irreducible

analytic subset of dimension n of $Y \times \mathbb{P}$. The maps ψ and ψ_0 are quasi-proper, q-fibering and holomorphic. The map χ is proper, surjective, holomorphic and has rank n and sheet number s. The set $P_\eta = F \cap (Y \times \{\infty\})$ is thin and analytic in F. The function f is algebraic over $\varphi^* \mathfrak{K}(Y)$ with $[f: \varphi^* \mathfrak{K}(Y)] \leqq s$. Take $W \neq \emptyset$ as an open subset of $V - P_g$, with $\overline{W}$ compact and $\overline{W} \subseteq V$. Define E, E', S_Y, S', S, F, F_0, Y_0, χ_0, N_2, N', $\tilde{N}$, P_η', S_X, R, y, H, $\mathfrak{B}$, $\lambda: \mathfrak{B} \to H$ as in part 5 of the proof of Lemma 9.1. Then λ is surjective and $\#H = s$. Moreover, f is holomorphic on $\varphi^{-1}(W)$ because $W \subseteq V - P_g$ and $f(x) = g(\varphi(x))$ if $x \in \varphi^{-1}(W)$. Hence $f|\varphi^{-1}(y) = g(y) \in \mathbb{C}$. Now, $\lambda(B)$ is the constant function value of $f|B - S_X$ if $B \in \mathfrak{B}$ is a branch of $\varphi^{-1}(y)$. Hence, $\lambda(B) = g(y)$ for all $B \in \mathfrak{B}$. Therefore, $H = \lambda(\mathfrak{B}) = \{g(y)\}$. Hence, $s = \#H = 1$, which implies $f \in \varphi^* \mathfrak{K}(Y)$, i.e., $f = \varphi^*(h)$ for some $h \in \mathfrak{K}(Y)$. Case 1 is proved.

2. Case: The map $\varphi: X \to Y$ is q-fibering, but X may not be irreducible: Let $\{X_\lambda\}_{\lambda \in \Lambda}$ be the family of branches of X with $X_\lambda \neq X_\mu$ if $\lambda \neq \mu$. By Proposition 8.10', $\varphi_\lambda = \varphi: X_\lambda \to Y$ is quasi-proper, q-fibering and holomorphic. Especially φ_λ has pure rank n. By Lemma 8.11, φ_λ is surjective. Especially, $U_\lambda = X_\lambda \cap U = \varphi_\lambda^{-1}(V) \neq \emptyset$. Define $\varphi_{\lambda_0} = \varphi_\lambda: U_\lambda \to V$. Then $f|U_\lambda = \varphi_{\lambda_0}^*(g)$. A meromorphic function $h_\lambda \in \mathfrak{K}(Y)$ exists such that $f|X_\lambda = \varphi_\lambda^*(h_\lambda)$. Then $g = h_\lambda|V$. Because Y is irreducible $h_\lambda = h$ is independent of $\lambda \in \Lambda$. $f|X_\lambda = \varphi_\lambda^*(h) = \varphi^*(h)|X_\lambda$ for each $\lambda \in \Lambda$. Hence, $f = \varphi^*(h)$ on X.

3. Case. The general situation is considered. Because φ has strict rank n, $D = \{x \in X | \operatorname{rank}_x \varphi \leqq n-1\}$ is thin and analytic. By Lemma 8.4, $D' = \varphi(D)$ is analytic with $\dim \varphi(D) \leqq n - 2$. By

Lemma 1.25, $T = \varphi^{-1}(D')$ is analytic and thin in X. Then $X_1 = X - T$ and $Y_1 = Y - D'$ are open and dense in X and Y respectively. The complex space X_1 is pure m-dimensional. The complex space Y_1 is irreducible and n-dimensional. The map $\varphi_1 = \varphi\colon X_1 \to Y_1$ is quasi-proper, q-fibering and holomorphic. The set $V_1 = V - D'$ is open and not empty in Y_1. Define $U_1 = \varphi_1^{-1}(V_1) = U - T \neq \emptyset$. Define $\varphi_2 = \varphi_1\colon U_1 \to V_1$. Then $f|U_1 = \varphi_0^*(g)|U_1 = \varphi_2^*(g|V_1)$. A meromorphic function $h_1 \in \mathfrak{K}(Y_1)$ exists such that $f|X_1 = \varphi_1^*(h_1)$. By Lemma 3.11, a meromorphic function $h \in \mathfrak{K}(Y)$ exists such that $h|Y_1 = h_1$. Then $f|X_1 = \varphi_1^*(h|Y_1) = \varphi^*(h)|X_1$ which implies $f = \varphi^*(h)$; q.e.d.

<u>Proposition 9.4.</u> Let X and Y be irreducible complex spaces of dimension n. Let $\varphi\colon X \to Y$ be a quasi-proper, holomorphic map of rank n. Let $V \neq \emptyset$ be an open subset of Y. Define $U = \varphi^{-1}(V)$. Suppose that $\varphi_0 = \varphi\colon U \to V$ is biholomorphic. Then $\varphi^*\colon \mathfrak{K}(Y) \to \mathfrak{K}(X)$ exists and is an isomorphism.

<u>Proof.</u> By Remark 3.5.3, $\varphi^*\colon \mathfrak{K}(Y) \to \mathfrak{K}(X)$ exists and is an injective homomorphism. Take $f \in \mathfrak{K}(X)$. Define $g = (\varphi_0^{-1})^*(f|U) \in \mathfrak{K}(V)$. Then $\varphi_0^*(g) = f|U$. By Theorem 9.3, $h \in \mathfrak{K}(Y)$ exists such that $f = \varphi^*(h)$. Hence, φ^* is surjective. The injective homomorphism φ^* is an isomorphism; q.e.d.

<u>Proposition 9.5.</u> Let X be a complex space of pure dimension m. Let Y be an irreducible complex space of dimension n with $m - n = q \geqq 0$. Let $\varphi\colon X \to Y$ be a semi-proper, homomorphic map of strict rank n. Let V be a dense open subset of Y. Define $U = \varphi^{-1}(V)$. Then U is open and dense in X. Define $\varphi_0 = \varphi\colon U \to V$.

Suppose that meromorphic functions $f \in \mathfrak{K}(X)$ and $g \in \tilde{\mathfrak{K}}(V)$ are given such that $\varphi_0^*(g) = f|U$. Then a meromorphic function $h \in \tilde{\mathfrak{K}}(Y)$ exists on Y such that $f = \varphi^*(h)$. Moreover, g and h are unique and $h|V = g$.

Proof. By Proposition 1.24, $D = \{x \in X | \operatorname{rank}_x \varphi \leqq n-1\}$ is analytic and thin in X and $D' = \varphi(D)$ is almost thin of dimension $n - 2$ in Y. By Lemma 7.6, $\varphi(X)$ is analytic and pure n-dimensional in Y. Hence, $\varphi(X) = Y$. The map φ is surjective. Hence, $U \neq \emptyset$. Let S_Y be the set of non-simple points of Y. By Lemma 1.25, $\tilde{S}_Y = \varphi^{-1}(S_Y)$ is a thin analytic subset of X. Let $W \neq \emptyset$ be open in X. Then $W_1 = W - (\tilde{S}_Y \cup D) \neq \emptyset$ is open in X and $\varphi_1 = \varphi: W_1 \to Y - S_Y$ is a holomorphic map of pure rank n. By Theorem 1.22, φ_1 is an open map. Hence, $\varphi(W_1) = \varphi_1(W_1)$ is open. Therefore, $V \cap \varphi(W_1) \neq \emptyset$ and $U \cap W_1 \neq \emptyset$. Hence $U \cap W \neq \emptyset$. The open set U is dense in X.

Define $S = Y - V$ and $\varphi_0 = \varphi: U \to V$. Suppose that $f \in \mathfrak{K}(X)$ and $g \in \tilde{\mathfrak{K}}(V)$ with $f|U = \varphi_0^*(g)$ are given. Let S_0 be the set of all points $a \in S$, such that an open neighborhood N of a and a meromorphic function $h \in \mathfrak{K}(N)$ exists such that $h|V \cap N = g|V \cap N$. Observe that $V \cap N \neq \emptyset$. Obviously, S_0 is open in S and $S_1 = S - S_0$ is closed in S and in Y.

Take $b \in S - D'$. An open neighborhood W of b exists such that $W = W_1 \cup \ldots \cup W_p$ where $W_\mu \neq W_\nu$ if $\mu \neq \nu$ and where $W_1, \ldots, W_p$ are the branches of W and where each W_μ contains b and is locally irreducible at b. Define $Z = \varphi^{-1}(V)$ and $\psi = \varphi: Z \to W$. Then Z and W are complex spaces of pure dimension m and n respectively and ψ is a semi-proper, surjective, holomorphic map of strict rank n. Take any branch, W_1. By Lemma 7.11, a branch C of Z and a point

$a \in C$ with $\psi(a) = b$ and $\psi(C) \subseteq W_i$ exists such that $\text{rank}_x \psi_i \geqq \text{rank}_x \psi$ if $x \in C$ and if $\psi_i = \psi\colon C \to W_i$. Because Z is open in X, $\text{rank}_x \psi = \text{rank}_x \varphi$ for all $x \in Z$. Now, $\varphi^{-1}(b) \subseteq X - D$. Hence, $\text{rank}_a \psi_i \geqq \text{rank}_a \varphi = n$. Therefore, rank $\psi_i = n$. An open neighborhood A of a in C exists such that $\text{rank}_x \psi_i = n$ for all $x \in A$. Then, $\psi_{i0} = \psi_i\colon A \to W_i$ is q-fibering, because dim $C = m$. By Proposition 2.4, a product representation (α, β, π) of ψ_{i0} centered at a and over b exists with $U_\alpha \subseteq A$. By Lemma 2.5.8, $\varphi(U_\alpha) = \psi_{i0}(U_\alpha) = U_\beta$. By Lemma 8.16, $\psi_{i1} = \psi_{i0}\colon U_\alpha \to U_\beta$ is quasi-proper. Moreover, ψ_{i1} is q-fibering and has pure rank n by Lemma 2.5.4.

Let $\tilde{j}\colon U_\alpha \cap U \to U$ and $j\colon U_\beta \cap V \to V$ be the inclusion maps. Define $\psi_{i2} = \psi_{i1}\colon U_\alpha \cap U \to U_\beta \cap V$. Then

$$j \circ \psi_{i2} = \varphi_0 \circ \tilde{j}.$$

By Lemma 3.4.4 and 3.4.5, $j^*\colon \mathfrak{K}(V) \to \mathfrak{K}(U_\beta \cap V)$, $\tilde{j}^*\colon \mathfrak{K}(U) \to \mathfrak{K}(U_\alpha \cap U)$, $\psi_{i2}^*\colon \mathfrak{K}(U_\beta \cap V) \to \mathfrak{K}(U_\alpha \cap U)$ and $\varphi_0^*\colon \mathfrak{K}(V) \to \mathfrak{K}(U)$ and $(j \circ \psi_{i2})^* = (\varphi_0 \circ \tilde{j})^*\colon \mathfrak{K}(V) \to \mathfrak{K}(U_\alpha \cap U)$ exist and are injective with $\psi_{i2}^* \circ j^* = \tilde{j}^* \circ \varphi_0^*$. Hence,

$$f|U_\alpha \cap U = \tilde{j}^*(f|U) = \tilde{j}^*(\varphi_0^*(g)) = \psi_{i2}^*(j^*(g)) = \psi_{i2}^*(g|V \cap U_\beta).$$

By Theorem 9.3, a meromorphic function $h_i \in \mathfrak{K}(U_\beta)$ exists such that $h_i|U_\beta \cap V = g|V \cap U_\beta$. Recall that U_β is an open neighborhood of b in W_i. Write $M_i = U_\beta$. An open neighborhood N of b exists with $N \subseteq M_1 \cup \ldots \cup M_p$. Then, $N_i = W_i \cap N = M_i \cap N$ is open in W_i and M_i and $h_i|N_i$ is meromorphic on N_i with $h_i|N_i \cap V = g|V \cap N_i$. Here N_i is a union of branches of N, and $N_i \cap N_j$ is thin in N if $i \neq j$. By Lemma 3.10, a meromorphic function h on N exists such that $h|N_i = h_i|N_i$. Then $h|N_i \cap V = h_i|N_i \cap V = g|N_i \cap V$ for $i = 1,\ldots,p$. Hence, $h|N \cap V = g|N \cap V$, because V is dense in N. Therefore, $b \in S_0$. Hence, $S - D' \subseteq S_0$ and $S_1 = S - S_0 \subseteq D'$. The closed set S_1 is almost thin of dimension $n - 2$.

For each $a \in S_0$, take an open neighborhood N_a of a and a meromorphic function $h_a \in \mathfrak{K}(N_a)$ such that $h_a|N_a \cap V = g|V \cap N$. Observe that $Y - S_1 = Y_0$ is open in Y and $Y_0 - S_0 = V$. By Lemma 3.10, one and only one meromorphic function $\tilde{h} \in \mathfrak{K}(Y_0)$ exists such that $\tilde{h}|V = g$. By Proposition 7.10, one and only one meromorphic function $h \in \mathfrak{K}(Y)$ exists such that $h|Y_0 = \tilde{h}$. Hence, $h|V = \tilde{h}|V = g$. Now, $f|U = \varphi_0^*(g) = \varphi_0^*(h|V) = \varphi^*(h)|U$ which implies $f = \varphi^*(h)$. Since φ_0^* and φ^* are injective, g and h are uniquely determined;

q.e.d.

Proposition 9.6. Let X be a complex space of pure dimension m. Let Y be an irreducible, locally irredicuble complex space of dimension n with $m - n = q \geqq 0$. Let $\varphi: X \to Y$ be a surjective, holomorphic map of strict rank n. Let V be open and dense in Y. Then $U = \varphi^{-1}(V)$ is open and dense in X. Define $\varphi_0 = \varphi: U \to V$. Suppose that meromorphic functions $f \in \mathfrak{K}(X)$ and $g \in \mathfrak{K}(V)$ are given such that $\varphi_0^*(g) = f|U$. Then g is unique. Moreover, one and only one meromorphic function $h \in \mathfrak{K}(Y)$ exists such that $f = \varphi^*(h)$. Moreover, $h|V = g$.

Proof. As in the proof of Proposition 9.5, it is shown that U is dense. The analytic set $D = \{x \in X | \text{rank}_x \varphi \leqq n-1\}$ is thin and $D' = \varphi(D)$ is almost thin of dimension n - 2 in Y. Define S = Y - V. Let S_0 be the set of all points a ∈ S, such that an open neighborhood N of a and a meromorphic function $h \in \mathfrak{K}(N)$ exists such that $h|V \cap N = g|V \cap N$. Then S_0 is open in S and $S = S = S_0$ is closed in S and in Y.

Take $b \in S - D'$. Then $a \in X - D = Z$ with $\varphi(a) = b$ exist. The map $\psi = \varphi: Z \to Y$ is q-fibering and holomorphic. By Proposition 2.4, a product representation (α, β, π) of ψ centered at a and over b exists. By Lemma 2.5.8, $\varphi(U_\alpha) = \psi(U_\alpha) = U_\beta$. The map $\psi_0 = \psi: U_\alpha \to U_\beta$ is quasi-proper by Lemma 8.16 and q-fibering by Lemma 2.5.5. Hence, ψ_0 has pure rank n. The open neighborhood U_β of b is connected, because β is centered at b. Define $\psi_1 = \psi: U_\alpha \cap U \to U_\beta \cap V$. Then, $\psi_1 = \varphi_0: U_\alpha \cap U \to U_\beta \cap V$. Hence,

$$f|U_\alpha \cap U = \varphi_0^*(g)|U_\alpha \cap U = \psi_1^*(g|U_\beta \cap V).$$

By Theorem 9.3, a meromorphic function $h \in \mathfrak{K}(U_\beta)$ exists such that $h|U_\beta \cap V = g|U_\beta \cap V$. Therefore, $b \in S_0$. Hence, $S - D' \subseteq S_0$ and $S_1 = S - S_0 \subseteq D'$. The closed set S_1 is almost thin of dimension n - 2. Now, the proof is completed as in the proof of Proposition 9.5; q.e.d.

Proposition 9.7. Let X be a complex space. Let Y be an irreducible complex space do dimension n. Let $\varphi: X \to Y$ be a quasi-proper holomorphic map. Define $D = \{x | \text{rank}_x \varphi < n\}$. Suppose that $X - D \neq \emptyset$ is pure m-dimensional with $m - n = q \geqq 0$. Let S be an analytic subset of X such that $A = X - S \neq \emptyset$. Define $X_0 = \overline{A}$. Then X_0 is analytic. Define $\varphi_0 = \varphi: X_0 \to Y$. Suppose that φ_0 has strict rank n. Let $V \neq \emptyset$ be an open subset of Y. Define $U = \varphi_0^{-1}(V)$ and $\psi = \varphi_0: U \to V$. Suppose that meromorphic functions $g \in \mathfrak{K}(V)$ and $f \in \mathfrak{K}(X_0)$ are given such that $f|U = \psi^*(g)$. Then one and only one meromorphic function $h \in \mathfrak{K}(Y)$ exists such that $\varphi_0^*(h) = f$. Moreover, $h|V = g$.

<u>Proof.</u> Because φ_0 has strict rank n, $\varphi_0^*: \mathfrak{K}(Y) \to \mathfrak{K}(X_0)$ is injective. Hence, h is unique. Let $\mathfrak{B}$ be the set of all branches of X. Define

$$\mathfrak{B}_0 = \{B \in \mathfrak{B} \,|\, B \cap S \neq B\}$$
$$\mathfrak{B}_1 = \{B \in \mathfrak{B} \,|\, \text{rank } \varphi|B = n\}$$
$$\mathfrak{B}_2 = \{B \in \mathfrak{B} \,|\, \text{rank } \varphi|B < n\}.$$

Then $\mathfrak{B}_1 \cap \mathfrak{B}_2 = \emptyset$. Because Y is n-dimensional, $\mathfrak{B} = \mathfrak{B}_1 \cup \mathfrak{B}_2$.

By Lemma 7.3, $X_0 = \bigcup_{B \in \mathfrak{B}_0} B$ and $\mathfrak{B}_0$ is the set of branches of X_0. Especially, $\mathfrak{B}_0 \neq \emptyset$. Because φ_0 has strict rank n, rank $\varphi|B$ = rank $\varphi_0|B = n$ if $B \in \mathfrak{B}_0$ by Lemma 1.16. Hence, $\mathfrak{B}_0 \subseteq \mathfrak{B}_1$. Take $B \in \mathfrak{B}_1$. Let S_X be the set of non-simple points of X. Then $x \in B - S_X$ exists such that $\text{rank}_x \varphi_0|B = n$. Hence, $\text{rank}_x \varphi = \text{rank}_x \varphi|B = \text{rank}_x \varphi_0|B = n$. Therefore, $x \in X - D$. Hence, dim B = $\dim_x X = m$. Therefore, $X_1 = \bigcup_{B \in \mathfrak{B}_1} B \neq \emptyset$ is a pure m-dimensional analytic subset of X. Define $\varphi_1 = \varphi: X_1 \to Y$. Because $\mathfrak{B}_1$ is the set of branches of X_1 and because rank $\varphi_1|B$ = rank $\varphi|B = n$, the map φ_1 has strict rank n by Lemma 1.16. If $B \in \mathcal{L}_2$, take any $x \in B - S_X$, then $\operatorname{rank}_x \varphi = \operatorname{rank}_x \varphi|B \leqq n - 1$. Hence, $x \in D$. Therefore, $B - S_X \subseteq D$, which implies $B \subseteq D$. Therefore, $X_2 = \bigcup_{B \in \mathcal{L}_2} B$ is an analytic subset of X contained in D and $\mathcal{L}_2$ is the set of branches of X_2.

If $X_2 \neq \emptyset$, define $\varphi_2 = \varphi: X_2 \to Y$. Because rank $\varphi_2|B$ = rank $\varphi|B \leqq n - 1$ for all $B \in \mathcal{L}$, rank $\varphi_2 \leqq n - 1$ by Lemma 1.17. The map φ_2 is semi-proper by Proposition 8.9'. Therefore, $E' = \varphi_2(X_2) = \varphi(X_2)$ is analytic in Y with $\dim E' \leqq n - 1$ by Theorem 7.9, and by Proposition 1.23. Therefore, E' is thin in Y. Because $\varphi_1: X_1 \to Y$ has strict rank n, the set $E_1 = \varphi_1^{-1}(E')$ is thin and analytic in X_1. The sets $\tilde{X}_1 = X_1 - E_1$ and $\tilde{Y}_1 = Y - E'$ are open and dense in X_1 and Y respectively and $\tilde{\varphi}_1 = \varphi_1: \tilde{X}_1 \to Y_1$ is holomorphic and has strict rank n. The set $E = \varphi^{-1}(E')$ is analytic in X with $E \supseteq X_2$ and $E \cap X_1 = E_1$. If $x \in X - E$, then $x \in B$ with $B \in \mathcal{L} - \mathcal{L}_2 = \mathcal{L}_1$. Hence, $x \in X_1$ and $x \in X_1 - E_1$. Therefore, $X - E \subseteq X_1 - E_1$. Also, $X_1 - E_1 = X_1 - E \subseteq X - E$. Hence, $X - E = X_1 - E_1 = \tilde{X}_1$ is open in X, with $\tilde{\varphi}_1 = \varphi: \tilde{X}_1 \to \tilde{Y}_1$ and $\tilde{X}_1 = \varphi^{-1}(\tilde{Y}_1)$. By Lemma 8.5 b), the map $\tilde{\varphi}_1$ is quasi-proper. Observe, that $\tilde{Y}_1 = Y - E'$ is an irreducible complex space of dimension n and that $\tilde{X}_1$ has pure dimension m. Define

$$D_1 = \{x \in \tilde{X}_1 \mid \mathrm{rank}_x \tilde{\varphi}_1 \leqq n-1\}.$$

$$D_1' = \tilde{\varphi}_1(D_1) = \varphi(D_1)$$

$$\tilde{D}_1 = \tilde{\varphi}_1^{-1}(D_1') \subseteq \tilde{X}_1 \text{ with } D_1 \subseteq \tilde{D}_1.$$

Because $\tilde{\varphi}_1$ has strict rank n, the set D_1 is analytic and thin in $\tilde{X}_1$. By Proposition 8.4, D_1' is analytic in $\tilde{Y}_1$ with $\dim D_1' \leqq n - 2$. Hence, D_1' is thin in $\tilde{Y}_1$. Because $\tilde{\varphi}_1$ has strict rank n, the set $\tilde{D}_1$ is thin and analytic in $\tilde{X}_1$ by Lemma 1.25. The sets $\hat{X}_1 = \tilde{X}_1 - \tilde{D}_1$ and $\hat{Y}_1 = \tilde{Y}_1 - D_1'$ are open and dense in $\tilde{X}_1$ and $\tilde{Y}_1$ respectively, with $\tilde{\varphi}_1^{-1}(\hat{Y}_1) = \hat{X}_1$. Hence $\hat{\varphi}_1 = \tilde{\varphi}_1 : \hat{X}_1 \to \hat{Y}_1$ is a quasi-proper holomorphic map of strict rank n. Here $\hat{X}_1$ is a complex space of pure dimension m and $\hat{Y}_1$ is an irreducible complex space of dimension n. If $x \in \hat{X}_1$, then $x \in \tilde{X}_1 - D_1$, hence, $n = \mathrm{rank}_x \tilde{\varphi}_1 = \mathrm{rank}_x \hat{\varphi}_1 = m - \dim_x \hat{\varphi}_1^{-1}(\hat{\varphi}_1(x))$ with $m - n = q$, because $\hat{X}_1$ is open in $\tilde{X}_1$. Hence, $\hat{\varphi}_1$ is q-fibering and has pure rank n.

If $B \in \mathfrak{B}_1$, define $B \cap \tilde{X}_1 = B - E_1 = B - E = \tilde{B}$. Then $\tilde{\mathfrak{B}}_1 = \{\tilde{B} \mid B \in \mathfrak{B}_1\}$ is the set of branches of $\tilde{X}_1$. Define $\hat{B} = \tilde{B} \cap \hat{X}_1 = \tilde{B} - \tilde{D}_1$. Then $\hat{\mathfrak{B}}_1 = \{\hat{B} \mid B \in \mathfrak{B}_1\}$ is the set of branches of $\hat{X}_1$. The set

$$\tilde{X}_0 = X_0 \cap \tilde{X}_1 = X_0 - E = \bigcup_{B \in \mathfrak{B}_0} \tilde{B} = \varphi_0^{-1}(\tilde{Y}_1)$$

is analytic in $\tilde{X}_1$ and $\tilde{\mathfrak{B}}_0 = \{\tilde{B} \mid B \in \mathfrak{B}_0\}$ is the set of branches of $\tilde{X}_0$. Define $\tilde{\varphi}_0 = \varphi_0 : \tilde{X}_0 \to \tilde{Y}_1$. Then $\tilde{\varphi}_0 = \tilde{\varphi}_1 : \tilde{X}_0 \to \tilde{Y}_1$. The set

$$\hat{X}_0 = \tilde{X}_0 \cap \hat{X}_1 = \tilde{X}_0 - D_1 = \bigcup_{B \in \mathcal{B}_0} \hat{B}$$

is analytic in $\hat{X}_0$ and $\hat{\mathcal{B}}_0 = \{\hat{B} | B \in \mathcal{B}_0\}$ is the set of branches of $\hat{X}_0$ with $\hat{\mathcal{B}}_0 \subseteq \hat{\mathcal{B}}_1$. Define $\hat{\varphi}_0 = \tilde{\varphi}_0 : \hat{X}_0 \to \hat{Y}_1$. Then $\hat{\varphi}_0 = \hat{\varphi}_1 : \hat{X}_0 \to \hat{Y}_1$. Hence, $\hat{\varphi}_0$ is q-fibering, quasi-proper and holomorphic by Proposition 8.10'. Because rank $\hat{\varphi}_0 | \hat{B}$ = rank $\hat{\varphi}_1 | \hat{B} = n$, the map $\hat{\varphi}_0$ has strict rank n.

The open subset $\hat{V} = V \cap \hat{Y}_1$ is not empty, because $\hat{Y}_1$ is open and dense in Y. Define $\hat{g} = g | \hat{V} \in \mathfrak{K}(\hat{V})$. By Lemma 8.11, $\hat{\varphi}_0$ is surjective. Hence,

$$\emptyset \neq \hat{U} = \hat{\varphi}_0^{-1}(\hat{V}) = \varphi_0^{-1}(V \cap \hat{Y}_1) \cap \hat{X}_0 = \varphi_0^{-1}(V) \cap \hat{X}_0 = U \cap \hat{X}_0.$$

Define $\hat{\psi} = \hat{\varphi}_0 : \hat{U} \to \hat{V}$. Then, $\hat{\psi} = \psi : \hat{U} \to \hat{V}$ and

$$\hat{\psi}^*(\hat{g}) = \hat{\psi}^*(g | \hat{V}) = \psi^*(g) | \hat{U} = f | U \cap \hat{U} = f | \hat{U}.$$

By Theorem 9.3, a meromorphic function $\hat{h} \in \mathfrak{K}(\hat{Y}_1)$ exists such that $\hat{\varphi}_0^*(\hat{h}) = f | \hat{X}_0$ and $\hat{h} | \hat{V} = \hat{g} = g | \hat{V}$. Because $\hat{Y}_1 = \tilde{Y}_1 - D_1'$ where D_1' is analytic in $\tilde{Y}_1$ with dim $D_1' \leqq n - 2$, one and only one meromorphic function $\tilde{h} \in \mathfrak{K}(\tilde{Y})$ exists such that $\tilde{h} | \hat{Y}_1 = \hat{h}$ (Lemma 3.7). Then

$$\tilde{\varphi}_0^*(\tilde{h}) | \hat{X}_0 = \hat{\varphi}_0^*(\tilde{h} | \hat{Y}_1) = \hat{\varphi}_0(\hat{h}) = f | \hat{X}_0 = (f | \tilde{X}_0) | \hat{X}_0.$$

Because $\hat{X}_0$ is open and dense in $\tilde{X}_0$, this implies $\tilde{\varphi}_0^*(\tilde{h}) = f | \tilde{X}_0$. By Proposition 8.9, the map $\varphi_0 : X_0 \to Y$ is semi-proper. By assumption,

φ_0 has strict rank n. The complex space X_0 has pure dimension m. The complex space Y is irreducible and has dimension n with $m - n = q \geqq 0$. The open subset $\tilde{Y}_1$ is dense in Y with $\tilde{X}_0 = \varphi_0^{-1}(\tilde{Y}_1)$. By Proposition 9.5 one and only one meromorphic function $h \in \mathfrak{K}(Y)$ exists such that $f = \varphi_0^*(h)$. Moreover, $h|\tilde{Y}_1 = \tilde{h}$. Especially, $h|\hat{V} = \tilde{h}|\hat{V} = \hat{h}|\hat{V} = g|\hat{V}$. Because V is dense in V, this implies $h|V = g$; q.e.d.

<u>Proposition 9.8.</u> Let X be a complex space. Let Y be a pure n-dimensional complex space. Let $\varphi: X \to Y$ be a quasi-proper, holomorphic map. Define $D = \{x \in X | \operatorname{rank}_x \varphi < n\}$. Suppose that $X - D \neq \emptyset$ is pure m-dimensional with $m - n = q \geqq 0$. Let B_0 be a branch of X with $B_0 - D \neq \emptyset$. Then $\varphi(B_0) = Y_0$ is a branch of Y. Let V be an open subset of Y such that $V_0 = V \cap Y_0 \neq \emptyset$. Define $U_0 = \varphi^{-1}(V) \cap B_0$ and $\psi = \varphi: U_0 \to V$. Suppose that meromorphic functions $g \in \mathfrak{K}(V)$ and $f \in \mathfrak{K}(B_0)$ are given such that $\psi^*(g) = f|U_0$. Define $\varphi_0 = \varphi: B_0 \to Y_0$. Then one and only one meromorphic function $h \in \mathfrak{K}(Y_0)$ exists such that $f = \varphi_0^*(h)$. Moreover, $h|V_0 = g|V_0$.

<u>Proof.</u> Let S_X be the set of non-simple points of X. By Propoistion 8.9', $\varphi_1 = \varphi: B_0 \to Y$ is semi-proper. If $x \in B_0 - (S_X \cup D)$, then $\operatorname{rank}_x \varphi_1 = \operatorname{rank}_x \varphi = n$. Hence, rank $\varphi_1 = n$. By Lemma 7.6 and by Lemma 1.27, $Y_0 = \varphi(B_0)$ and $\varphi_1(B_0)$ is an n-dimensional, irreducible, analytic subset of Y. Hence, Y_0 is a branch of Y. Let $\mathfrak{B}$ be the set of branches of X. Define

$$\mathfrak{T} = \{ B \in \mathfrak{B} | \varphi^{-1}(Y_0) \cap B \neq B\}.$$

Define $Z = \bigcup_{B \in \mathcal{L}} B$. If $Z \neq \emptyset$, then $\chi = \varphi\colon Z \to Y$ is semi-proper by Proposition 8.9'. Therefore, $Z' = \chi(Z)$ is analytic in Y by Theorem 7.9. For the same reason, $\chi(B)$ is analytic in Y if $B \in \mathcal{L}$. Then $\chi(B) \cap Y_0 = \varphi(B) \cap Y_0$ is a thin analytic subset of Y_0. Therefore,

$$Z' \cap Y_0 = \chi(Z) \cap Y_0 = \bigcup_{B \in} \chi(B) \cap Y_0$$

is almost thin in Y_0. Because Z' is analytic, $Z' \cap Y_0$ is thin in Y_0. If $B \in \mathcal{B} - \mathcal{L} = \tilde{\mathcal{B}}$, then $B \subseteq \varphi_0^{-1}(Y_0)$ or $\varphi(B) \subseteq Y_0$. Hence, $\tilde{X} = \bigcup_{B \in \tilde{\mathcal{B}}} B$ is an analytic subset of X with $\varphi(\tilde{X}) \subseteq Y_0$. Moreover, $\tilde{\mathcal{B}}$ is the set of branches of $\tilde{X}$ and $B_0 \subseteq \tilde{X}$ is a branch of $\tilde{X}$. Define $\tilde{\varphi} = \varphi\colon \tilde{X} \to Y_0$.

The set $Z'' = \varphi^{-1}(Z')$ is analytic in X with $Z \subseteq Z''$. Define $\tilde{Z} = \tilde{X} \cap Z'' = \tilde{\varphi}^{-1}(Z' \cap Y_0)$. If $x \in X - Z''$, then $x \in B$ with $B \in (\mathcal{B} - \mathcal{L}) = \tilde{\mathcal{B}}$. Hence, $x \in \tilde{X}$. Therefore, $X = Z'' \subseteq \tilde{X} - \tilde{Z}$. If $x \in \tilde{X} - \tilde{Z}$, then $\varphi(x) = \tilde{\varphi}(x) \in Y_0$ and $x \in \varphi^{-1}(Y_0)$. Therefore, $\tilde{X} - \tilde{Z} \subseteq \varphi^{-1}(Y_0) - Z''$. Then

$$\varphi^{-1}(Y_0) - Z'' \subseteq X - Z'' \subseteq \tilde{X} - \tilde{Z} \subseteq \varphi^{-1}(Y_0) - Z''.$$

Hence,

$$\hat{X} = \tilde{X} - \tilde{Z} = X - Z = \varphi^{-1}(Y_0) - Z''$$

is open in $\tilde{X}$, X and $\varphi^{-1}(Y_0)$. If $B \in \tilde{\mathcal{B}}$, then $\hat{B} = B - \tilde{Z}$ is either empty or a branch of $\hat{X}$ and $\hat{\mathcal{B}} = \{B \in \tilde{\mathcal{B}} \mid \hat{B} \neq \emptyset\}$ is the set of branches of $\hat{X}$. Because $\tilde{\varphi}(Z'') \subseteq Z' \cap Y_0$ and $\tilde{\varphi}(B_0) = \varphi(B_0) = Y_0$, and because $Y_0 \cap Z'$ is thin in Y_0, the set $\hat{B}_0 = B_0 - \tilde{\varphi}^{-1}(Y_0 \cap Z')$ is not empty. Hence, $\hat{X} \neq \emptyset$ and $\hat{B}_0$ is a branch of $\hat{X}$. Define $\hat{Y} = Y_0 - Z'$. Then $\hat{Y}$ is open and dense in Y_0 with $\varphi(\hat{X}) \subseteq \hat{Y}$. Moreover, $\hat{Y}$ is an irreducible complex space of dimension n. Then $\hat{\varphi} = \tilde{\varphi}: \hat{X} \to \hat{Y}$ is holomorphic with $\hat{\varphi} = \varphi: \hat{X} \to \hat{Y}$. Because φ is quasi-proper, the restriction $\varphi' = \varphi: \varphi^{-1}(Y_0) \to Y_0$ is quasi-proper by Lemma 8.5a). Now,

$$(\varphi')^{-1}(Y_0 \cap Z') = \varphi^{-1}(Y_0) \cap \varphi^{-1}(Y_0 \cap Z') = \varphi^{-1}(Y_0) \cap \varphi^{-1}(Z')$$

$$= \varphi^{-1}(Y_0) - Z'' = \hat{X}.$$

By Lemma 8.5 b), the restriction

$$\hat{\varphi}' = \varphi': \hat{X} \to Y_0 - Z' = \hat{Y}$$

is quasi-proper. Here $\hat{\varphi}' = \varphi: \hat{X} \to \hat{Y}$. Hence, $\hat{\varphi}' = \hat{\varphi}$. Therefore, $\hat{\varphi}: \hat{X} \to \hat{Y}$ is a quasi-proper map.

Because $\hat{X}$ is open in X,

$$\hat{D} = \{x \in \hat{X} \mid \mathrm{rank}_x \hat{\varphi} < n\} = \{x \in \hat{X} \mid \mathrm{rank}_x \varphi < n\} = \hat{X} \cap D.$$

Hence, $\hat{X} - \hat{D} = \hat{X} - D$ is pure m-dimensional or empty. The sets $B_0 - D$ and $\hat{B}_0 = B_0 \cap \hat{X}$ are open and dense on B_0. Hence, $\hat{B}_0 - D \neq \emptyset$. Therefore, $\hat{X} - D \neq \emptyset$. As shown, the map $\varphi_1 = \varphi: B_0 \to Y$ has rank n. Then $\hat{\varphi}_0 = \varphi_1: \hat{B}_0 \to \hat{Y}$ has rank n. Let $\hat{S}$ be the union of all branches $\hat{B} \neq \hat{B}_0$ of $\hat{X}$. Then, $\hat{S}$ is analytic in $\hat{X}$ and $\hat{A} = \hat{B}_0 - \hat{S} = \hat{X} - \hat{S} \neq \emptyset$. The closure of $\hat{A}$ in $\hat{X}$ is $\hat{B}_0$.

Because $\hat{Y}$ is dense in Y_0, the open subset $\hat{V} = V_0 \cap \hat{Y}$ of $\hat{Y}$ is not empty and $\hat{g} = g|\hat{V}$ is meromorphic on $\hat{Y}$. Because $\hat{B}_0$ is open and dense in B_0, the function $\hat{f} = f|\hat{B}_0$ is defined and meromorphic on $\hat{B}_0$. Because $\varphi(B_0) = Y_0$, the set

$$U_0 = \varphi^{-1}(V) \cap B_0 = \varphi^{-1}(V \cap Y_0) \cap B_0 = \varphi_0^{-1}(V_0)$$

is not empty. Because $\hat{B}_0$ is dense in B_0, the set

$$\hat{U} = \hat{\varphi}_0^{-1}(\hat{V}) = \varphi_0^{-1}(V_0 \cap \hat{Y}) \cap \hat{X} = \varphi_0^{-1}(V_0) \cap \hat{B}_0 = U_0 \cap \hat{B}_0$$

is not empty. Define $\hat{\psi} = \hat{\varphi}_0\colon \hat{U} \to \hat{V}$. Then $\hat{\psi} = \psi\colon \hat{U} \to \hat{V}$ and

$$\psi^*(\hat{g}) = \psi^*(g|\hat{V}) = \psi^*(g)|\hat{U} = (f|U_0)|\hat{U} = f|\hat{U}.$$

By Proposition 9.7, one and only one meromorphic function $\hat{h} \in \mathfrak{K}(\hat{Y})$ exists such that $\hat{\varphi}_0(\hat{h}) = f|\hat{B}_0$. Moreover, $\hat{h}|\hat{V} = \hat{g} = g|\hat{V}$.

As already shown, the map $\varphi_1\colon B_0 \to Y$ is semi-proper with $\varphi_1(B_0) = \varphi(B_0) = Y_0$. If K is a compact subset of Y_0, a compact subset K' of B_0 exists such that $\varphi_0(K') = \varphi_1(K') = K \cap \varphi_1(B_0) = K \cap Y_0 = K$. Hence, $\varphi_0 = \varphi_1\colon B_0 \to Y_0$ is semi-proper. The map φ_0 is also holomorphic with rank n and B_0 and Y_0 are irreducible complex spaces of dimension m and n respectively with $m - n = q \geqq 0$. Also $\hat{Y}$ is open and dense in Y_0 with $\hat{\varphi}_0^*(\hat{h}) = f|\hat{B}_0$ where

$$\hat{\varphi}_0^{-1}(\hat{Y}) = B_0 \cap \hat{\varphi}^{-1}(\hat{Y}) = B_0 \cap \hat{X} = \hat{B}_0 .$$

By Proposition 9.5, one and only one meromorphic function $h \in \mathfrak{K}(Y_0)$ exists such that $\varphi_0^*(h) = f$. Moreover, $h|\hat{B}_0 = \hat{h}$. Hence, $h|\hat{V} = \hat{h}|\hat{V} = g|\hat{V}$, which implies $h|V_0 = g|V_0$, because $\hat{V}$ is dense in V_0; q.e.d.

§10. Quasi-proper maps of codimension k

A holomorphic map $\varphi: X \to Y$ of a complex space X into a complex space Y is said to be quasi-proper of codimension k if and only if for every compact subset K of Y a compact subset K' of X exists such that the following property holds.

(P) Take $y \in \varphi(X) \cap K$. Let B be a branch of $\varphi^{-1}(y)$. Let C be an irreducible analytic subset of B with $\dim B - \dim C \leqq k$. Then $C \cap K' \neq \emptyset$.

Obviously, a quasi-proper map of codimension k is quasi-proper of codimension k' if $0 \leqq k' \leqq k$. Clearly, a proper map is quasi-proper of codimension k for any $k \geqq 0$.

Lemma 10.1. Let X and Y be complex spaces. Let $\varphi: X \to Y$ be a holomorphic map. Then, φ is quasi-proper of codimension 0 if and only if φ is quasi-proper. Especially, every holomorphic map which is quasi-proper of codimension k is quasi-proper.

Proof. a) Suppose that φ is quasi-proper. Let K be a compact subset of Y. A compact subset K' of X exists such that $B \cap K' \neq \emptyset$ is B is a branch of $\varphi^{-1}(y)$ and if $y \in K \cap \varphi(X)$. Take $y \in \varphi(X) \cap K$. Let B be a branch of $\varphi^{-1}(y)$. Let C be an irreducible analytic subset of B with $\dim B - \dim C \leqq 0$. Then $B = C$ and $C \cap K' = B \cap K' \neq \emptyset$. The map φ is quasi-proper of codimension 0.

b) Suppose that φ is quasi-proper of codimension 0. Let K be a compact subset of Y. Let K' be a compact subset of X such that (P) holds for $k = 0$. Take $y \in K \cap \varphi(X)$. Let B be a branch of $\varphi^{-1}(y)$. Then $\dim B - \dim B \leqq 0$. Hence, $B \cap K' \neq \emptyset$. The map φ

is quasi-proper; q.e.d.

Lemma 10.2. Let X and Y be complex spaces. Let $\varphi: X \to Y$ be a q-fibering holomorphic map. Let K be a compact subset of Y. Let K' be a compact subset of X. Then the following conditions are equivalent:

1) The condition (P) (see above).

2) The condition

(P') Take $y \in \varphi(X) \cap K$. Let C be an irreducible analytic of $\varphi^{-1}(y)$ with $\dim \varphi^{-1}(y) - \dim C \leqq k$. Then, $C \cap K' \neq \emptyset$.

3) The condition

(P") Take $y \in \varphi(X) \cap K$. Let C be an irreducible analytic subset of $\varphi^{-1}(y)$ with $\dim C \geqq q - k$. Then $C \cap K' \neq \emptyset$.

Proof. Because $\varphi^{-1}(y) \neq \emptyset$ is pure q-dimensional and because each irreducible analytic subset of $\varphi^{-1}(y)$ is contained in a branch of $\varphi^{-1}(y)$, this Lemma is trivial; q.e.d.

Lemma 10.3. Let X and Y be complex spaces. Let $\varphi: X \to Y$ be a q-fibering holomorphic map. Then φ is proper if and only if φ is quasi-proper of codimension q.

Proof. If φ is proper, then φ is quasi-proper of codimension q. Suppose that φ is quasi-proper of codimension q. Let K be a compact subset of Y. A compact subset K' of X exists such that (P") is satisfied with $k = q$. Take $x \in \varphi^{-1}(K)$. Then $y = \varphi(x) \in K \cap \varphi(X)$. Then $C = \{x\}$ is an irreducible analytic subset of $\varphi^{-1}(y)$ with $\dim C = 0 = q - q$. Hence, $C \cap K' \neq \emptyset$, which implies $x \in K'$.

Therefore, $\varphi^{-1}(K) \subseteq K'$. The set $\varphi^{-1}(K)$ is compact. The map φ is proper; q.e.d.

Obviously, if $\varphi: X \to Y$ is holomorphic, then φ is proper if and only if φ is quasi-proper of codimension k for all $k \geqq 0$. Therefore, the concept "quasi-proper of codimension k" provides a scale to measure the "properness" of a holomorphic map.

Lemma 10.4. Let X and Y be irreducible complex spaces of dimension m and n respectively with $q = m - n \geqq 0$. Let $\varphi: X \to Y$ be a q-fibering, holomorphic map. Assume that φ is quasi-proper of codimension k with $1 \leqq k \leqq q$. Let $f_1,\ldots,f_k$ be φ-independent meromorphic functions on X. Then $A = X - (P_{f_1} \cup \ldots \cup P_{f_k})$ is the largest open subset of X where $f_1,\ldots,f_k$ are holomorphic. Define $f: A \to \mathbb{C}^k$ by $f(x) = (f_1(x),\ldots,f_k(x))$. Define $A^* = \{(x,f(x)) | x \in A\}$. Let Γ be the closure of A^* in $X \times \mathbb{P}^k$. Let $\pi: \Gamma \to X$ and $\mathfrak{v}: \Gamma \to \mathbb{P}^k$ be the projections. Define $\psi: \Gamma \to Y \times \mathbb{P}^k$ by $\psi(x,z) = (\varphi(x),z) = (x,z) \in \Gamma$. Let $\chi: Y \times \mathbb{P}^k \to Y$ and $\eta: Y \times \mathbb{P}^k \to \mathbb{P}^k$ be the projections. Define $F = \{f_1,\ldots,f_k\}$. Then

1. The set Γ is an irreducible, m-dimensional, analytic subset of $X \times \mathbb{P}^k$. The projection π and $\mathfrak{v}$ are holomorphic and π is proper and surjective. Moreover, $\pi_0 = \pi: A^* \to A$ is biholomorphic.
2. The map $\psi: \Gamma \to Y \times \mathbb{P}^k$ is surjective, quasi-proper and holomorphic and has rank $n + k$.

3. The diagram

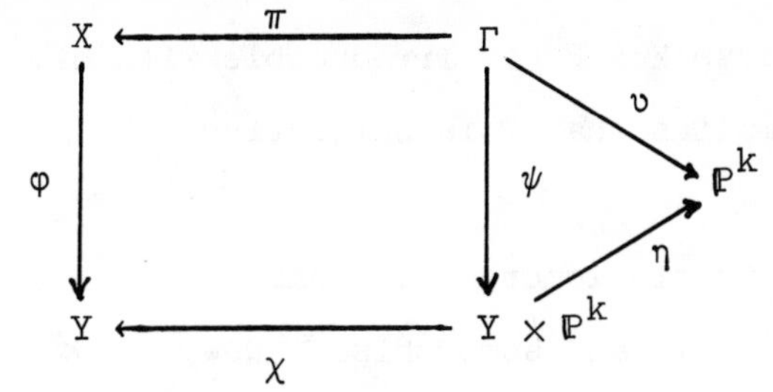

is commutative.

4. The homomorphism $\pi^*: \mathfrak{K}(X) \to \mathfrak{K}(\Gamma)$ is well defined and is an isomorphism and restricts to an isomorphism

$$\pi^*: \mathfrak{K}_\varphi(X;F) \longrightarrow \mathfrak{K}_\psi(\Gamma).$$

<u>Proof.</u> By Proposition 3.12, f is meromorphic. Hence, 1. holds. Obviously the diagram in 3. commutes.

2. Because $f_1,\ldots,f_k$ and φ-independent, ψ has rank $n + k$. Obviously, ψ is holomorphic. Let K be a compact subset of $Y \times \mathbb{P}^k$. Then $K_1 = \chi(K)$ is compact. A compact subset K' of X exists such that (P'') holds for K_1 in Y and K' in X and for k. Then $K'' = \pi^{-1}(K') = \Gamma \cap (K' \times \mathbb{P}^k)$ is compact in Γ. Take $(y,z) \in K \cap \psi(\Gamma)$. Let B be a branch of $\psi^{-1}(y,z) = (\varphi^{-1}(y) \times \{z\}) \cap \Gamma$. Then $B = B' \times \{z\}$ where B' is an irreducible analytic subset of $\varphi^{-1}(y)$ with $\dim B' = \dim B$. Obviously, $B' = \pi(B)$. Then $\dim B = \dim B' \leqq q$. Take a simple point (x,z) of $\psi^{-1}(y,z)$ with $(x,z) \in B$. Then

$$\begin{aligned} q \geqq \dim B' &= \dim_{(x,z)} B = \dim_{(x,z)} \psi^{-1}(y,z) = m - \operatorname{rank}_{(y,z)} \psi \\ &\geqq m - \operatorname{rank} \psi = m - n - k = q - k. \end{aligned}$$

Therefore, $B' \cap K' \neq \emptyset$. Take $x_0 \in B' \times K'$. Then $(x_0, z) \in B \cap K''$. Hence, $B \cap K'' \neq \emptyset$. The map ψ is quasi-proper. Because ψ has rank $n + k$, and because $Y \times \mathbb{P}^k$ is irreducible with $\dim Y \times \mathbb{P}^k = n + k$, Lemma 8.11 implies that ψ is surjective.

4. Because π_0 is biholomorphic, π has rank m. By Proposition 9.4, $\pi^*: \mathfrak{K}(X) \to \mathfrak{K}(\Gamma)$ is an isomorphism. Now, $\pi^*(\mathfrak{K}_\varphi(X;F)) = \mathfrak{K}_\psi(\Gamma)$ remains to be shown.

Take $h \in \mathfrak{K}_\varphi(X;F)$. Then $f_1, \ldots, f_k$, h are φ-dependent by Lemma 4.15. By Lemma 4.3, an open subset $U \neq \emptyset$ consisting of simple points of X with $U \subseteq A - P_h$ and a schlicht chart $\beta: U_\beta \to U'_\beta$ of Y with $\varphi(U) \subseteq U_\beta$ exists such that $\beta \circ \varphi | U = (g_1, \ldots, g_n)$ and such that

$$\omega = dg_1 \wedge \cdots \wedge dg_n \wedge df_1 \wedge \cdots \wedge df_k \text{ on } U$$

with $\omega(x) \neq 0$ if $x \in U$. By Lemma 4.5, $\omega \wedge dh = 0$ on U.

Define $U^* = \pi_0^{-1}(U) = \{(x, f(x)) | x \in U\}$. Then U is open in A^* and Γ. The map $\tilde{\beta} = \beta \times \mathrm{Id}: U_\beta \times \mathbb{C}^k \to U'_\beta \times \mathbb{C}^k$ is biholomorphic and $\tilde{\beta}$ is a schlicht chart of $Y \times \mathbb{P}^k$. The open subset U^* of Γ consists of simple points of U^* only and $\pi^*(h)$ is holomorphic on U^*. Moreover, $\psi(U^*) \subseteq U_\beta \times \mathbb{C}^k$. If $(x,z) \in U^*$, then $z = f(x)$ and $\psi(x,z) = (\varphi(x), f(z))$. Also

$$\tilde{\beta} \circ \psi | U^* = (g_1, \ldots, g_n, f_1 | U, \ldots, f_k | U) \circ \pi | U^* \tag{10.1}$$

$$0 = (\pi | U^*)^* (\omega \wedge dh) = dg_1 \circ \pi \wedge \cdots \wedge dg_h \circ \pi \wedge \cdots \wedge df_k \circ \pi \wedge d\pi^*(h) \tag{10.2}$$

on U^*. By Lemma 4.6, $\pi^*(h)$ is ψ-dependent, i.e., $\pi^*(h) \in \mathfrak{K}_\psi(\Gamma)$.

Take $\tilde{h} \in \mathfrak{K}_\psi(\Gamma)$. Then $\tilde{h} = \pi^*(h)$ for one and only one $h \in \mathfrak{K}(X)$. Now, U and β exist as before by Lemma 4.8. Define $\tilde{\beta}$, U^* and ω as before. Because of (10.1) and because $\tilde{h} = \pi^*(h) \in \mathfrak{K}_\psi(\Gamma)$, Lemma 4.5 implies (10.2) on U^*. Because $\pi|U^*: U^* \to U$ is biholomorphic, $\omega \wedge dh = 0$ on U. By Lemma 4.6, $f_1,\ldots,f_k,h$ are φ-dependent. By Lemma 4.15, $h \in \mathfrak{K}_\varphi(X;F)$. Therefore, $\pi^*(\mathfrak{K}_\varphi(X;F)) = \mathfrak{K}_\psi(\Gamma)$ and $\pi^*: \mathfrak{K}_\varphi(X;F) \to \mathfrak{K}_\psi(\Gamma)$ is an isomorphism; q.e.d.

<u>Theorem 10.5.</u> Let X and Y be irreducible complex spaces of dimension m and n respectively with $m - n = q \geqq 0$. Let $\varphi: X \to Y$ be a holomorphic map of rank n. Suppose that φ is quasi-proper of codimension k with $1 \leqq k \leqq q$. Let $f_1,\ldots,f_k$ be φ-independent meromorphic functions on X. Define $F = \{f_1,\ldots,f_k\}$. Then $\mathfrak{K}_\varphi(X;F)$ is an algebraic function field over $\varphi^*\mathfrak{K}(Y)$ of transcendence degree k. Moreover, $\varphi^*\mathfrak{K}(Y)(f_1,\ldots,f_k)$ is a pure transcendental extension of degree k of $\varphi^*\mathfrak{K}(Y)$ and $\mathfrak{K}_\varphi(X;F)$ is a finite algebraic extension of $\varphi^*\mathfrak{K}(Y)(f_1,\ldots,f_k)$.

<u>Remark.</u> The case $k = 0$ is given by Theorem 9.2.

<u>Proof.</u> Suppose, that $f_1,\ldots,f_k$ are algebraically dependent over $\varphi^*\mathfrak{K}(Y)$. By Proposition 4.25, $f_1,\ldots,f_k$ are φ-dependent, which is wrong. Hence, $f_1,\ldots,f_k$ are algebraically independent over $\varphi^*\mathfrak{K}(Y)$. Therefore, $\varphi^*\mathfrak{K}(Y)(f_1,\ldots,f_k)$ is a pure transcendental extension of degree k of $\varphi^*\mathfrak{K}(Y)$. It remains to be shown that $\mathfrak{K}_\varphi(X;F)$ is a finite algebraic extension of $\varphi^*\mathfrak{K}(Y)(f_1,\ldots,f_k)$.

At first, consider the case, where φ is q-fibering. Then the situation of Lemma 10.4 is given. Adopt the notation there. On $\mathbb{P}^k$, meromorphic function $\zeta_1,\ldots,\zeta_k$ exist which are holomorphic on $\mathbb{C}^k$ with $\zeta_\mu(z_1,\ldots,z_k) = z_\mu$ for $(z_1,\ldots,z_k) \in \mathbb{C}^k$. The maps $\psi\colon \Gamma \to Y \times \mathbb{P}^k$ and $\eta\colon Y \times \mathbb{P}^k \to \mathbb{P}^k$ have rank $n + k$ and k respectively with $\vartheta = \eta\circ\psi\colon \Gamma \to \mathbb{P}^k$. Because ψ and η are surjective, ϑ is surjective. Hence, ϑ has rank k. Therefore, $\vartheta^*(\zeta_\mu) = \psi^*\eta^*(\zeta_\mu)$ for $\mu = 1,\ldots,k$. Define $\tilde{\zeta}_\mu = \eta^*(\zeta_\mu)$, then $\vartheta^*(\zeta_\mu) = \psi^*(\tilde{\zeta}_\mu)$. If $u = (x,z) \in A^*$, then $x = \pi(u)$ and $z = \vartheta(u)$. Moreover, $z = f(x) = f\circ\pi(u) \in \mathbb{C}^k$, which implies $\zeta_\mu\circ\vartheta(u) = \zeta_\mu(z) = f_\mu\circ\pi(u)$. Hence, $\vartheta^*(\zeta_\mu)|A^* = \pi^*(f_\mu)|A^*$. Because A^* is open and dense in Γ, also $\vartheta^*(\tilde{\zeta}_\mu) = \pi^*(f_\mu)$ holds, which implies $\psi^*(\tilde{\zeta}_\mu) = \pi^*(f_\mu)$ for $\mu = 1,\ldots,k$.

By Theorem 6.9

$$\mathfrak{K}(Y \times \mathbb{P}^k) = \chi^*\mathfrak{K}(Y)(\tilde{\zeta}_1,\ldots,\tilde{\zeta}_k)$$

is a pure transcendental extension of $\chi^*\mathfrak{K}(Y)$. By Theorem 9.2 $\mathfrak{K}_\psi(\Gamma)$ is a finite algebraic extension of $\psi^*(\mathfrak{K}(Y \times \mathbb{P}^k))$ since ψ is quasi-proper and has rank $n + k$. Because $\pi^*\colon \mathfrak{K}_\varphi(X;F) \to \mathfrak{K}_\psi(\Gamma)$ is an isomorphism, $\pi^*(\mathfrak{K}_\varphi(X;F))$ is a finite algebraic extension of

$$\begin{aligned}\psi^*(\mathfrak{K}(Y \times \mathbb{P}^k)) &= \psi^*\chi^*\mathfrak{K}(Y)(\psi^*(\tilde{\zeta}_1),\ldots,\psi^*(\tilde{\zeta}_k))\\ &= \pi^*\varphi^*\mathfrak{K}(Y)(\pi^*(f_1),\ldots,\pi^*(f_k))\\ &= \pi^*(\varphi^*\mathfrak{K}(Y)(f_1,\ldots,f_k)).\end{aligned}$$

Hence, $\mathfrak{K}_\varphi(X;F)$ is a finite algebraic extension of $\varphi^*\mathfrak{K}(Y)(f_1,\ldots,f_k)$.

Here the equality $\psi^* \circ \chi^* = \pi^* \circ \varphi^*$ was used, which holds by Lemma 3.4.4 and Remark 3.5.3 because the maps ψ, χ, φ, π and $\chi \circ \psi = \varphi \circ \pi$ have rank $n + k$, n, n, m and n respectively. The map $\chi \circ \psi$ is surjective; therefore, it has rank n.

<u>Now, consider the general situation.</u> By Proposition 1.24 and by Proposition 8.4, the analytic set $D = \{x \in X \mid \mathrm{rank}_x \varphi \leqq n-1\}$ is thin and $D' = \varphi(D)$ is analytic with $\dim D' \leqq n - 2$. By Lemma 1.25, $T = \varphi^{-1}(D')$ is a thin analytic subset of X. Then $X_1 = X - T$ and $Y_1 = Y - D'$ are open and dense in X and Y respectively and are irreducible complex spaces of dimension m and n respectively with $\varphi(X_1) = Y_1$. The map $\varphi_1 \colon X_1 \to Y_1$ is q-fibering, holomorphic and quasi-proper of codimension k. By Lemma 4.14, the meromorphic functions $f_1|X_1, \ldots, f_k|X_1$ are φ_1-independent. Define $F_1 = \{f_1|X_1, \ldots, f_k|X_1\}$. Then $\mathfrak{K}_{\varphi_1}(X_1, F_1)$ is a finite algebraic extension of $\varphi_1^* \mathfrak{K}(Y_1)(f_1|X_1, \ldots, f_k|X_1)$. Define

$$r = [\mathfrak{K}_{\varphi_1}(X_1, F_1) : \varphi_1^* \mathfrak{K}(Y_1)(f_1|X_1, \ldots, f_k|X_1)].$$

Take $g \in \mathfrak{K}_\varphi(X, F)$. By Lemma 4.15, $f_1, \ldots, f_k, g$ are φ-dependent. By Lemma 4.14, $f_1|X_1, \ldots, f_k|X_1, g|X_1$ are φ_1-dependent. By Lemma 4.15, $g|X_1 \in \mathfrak{K}_{\varphi_1}(X, F_1)$. A polynomial

$$\sum_{\nu=0}^{s} Q_\nu \xi^\nu = Q \in \mathfrak{K}(Y_1)[\xi_1, \ldots, \xi_k, \xi]$$

with $s \leqq r$ and with

$$\sum_{\mu_1,\ldots,\mu_k=0}^{t} a_{\nu\mu_1\ldots\mu_k}\xi_1^{\mu_1}\ldots\xi_k^{\mu_k} = Q_\nu \in \widetilde{\mathfrak{K}}(Y_1)[\xi_1,\ldots,\xi_k]$$

with $Q_s \not\equiv 0$ exists such that $\varphi^*Q(f_1|X_1,\ldots,f_k|X_1,g|X_1) = 0$ on X_1. Because $\dim D' \leqq n-2$, the meromorphic functions $a_{\nu\mu_1\ldots\mu_p}$ on $Y_1 = Y - D'$ extend to meromorphic functions $\tilde{a}_{\nu\mu_1\ldots\mu_p}$ on Y. Hence, Q_ν extends to

$$\sum_{\mu_1,\ldots,\mu_k=0}^{t} \tilde{a}_{\nu\mu_1\ldots\mu_k}\xi_1^{\mu_1}\ldots\xi_k^{\mu_k} = \tilde{Q}_\nu \in \mathfrak{K}(Y)[\xi_1,\ldots,\xi_k]$$

with $\tilde{Q}_s \not\equiv 0$ and $s \leqq r$. Also Q extends to

$$\tilde{Q} = \sum_{\nu=0}^{s} \tilde{Q}_\nu \xi^\nu \in \mathfrak{K}(Y)[\xi_1,\ldots,\xi_p,\xi]$$

with $\varphi^*\tilde{Q}(f_1,\ldots,f_k,g) = 0$ on X, because X_1 is open and dense in X. Since $f_1,\ldots,f_k$ are algebraically independent over $\varphi^*\mathfrak{K}(Y)$ and since $\varphi^*\tilde{Q}_s(\xi_1,\ldots,\xi_k) \not\equiv 0$, also $\varphi^*\tilde{Q}_s(f_1,\ldots,f_k) \not\equiv 0$ on X. Hence, g is algebraic over $\varphi^*\mathfrak{K}(Y)(f_1,\ldots,f_k)$ with

$$[g\colon \varphi^*\mathfrak{K}(Y)(f_1,\ldots,f_k)] \leqq s \leqq r.$$

By Lemma 4.22, $\mathfrak{K}_\varphi(X;F)$ is a finite algebraic extension of $\varphi^*\mathfrak{K}(Y)(f_1,\ldots,f_k)$ which is a pure transcendental extension of degree k of $\varphi^*\mathfrak{K}(Y)$. Hence, $\mathfrak{K}_\varphi(X;F)$ is an algebraic function field of transcendence degree k over $\varphi^*\mathfrak{K}(Y)$; q.e.d.

Recall the definition of φ-dim F and dim φ introduced in §4.

Theorem 10.6. Let X and Y be irreducible complex spaces of dimension m and n respectively with $m - n = q$. Let $F \subseteq \mathfrak{K}(X)$ be a set of meromorphic functions on X. Let $\varphi: X \to Y$ be a holomorphic map of rank n which is quasi-proper of codimension k with $k = \varphi\text{-dim } F$. Then $\mathfrak{K}_\varphi(X,F)$ is an algebraic field of transcendence degree k over $\varphi^* \mathfrak{K}(Y)$.

Proof. If $k = 0$, then $\mathfrak{K}_\varphi(X,F) = \mathfrak{K}_\varphi(X)$. Then Lemma 10.1 and Theorem 9.2 imply Theorem 10.6. If $k > 0$, then φ-independent meromorphic functions $f_1,\ldots,f_k$ on X exist such that $\mathfrak{K}_\varphi(X,F) = \mathfrak{K}_\varphi(X,F')$ where $F' = \{f_1,\ldots,f_k\}$. Here $1 \leqq k \leqq q$ by Lemma 4.1. Hence, Theorem 10.5 implies Theorem 10.6; q.e.d.

If $F = \mathfrak{K}(X)$, then $\varphi\text{-dim } F = \dim \varphi$. Therefore, Theorem 10.6 implies

Theorem 10.7. Let X and Y be irreducible complex spaces of dimension m and n respectively with $m - n = q \geqq 0$. Let $\varphi: X \to Y$ be a holomorphic map of rank n which is quasi-proper of codimension k with $k = \dim \varphi$. Then $\mathfrak{K}(X)$ is an algebraic function field of transcendence degree k over $\varphi^* \mathfrak{K}(Y)$.

Because $\dim \varphi \leqq q$ and because any map which is quasi-proper of codimension q is also quasi-proper of codimension k if $0 \leqq k \leqq q$, Theorem 10.7 implies

Theorem 10.8. Let X and Y be irreducible complex spaces of dimension m and n respectively with $m - n = q \geqq 0$. Let $\varphi: X \to Y$ be a holomorphic map of rank n which is quasi-proper of codimension q. Then $\mathfrak{K}(X)$ is an algebraic function field over $\varphi^* \mathfrak{K}(Y)$ with transcendence degree at most q.

A proper holomorphic map is quasi-proper of codimension q. Therefore, Theorem 10.8 implies

Theorem 10.9. Let X and Y be irreducible complex spaces of dimension m and n respectively with $m - n = q \geqq 0$. Let $\varphi: X \to Y$ be a proper, holomorphic map of rank n. Then $\mathfrak{K}(X)$ is an algebraic function field over $\varphi^* \mathfrak{K}(Y)$ with transcendence degree $\leqq q$.

Theorem 9.2, Theorem 10.5 to Theorem 10.9 comprise the first main result of these investigations, which was based on the methods of Remmert, who considered the case where Y is a point.

§11. Full holomorphic maps

A holomorphic map $\varphi: X \to Y$ of a complex space X into a complex space Y is said to be <u>full</u>, if and only if every point $b \in Y$ has an open neighborhood V such that the following condition (F) is satisfied:

(F) Open subsets U_ν in X exist for each $\nu \in \mathbb{N}$ such that

$$\varphi^{-1}(V) = \bigcup_{\nu=1}^{\infty} U_\nu$$

and such that $\varphi_\nu = \varphi: U_\nu \to V$ is quasi-proper if $U_\nu \neq \emptyset$.

Here, $U_\nu = U_\mu$ for $\nu \neq \mu$ is permitted. The set $\{U_\nu | \nu \in \mathbb{N}\}$ is finite or countable. Hence, every quasi-proper map is full. Obviously, it suffices to require (F) for $b \in \overline{\varphi(X)}$ only, in which case, $U_\nu \neq \emptyset$ for each $\nu \in \mathbb{N}$ can be assumed.

<u>Lemma 11.1.</u> Let X and Y be complex spaces. Let $\varphi: X \to Y$ be a full holomorphic map. Let Y_0 be open in Y such that $\varphi^{-1}(Y_0) = X_0 \neq \emptyset$. Define $\varphi_0 = \varphi: X_\varphi \to Y_0$. Then φ_0 is full.

<u>Proof.</u> Take $b \in Y_0$. An open neighborhood V of Y exists such that (F) holds for X, V and φ. Then $\varphi_0^{-1}(V \cap Y_0) = \bigcup_{\nu=1}^{\infty} (U_\nu \cap X_0)$. If $U_\nu \cap X_0 \neq \emptyset$, then $U_\nu \neq \emptyset$ and $\varphi_\nu = \varphi: U_\nu \to V$ is quasi-proper. Define $\varphi_{0\nu} = \varphi_0: U_\nu \cap X_0 \to V \cap Y_0$. Then $\varphi_{0\nu} = \varphi_\nu: U_\nu \cap X_0 \to V \cap Y_0$ with

$$U_\nu \cap X_0 = U_\nu \cap \varphi^{-1}(Y_0) = U_\nu \cap \varphi^{-1}(Y_0 \cap V) = \varphi_\nu^{-1}(Y_0 \cap V).$$

By Lemma 8.5 b) $\varphi_{0\nu}$ is quasi-proper. Hence, φ_0 is full; q.e.d.

<u>Lemma 11.2.</u> Let X and Y be complex spaces. Let $\varphi: X \to Y$ be a full, holomorphic map. Let T' be an analytic subset of Y. Suppose that $T = \varphi^{-1}(T') \neq \emptyset$. Then $\varphi_0 = \varphi: T \to T'$ is full.

<u>Proof.</u> Take $b \in T'$. An open neighborhood V of b exists such that (F) is satisfied for φ and V. Then $\varphi_0^{-1}(V \cap T') = \bigcup_{\nu=1}^{\infty} U_\nu \cap T$. If $U_\nu \cap T \neq \emptyset$, then $U_\nu \neq \emptyset$ and $\varphi_\nu = \varphi: U_\nu \to V$ is quasi-proper. Define $\varphi_{0\nu} = \varphi_0: U_\nu \cap T \to V \cap T'$. Then $\varphi_{0\nu} = \varphi_\nu: U_\nu \cap T \to V \cap T'$ with

$$U_\nu \cap T = U_\nu \cap \varphi^{-1}(T') = U_\nu \cap \varphi^{-1}(T' \cap V) = \varphi_\nu^{-1}(T' \cap V).$$

By Lemma 8.5 a), $\varphi_{0\nu}$ is quasi-proper. Therefore, φ_0 is full; q.e.d.

<u>Lemma 11.3.</u> Let X and Y be complex spaces. Let $\varphi: X \to Y$ be a full holomorphic map. Let $S \neq \emptyset$ be a φ-saturated analytic subset of X. Then $\varphi_0 = \varphi: S \to Y$ is a full holomorphic map.

<u>Proof.</u> Take $b \in Y$. An open neighborhood V of b exists such that condition (F) is satisfied for φ and V. Then

$$\varphi_0^{-1}(V) = \varphi^{-1}(V) \cap S = \bigcup_{\nu=1}^{\infty} U_\nu \cap S.$$

If $S_\nu = U_\nu \cap S \neq \emptyset$, then $U_\nu \neq \emptyset$ and $\varphi_\nu = \varphi: U_\nu \to V$ is defined and quasi-proper. Take $x \in S_\nu$. Let B be a branch of $\varphi_\nu^{-1}(\varphi_\nu(x)) \cap S_\nu = \varphi^{-1}(\varphi(x)) \cap S \cap U_\nu$. One and only one branch B' of $\varphi^{-1}(\varphi(x)) \cap S$ with $B' \supseteq B$ exists and B is a branch of $B' \cap U_\nu$. Moreover, dim B' = dim B. Since S is φ-saturated, B' is a branch of $\varphi^{-1}(\varphi(x))$. Hence, each branch of $B' \cap U_\nu$ is a branch of $\varphi^{-1}(\varphi(x)) \cap U_\nu$. Hence, B is a branch of $\varphi^{-1}(\varphi(x)) \cap U_\nu = \varphi_\nu^{-1}(\varphi_\nu(x))$. Therefore, S_ν if φ_ν-saturated. By Lemma 8.3, $\varphi_{0\nu} = \varphi_\nu: S_\nu \to V$ is quasi-proper where $\varphi_{0\nu} = \varphi_0|S_\nu$. Hence, φ_0 is full; q.e.d.

Lemma 11.4. Let X and Y be complex spaces. Let $\varphi: X \to Y$ be a full holomorphic map. Let T' be an analytic subset of Y such that $T = \varphi^{-1}(T') \neq \emptyset$. Then $\varphi_0 = \varphi: T \to Y$ is a full holomorphic map.

Proof. The set T is φ-saturated.

Lemma 11.5. Let X and Y be complex spaces. Suppose that X has pure dimension m. Let $\varphi: X \to Y$ be a full, holomorphic map. Let p be a non-negative integer. Define

$$E = \{x \in X \mid \operatorname{rank}_x \varphi \leqq p\}.$$

Then E is analytic and $\varphi_0 = \varphi: E \to Y$ is a full holomorphic map.

Proof. By Lemma 1.19, E is φ-saturated. Therefore, φ_0 is full by Lemma 11.3; q.e.d.

Lemma 11.6. Let $X_1,\ldots,X_p,Y_1,\ldots,Y_p$ be complex spaces. Let $\varphi_\nu: X_\nu \to Y_\nu$ be full holomorphic maps. Define $X = X_1 \times \ldots \times X_p$

and $Y = Y_1 \times \ldots \times Y_p$. Then

$$\varphi = \varphi_1 \times \ldots \times \varphi_p : X \to Y$$

is a full, holomorphic map.

Proof. Take $b = (b_1,\ldots,b_p) \in Y$. Then $b_\mu \in Y_\mu$ for $\mu = 1,\ldots,p$. An open neighborhood V_μ of b_μ in Y_μ and open subsets $U_{\mu\nu}$ of X_μ for $\nu \in \mathbb{N}$ exist such that $\varphi_\mu^{-1}(V_\mu) = \bigcup_{\nu=1}^{\infty} U_{\mu\nu}$ and such that $\varphi_{\mu\nu} = \varphi_\mu : U_{\mu\nu} \to V_\mu$ is quasi-proper. Define $V = V_1 \times \ldots \times V_p$ and

$$U_{\nu_1\ldots\nu_p} = U_{1\nu_1} \times \ldots \times U_{p\nu_p}$$

if $(\nu_1,\ldots,\nu_p) \in \mathbb{N}^p$. Then

$$\varphi^{-1}(V) = \bigcup_{\nu_1,\ldots,\nu_p=1}^{\infty} U_{\nu_1\ldots\nu_p}.$$

If $U_{\nu_1\ldots\nu_p} \neq \emptyset$, then $U_{\mu\nu_\mu} \neq \emptyset$ for $\mu = 1,\ldots,p$ and $\varphi_{\mu\nu_\mu}$ is quasi-proper. By Lemma 8.6

$$\varphi | U_{\nu_1\ldots\nu_p} = \varphi_{1\nu_1} \times \ldots \times \varphi_{p\nu_p} : U_{\nu_1\ldots\nu_p} \to V$$

is quasi-proper. Therefore, φ is full; q.e.d.

For the next Lemma, recall the notation in and before Lemma 1.28:

Lemma 11.7. Let X and Y be complex spaces. Let $\varphi: X \to Y$ be a full, holomorphic map. Let p be a positive integer. Then $\delta_\varphi: X_\varphi^p \to Y$ is full.

Proof. By Lemma 11.6, $\varphi_p: X^p \to Y^p$ is full. By Lemma 11.2 the restriction $\varphi_p: X_\varphi^p \to \Delta_Y$ if full. Because $\delta_Y: \Delta_Y \to Y$ is biholomorphic, the map $\delta_\varphi = \delta_Y \circ \varphi_p: X_\varphi^p \to Y$ is full; q.e.d.

Lemma 11.8. Let X and Y be complex spaces. Let $\varphi: X \to Y$ be a full holomorphic map. Let C be a connectivity component of X. Then $\varphi_0 = \varphi: C \to Y$ is full.

Proof. By Lemma 8.8, C is φ-saturated. Hence, φ_0 is full by Lemma 11.3; q.e.d.

Lemma 11.9. Let X and Y be complex spaces. Let $\varphi: X \to Y$ be a q-fibering, full, holomorphic map. Let $S \neq X$ be an analytic subset of X. Define $A = X - S$. Then $X_0 = \overline{A}$ is analytic in X and X_0 is a union of branches of X. The map $\varphi_0 = \varphi: X_0 \to Y$ is q-fibering, full and holomorphic.

Proof. By Proposition 8.10, X_0 is analytic and φ-saturated. Therefore φ_0 is full and holomorphic by Lemma 11.4. By Proposition 8.10, φ_0 is q-fibering. By Lemma 7.3, X_0 is the union of all branches of X which are not contained in S; q.e.d.

In accordance with the considerations after Proposition 8.10 and before Proposition 8.9', Lemma 11.9 is equivalent to:

<u>Lemma 11.9'.</u> Let X and Y be complex spaces. Let $\varphi: X \to Y$ be a q-fibering, full, holomorphic map. Let $\mathcal{L}_0 \neq \emptyset$ be a subset of the set $\mathcal{L}$ of branches of X. Then $X_0 = \bigcup_{B \in \mathcal{L}_0} B$ is analytic in X. The map $\varphi_0 = \varphi: X_0 \to Y$ is q-fibering, full and holomorphic.

<u>Lemma 11.10.</u> Let X and Y be complex spaces. Suppose that Y is irreducible with dim Y = n. Let S be an analytic subset of X with $A = X - S \neq \emptyset$. Then $X_0 = \overline{A}$ is analytic. Let $\varphi: X \to Y$ be a full, holomorphic map such that $\varphi_0 = \varphi: X_0 \to Y$ has strict rank n. Then φ_0 is surjective.

<u>Proof.</u> By Lemma 7.3, X_0 is analytic. Take $b \in \overline{\varphi_0(X_0)}$. Then it is claimed that $b \in \varphi_0(X_0) = \varphi(X_0)$ and that b is an interior point of $\varphi(X_0)$ if Y is locally irreducible at b. An open neighborhood V of b exists such that condition (F) is satisfied. An open neighborhood W of b in V exists such that N consists of finitely many branches $W_1,\ldots,W_p$ with $b \in W_\mu$ for each μ and such that each W_μ is locally irreducible at b. If Y is locally irreducible at b, then p = 1 and $W = W_1$ is irreducible. Define $U_\nu' = \varphi^{-1}(W) \cap U_\nu = \varphi_\nu^{-1}(W)$ for $\nu \in \mathbb{N}$. Define $\tilde{U}_\nu = U_\nu' \cap X_0$. Then

$$\varphi_0^{-1}(W) = \varphi^{-1}(W) \cap X_0 = \bigcup_{\nu=1}^{\infty} \tilde{U}_\nu.$$

Because $b \in \overline{\varphi_0(X_0)}$, the intersection $W \cap \varphi_0(X_0)$ is not empty. Hence, $\tilde{U}_\nu \neq \emptyset$ for some ν. Then $U_\nu' \neq \emptyset \neq U_\nu$ and $\varphi_\nu = \varphi: U_\nu \to V$ is quasi-proper. By Lemma 8.5, $\varphi_\nu' = \varphi_\nu: U_\nu' \to W$ is quasi-proper. Now, $S \cap U_\nu'$ is analytic and $A \cap U_\nu' = U_\nu' - S$ is open in U_ν' and has the closure $\overline{A} \cap U'_\nu = X_0 \cap U'_\nu$ in U'_ν. By Proposition 8.9, $\varphi'_{\nu 0} = \varphi'_\nu: U'_\nu \cap X_0 \to W$ is semi-proper. Observe that $\varphi'_{\nu 0} = \varphi_0: U' \cap X_0 \to W$. By Lemma 1.16, $D = \{x \in X_0 | \operatorname{rank}_x \varphi_0 < n\}$ is thin in X_0. If $x \in (U'_\nu \cap X_0) - D$, then $\operatorname{rank}_x \varphi'_{\nu 0} = \operatorname{rank}_x \varphi_0 = n$. Hence, $\varphi'_{\nu 0}$ has strict rank n. By Lemma 7.6, $\varphi'_{\nu 0}(U'_\nu \cap X_0) = \varphi'_{\nu 0}(\tilde{U}_\nu)$ is a pure n-dimensional analytic subset of W which is pure n-dimensional. Hence, $W_{\mu_1}, \ldots, W_{\mu_r}$ with $1 \leqq r \leqq p$ exist such that

$$\varphi'_{\nu 0}(\tilde{U}_\nu) = W_{\mu_1} \cup \ldots \cup W_{\mu_r}.$$

Especially, $b \in \varphi'_{\nu 0}(\tilde{U}_\nu) \subseteq \varphi_0(X_0)$. If Y is locally irreducible at b, then $p = r = 1$ and $\varphi'_{\nu 0}(\tilde{U}_\nu) = W$. Hence, $b \in W \subseteq \varphi_0(X_0)$ is an interior point of $\varphi(X_0)$. The image set $\varphi(X_0)$ is closed. Let S_Y be the set of non-simple points of Y. Then $Y - S_Y$ is connected and $\varphi(X_0) - S_Y$ is open and closed in $Y - S_Y$. Therefore, $\varphi(X_0) - S_Y = Y - S_Y$, which implies $\varphi(X_0) = Y$ since $\varphi(X_0)$ is closed; q.e.d.

Corollary 11.11. Let X be a complex space. Let Y be an irreducible complex space of dimension n. Let $\varphi: X \to Y$ be a full, holomorphic map of strict rank n. Then φ is surjective.

Proof. Take $S = \emptyset$ in Lemma 11.10. Then $X_0 = A = X$ and $\varphi_0 = \varphi$. Hence, φ is surjective; q.e.d.

Lemma 11.12. Let X be a complex space. Let Y be an irreducible complex space with $\dim Y = n$. Let $\varphi: X \to Y$ be a full holomorphic map of rank n. Define $D = \{x \in X \mid \text{rank}_x \varphi < n\}$ and $A = X - D$. Then $A \neq \emptyset$ is open in X and $X_0 = \overline{A}$ is analytic. Moreover, $\varphi(X_0) = Y$.

<u>Proof.</u> Because φ has rank n, $A \neq \emptyset$. Because D is analytic, A is open. By Lemma 7.7. $\varphi_0 = \varphi: X_0 \to Y$ has strict rank n. By Lemma 11.10, φ_0 is surjective. Hence, $\varphi(X_0) = \varphi_0(X_0) = Y$; q.e.d.

<u>Lemma 11.13.</u> Let X be a complex space. Let $\mathfrak{B}_0 \neq \emptyset$ be a subset of the set $\mathfrak{B}$ of branches of X. Suppose that each $B \in \mathfrak{B}_0$ has dimension m. Define $X_0 = \bigcup_{B \in \mathfrak{B}_0} B$. Then X_0 is analytic in X. Let Y be an irreducible complex space of dimension n with $m - n = q \geqq 0$. Let $\varphi: X \to Y$ be a q-fibering, full holomorphic map. Then $\varphi(X_0) = Y$.

<u>Proof.</u> By Lemma 11.9', $\varphi_0 = \varphi: X_0 \to Y$ is holomorphic, full and q-fibering. Because X_0 is pure m-dimensional, φ_0 has pure rank n. By Corollary 11.12, φ_0 is surjective. Hence, $\varphi(X_0) = \varphi_0(X_0) = Y$; q.e.d.

<u>Lemma 11.14.</u> Let X be a complex space. Let Y be a complex space of pure dimension n. Let $\varphi: X \to Y$ be a surjective, holomorphic map of strict rank n. Let f be a meromorphic function on X. Let H be the set of all points $y \in Y$ such that an open neighborhood N of y and a meromorphic function g on N exist such that $\tilde{\varphi}^*(g) = f|M$ where $M = \varphi^{-1}(N)$ and $\tilde{\varphi} = \varphi: M \to N$. Suppose that $H \neq \emptyset$. Define $G = \varphi^{-1}(H)$ and $\hat{\varphi} = \varphi: G \to H$. Then one and only one meromorphic function h on H exists such that $\hat{\varphi}^*(h) = f|G$.

<u>Proof.</u> Obviously, $H \neq \emptyset$ and $G \neq \emptyset$ are open. Hence, $\hat{\varphi}$ is defined. For each $y \in H$, choose an open neighborhood N_y and a meromorphic function $g_y \in \mathfrak{K}(N_y)$ such that $\varphi_y^*(g_y) = f|M_y$ if $M_y = \varphi^{-1}(N_y)$ and $\varphi_y = \varphi: M_y \to N_y$. Then

$$G = \bigcup_{y \in H} M_y \qquad\qquad H = \bigcup_{y \in H} N_y.$$

For $(y,z) \in H^2 = H \times H$, define $N_{yz} = N_y \cap N_z$ and $M_{yz} = M_y \cap M_z = \varphi^{-1}(N_y \cap N_z)$. Define $H_1^2 = \{(y,z) \in H^2 | N_{yz} \neq \emptyset\}$. Then $M_{yz} \neq \emptyset$ and $\varphi_{yz} = \varphi: M_{yz} \to N_{yz}$ are defined if $(y,z) \in H_1^2$. Because φ has strict rank n, the maps φ_y if $y \in H$ and φ_{yz} if $(y,z) \in H_1^2$ and $\hat{\varphi}$ have strict rank n by Lemma 1.16 and Lemma 1.3. Because φ is surjective, they are surjective. By Remark 3.5.1

$$\varphi_y^*: \mathfrak{K}(N_y) \to \mathfrak{K}(M_y) \qquad \text{if } y \in H$$

$$\varphi_{y2}^*: \mathfrak{K}(N_{yz}) \to \mathfrak{K}(M_{yz}) \qquad \text{if } (y,z) \in H_1^2$$

$$\hat{\varphi}^*: \mathfrak{K}(H) \to \mathfrak{K}(G)$$

are injective homomorphisms.

If $(y,z) \in H_1^2$, then $g_y|N_{yz} = g_z|N_{yz} \in \mathfrak{K}(N_{yz})$ and

$$\varphi_{yz}^*(g_y|N_{yz} - g_z|N_{yz}) = \varphi_y^*(g_y)|M_{yz} - \varphi_z^*(g_z)|M_{yz}$$

$$= f|M_{yz} = f|M_{yz} = 0.$$

Therefore, $g_y|N_{yz} = g_z|N_{yz}$. One and only one meromorphic function h on H exists such that $h|N_y = g_y$ if $y \in H$.

If $y \in H$, then $f|M_y = \varphi_y^*(h|N_y) = \hat{\varphi}^*(h)|M_y$. Hence, $f|G = \hat{\varphi}^*(h)$. If $\tilde{h} \in \mathfrak{K}(H)$ with $f|G = \hat{\varphi}^*(\tilde{h})$, then $\hat{\varphi}^*(\tilde{h}) = \varphi^*(h)$, which implies $\tilde{h} = h$, because $\hat{\varphi}^*$ is injective; q.e.d.

Lemma 11.15. Let X be a complex space. Let Y be a complex space of pure dimension n. Let $\varphi: X \to Y$ be a surjective, full, holomorphic map of strict rank n. Then φ is semi-proper.

Proof. Take $b \in Y$. An open neighborhood V of b and open subsets $U_\nu \neq \emptyset$ of X exist such that $\varphi^{-1}(V) = \bigcup_{\nu=1}^{\infty} U_\nu$ and such that $\varphi_\nu = \varphi: U_\nu \to V$ is quasi-proper. An open subset W of V exists such that W consists of finitely many branches $W_1, \ldots, W_p$ with $b \in W_\mu$ for $\mu = 1, \ldots, p$ and where each W_μ is irreducible at b. Define $\tilde{U}_\nu = \varphi^{-1}(W) \cap U_\nu = \varphi_\nu^{-1}(W)$. If $\tilde{U}_\nu \neq \emptyset$, then $\tilde{\varphi}_\nu = \varphi_\nu: \tilde{U}_\nu \to W$ is quasi-proper by Lemma 8.5 b). Then $\tilde{\varphi}_\nu$ has strict rank n. By Lemma 8.1, $\tilde{\varphi}_\nu(\tilde{U}_\nu)$ is analytic and pure n-dimensional in W. Hence, $\tilde{\varphi}_\nu(\tilde{U}_\nu)$ is a union of branches of W.

For each $\mu = 1, \ldots, p$, pick a point $y_\mu \in W_\mu$ which is a simple point of W. Hence, $y_\mu \notin W_\nu$ if $\nu \neq \mu$. Because φ is surjective $x_\mu \in X$ with $y_\mu = \varphi(x_\mu)$ exists. Then $x_\mu \in \varphi^{-1}(V)$. Hence, $x_\mu \in U_{\nu_\mu}$ for some index ν_μ. Now, $\varphi(x_\mu) = y_\mu \in W$. Hence, $x_\mu \in U_{\nu_\mu} \cap \varphi^{-1}(W) = \tilde{U}_{\nu_\mu}$ which implies $y_\mu \in \tilde{\varphi}_{\nu_\mu}(\tilde{U}_{\nu_\mu})$. Hence, y_μ is contained in a branch of $\tilde{\varphi}_{\nu_\mu}(\tilde{U}_{\nu_\mu})$ which is some W_λ. But $y_\mu \in W_\lambda$ implies $\lambda = \mu$.

Therefore, $W_\mu \subseteq \tilde{\varphi}_{\nu_\mu}(\tilde{U}_{\nu_\mu})$. Define $\tilde{U}_{\nu_\mu} \cap \varphi_{\nu_\mu}^{-1}(W_\mu) = Z_\mu$. Then Z_μ is analytic in $\tilde{U}_{\nu_\mu}$ and $\hat{\varphi}_\mu = \tilde{\varphi}_{\nu_\mu} : Z_\mu \to W_\mu$ is surjective and quasi-proper by Lemma 8.5. Let K be a compact subset of W. Then $K \cap W_\mu$ is compact. A compact subset K_μ' of Z_μ exsits such that each branch of $\hat{\varphi}_\mu^{-1}(y)$ intersects K_μ' if $y \in K \cap W_\mu$. The set $K' = K_1' \cup \ldots \cup K_p'$ is compact in $\varphi^{-1}(W) = \tilde{W}$. If $y \in K$ then $y \in K \cap W_\mu$ for some μ. Hence, $B \cap K_\mu' \neq \emptyset$ for some branch B of $\hat{\varphi}_\mu^{-1}(y)$. Take $x \in B \cap K_\mu'$. Then $y = \hat{\varphi}_\mu(x) = \varphi(x)$ and $x \in \varphi^{-1}(y) \cap K'$. The surjective map $\varphi: \tilde{W} \to W$ is semi-proper. By Lemma 1.12 $\varphi: X \to Y$ is semi-proper; q.e.d.

<u>Theorem 11.16.</u> Let X be a complex space. Let Y be an irreducible complex space of dimension n. Let $\varphi: X \to Y$ be a full, holomorphic map. Define

$$D = \{x \in X \mid \mathrm{rank}_x \varphi \leqq n-1\}.$$

Suppose that $X - D \neq \emptyset$ is pure m-dimensional with $m - n = q \geqq 0$. Let X_0 be a pure m-dimensional analytic subset of X such that $X_0 \cap D$ is a thin analytic subset of X_0. Then $\varphi_0 = \varphi: X_0 \to Y$ is a surjective, holomorphic map of strict rank n. Let $W \neq \emptyset$ be an open subset of Y. Define $U = \varphi_0^{-1}(W)$ and $\psi = \varphi_0: U \to W$. Let $f \in \mathfrak{K}(X_0)$ and $g \in \mathfrak{K}(W)$ be meromorphic functions such that $f|U = \psi^*(g)$. Then g is unique. One and only one meromorphic function $h \in \mathfrak{K}(Y)$ exists such that $f = \varphi_0^*(h)$. Moreover, $h|W = g$.

<u>Proof.</u> Let S_X and S_Y be the set of non-simple points of X and

Y respectively. Then $X_0 - D$ is a pure m-dimensional analytic subset of the pure m-dimensional complex space $X - D$. Therefore, $X_0 - D$ is a union of branches of $X - D$. Hence, $X_0 - (D \cup S_X)$ is open in $X - D$. Therefore, $\operatorname{rank}_x \varphi_0 = \operatorname{rank}_x \varphi = n$ if $x \in X_0 - (D \cup S_X)$. Because $X_0 - (D \cup S_X)$ is open and dense on X_0, the map φ_0 has strict rank n. Let $\mathfrak{B}$ and $\mathfrak{B}_0$ be the sets of branches of X and X_0 respectively. If $B \in \mathfrak{B}_0$, then $B - D \neq \emptyset$ is a branch of $X_0 - D$, hence, a branch of $X - D$. Then B is a branch of X, i.e., $B \in \mathfrak{B}$. Therefore, $\mathfrak{B}_0 \subseteq \mathfrak{B}$. Define $\mathfrak{B}_1 = \mathfrak{B} - \mathfrak{B}_0$. Then $S = \bigcup_{B \in \mathfrak{B}_1} B$ is an analytic subset of X and $\mathfrak{B}_1$ is the set of branches of S. The analytic set $X_0 \cap S$ is thin. Define $A = X - S$. If $x \in A$, then $x \in B \in \mathfrak{B} - \mathfrak{B}_1 = \mathfrak{B}_0$. Hence, $x \in X_0 - S$. Therefore, $X - S \subseteq X_0 - S$. Obviously, $X - S \supseteq X_0 - S$, which implies $A = X - S = X_0 - S$. Moreover, $\overline{A} = X_0$. By Lemma 11.10, φ_0 is surjective.

Let H be the set of all points $y \in Y$ such that an open neighborhood N of y and a meromorphic function k on N exists such that $\varphi_N^*(k) = f|M$ where $M = \varphi_0^{-1}(N) = \varphi^{-1}(N) \cap X_0$ and where $\varphi_N = \varphi_0 \colon M \to N$. Then $H \neq \emptyset$ and $G = \varphi_0^{-1}(H) \neq \emptyset$ in X_0. Define $\hat{\varphi}_0 = \varphi_0 \colon G \to H$. By Lemma 11.14 one and only one meromorphic function h on H exists such that $\hat{\varphi}_0^*(h) = f|G$. Obviously, $W \subseteq H$ and $U \subseteq G$ with $\psi = \hat{\varphi}_0 \colon U \to W$. Hence,

$$\psi^*(g) = f|U = \hat{\varphi}_0^*(h)|U = \psi^*(h|N).$$

Because φ_0 has strict rank n, also ψ has strict rank n and ψ^* is injective. Therefore, $g = h|W$. For the same reason g is uniquely determined by the equation $f|U = \psi^*(g)$.

Now, $H = Y$ shall be proved. If this is shown, then $G = X_0$ and $\hat{\varphi}_0 = \varphi_0$ follows and h is a solution to the equation $\varphi_0^*(h) = f$. Because φ_0 has strict rank n, this solution is unique. Hence, it remains to prove $H = Y$.

Take $b \in \overline{H}$. An open neighborhood V of b and open subsets $\tilde{U}_\nu \neq \emptyset$ of X exists such that $\varphi^{-1}(V) = \bigcup_{\nu=1}^{\infty} \tilde{U}_\nu$ and such that $\tilde{\varphi}_\nu = \varphi\colon \tilde{U}_\nu \to V$ is quasi-proper. Take an open neighborhood N of b such that $N \subseteq V$ and such that N consists of finitely many branches, $N_1, \ldots, N_p$ with $b \in N_\mu$ for $\mu = 1, \ldots, p$ and such that N_μ is irreducible at b. Of course, $N_\mu \neq N_\nu$ if $\mu \neq \nu$ can be assumed. The set $U_\nu = \tilde{\varphi}_\nu^{-1}(N) = \varphi^{-1}(N) \cap \tilde{U}_\nu$ is open in $\tilde{U}_\nu$ and X. If $U_\nu \neq \emptyset$, then $\varphi_\nu = \tilde{\varphi}_\nu\colon U_\nu \to N$ is quasi-proper. Moreover, $\varphi_\nu = \varphi\colon U_\nu \to N$. The set $U_{\nu 0} = U_\nu \cap X_0$ is open in X_0. Define $I = \{\nu \in \mathbb{N} \mid U_{\nu_0} \neq \emptyset\}$. Then $\varphi^{-1}(N) = \bigcup_{\nu=1}^{\infty} \tilde{U}_\nu$ and

$$M = \varphi_0^{-1}(N) = \varphi^{-1}(N) \cap X_0 = \bigcup_{\nu \in I} U_{\nu 0}.$$

The set $M \neq \emptyset$ is open in X_0. Then $\varphi_N = \varphi_0\colon M \to N$ has strict rank n. If $\nu \in I$, then $\varphi_{\nu 0} = \varphi_\nu\colon U_{\nu 0} \to N$ is defined with $\varphi_{\nu 0} = \varphi_0\colon U_{\nu 0} \to N$. Hence, $\varphi_{\nu 0}$ has strict rank n. Moreover, $\varphi_{\nu 0} = \varphi_N\colon U_{\nu 0} \to N$. Define $D_\nu = \{x \in U_\nu \mid \operatorname{rank}_x \varphi_\nu < n\}$. Because U_ν is open in X, and because $\varphi_\nu = \varphi\colon U_\nu \to N$, it follows that $D_\nu = U_\nu \cap D$.

Now, the case $b \in Y - S_Y$ shall be considered at first. Then Y is irreducible at b, which implies $p = 1$ and $N = N_1$. By Lemma 8.11, the map $\varphi_{\nu 0}\colon U_{\nu 0} \to N$ is surjective if $\nu \in I$. Now,

Proposition 9.7 shall be applied if $\nu \in I$:

Prop. 9.7	X	Y	n	φ	D	m	q	S	A	X_0	φ_0	V
Here	U_ν	N	n	φ_ν	D_ν	m	q	S_ν	A_ν	$U_{\nu 0}$	$\varphi_{\nu 0}$	Q

Prop. 9.7	U	ψ	g	f	h
Here	R_ν	ψ_ν	$h\|Q$	$f\|U_{\nu 0}$	k_ν

Here $U_\nu - D \supseteq U_{\nu 0} - D \neq \emptyset$ and $U_\nu - D$ is open in $X - D$. Hence, $U_\nu - D \neq \emptyset$ is pure m-dimensional with $m - n = q \geqq 0$. The set $S_\nu = U_\nu \cap S$ is analytic in U_ν with $A_\nu = U_\nu \cap A = U_\nu - S$. The closure of A_ν in U_ν is $\overline{A} \cap U_\nu = X_0 \cap U_\nu = U_{\nu 0}$. Because $b \in \overline{H}$, the open subset $Q = N \cap H$ of N is not empty and $h|Q$ is defined as a meromorphic function on Q. Define

$$R_\nu = \varphi_{\nu 0}^{-1}(Q) = \varphi_\nu^{-1}(H \cap N) \cap X_0 = \varphi^{-1}(H \cap N) \cap U_\nu \cap X_0$$
$$= \varphi^{-1}(H) \cap U_{\nu 0} = G \cap U_{\nu 0} \subseteq X_0.$$

Because $\varphi_{\nu 0}$ is surjective, $R_\nu \neq \emptyset$. Define $\psi_\nu = \varphi_{\nu 0}: R_\nu \to Q$. Then $\psi_\nu = \hat{\varphi}_0: R_\nu \to Q$. Because R_ν is open in X_0, the meromorphic function $f|R_\nu = (f|U_{\nu 0})|R_\nu$ is well defined, and

$$f|R_\nu = f|U_{\nu 0} \cap G = \hat{\varphi}_0^*(h)|U_{\nu 0} \cap G = \hat{\varphi}_0^*(h)|R_\nu$$
$$= \psi_\nu^*(h|Q).$$

The assumptions of Proposition 9.7 are satisfied. Therefore, one and only one meromorphic function $k_\nu \in \mathfrak{K}(N)$ exists such that

$\varphi^*_{\nu 0}(k_\nu) = f|U_{\nu 0}$. Moreover, $k_\nu|Q = h|Q$. This is true for each $\nu \in I$. Because $Q \neq \emptyset$ is an open subset of the irreducible complex space N, the function $k_\nu \in \mathfrak{F}(N)$ does not depend on $\nu \in I$. Hence, $k \in \mathfrak{F}(N)$ exists such that $\varphi^*_{\nu 0}(k) = f|U_{\nu 0}$ and $k|Q = h|Q$. Now, $\varphi_{\nu 0} = \varphi_N|U_{\nu 0}$ implies

$$\varphi^*_N(k)|U_{\nu 0} = \varphi^*_{\nu 0}(k) = f|U_{\nu 0} = (f|M)|U_{\nu 0}.$$

Because $M = \bigcup_{\nu \in I} U_{\nu 0}$, this implies $\varphi^*_N(k) = f|M$. Therefore, $b \in H$ by the definition of the set H.

Therefore $(Y - S_Y) \cap \overline{H} = (Y - S_Y) \cap H \neq \emptyset$ is open and closed in the connected complex manifold $Y - S_Y$. Therefore $Y - S_Y \subseteq H$ and $\overline{H} = Y$.

Now, the case $b \in S_Y$ shall be considered. If $\nu \in I$, let $\mathfrak{B}_\nu$ be the set of branches of $U_{\nu 0}$. Because $U_{\nu 0}$ is open in X_0, each branch $B \in \mathfrak{B}_\nu$ has dimension m. Take $\nu \in I$ and $B \in \mathfrak{B}_\nu$. For this selection, Proposition 9.8 shall be applied:

Prop. 9.8	X	Y	n	φ	D	m	B_0	Y_0	V	V_0	U_0
Here	U_ν	N	n	φ_ν	D_ν	m	B	N_μ	Q	Q_μ	$R_{\nu B}$

Prop. 9.8	ψ	g	f	h	φ_0
Here	$\psi_{\nu B}$	$h\|Q$	$f\|B$	$k_{\nu B}$	$\varphi_{\nu B}$

As stated, U_ν is open in X and N is open in Y. Hence, N has pure dimension n. The holomorphic map $\varphi_\nu: U_\nu \to N$ is quasi-proper with $D_\nu = \{x \in U_\nu | \text{rank } \varphi_\nu < n\}$ where $U_\nu - D_\nu = U_\nu \cap (X - D)$ is pure m-dimensional with $m - n = q \geqq 0$. Moreover, $U_{\nu 0} - D_\nu$ is open and dense in $U_{\nu 0}$. Also $U_{\nu 0} - D_\nu$ is a pure m-dimensional analytic subset of $U_\nu - D_\nu$, hence, $U_{\nu 0} - D_\nu \neq \emptyset$ is a union of branches of $U_\nu - D_\nu$.

Therefore, $B - D_\nu = B - D$ is a branch of $U_\nu - D_\nu$. Hence, $B = \overline{B - D_\nu}$ is a branch of U_ν. By Proposition 9.8, $\varphi_\nu(B)$ is a branch of N. One and only one index $\mu = \mu(\nu, B)$ exists such that $\varphi_\nu(B) = N_\mu$.

Because H is dense in Y, the open subset $Q = H \cap N$ of N and the open subset $Q_\mu = Q \cap N_\mu = H \cap N_\mu$ of N_μ are not empty. Define $\varphi_{\nu B} = \varphi_\nu: B \to N_\mu$. Then $\varphi_{\nu B}$ is holomorphic and surjective and has rank n. Therefore, $R_{\nu B} = \varphi_{\nu B}^{-1}(Q_\mu) = \varphi_\nu^{-1}(Q \cap N_\mu) \cap B = \varphi_\nu^{-1}(Q) \cap B$ is open and not empty in B. The meromorphic function $h|Q \in \mathfrak{K}(Q)$ is defined. The function f is meromorphic on X_0. Hence, $f|U_{\nu 0} \in \mathfrak{K}(U_{\nu 0})$ is defined, since $U_{\nu 0}$ is open and not empty in X_0. Because B is a branch of $U_{\nu 0}$, also $f|B \in \mathfrak{K}(B)$ is well defined. Moreover, $\hat{\varphi}_0^*(h) = f|G$ was already proved. Define $\psi_{\nu B} = \varphi_\nu: R_{\nu B} \to Q$. Then $\text{rank}_x \psi_{\nu B} = \text{rank}_x \varphi_{\nu B}$ for each $x \in R_{\nu B}$ because $R_{\nu B}$ is open in B. Because $\varphi_{\nu B}$ has rank n on the irreducible complex space B, the map $\psi_{\nu B}$ has strict rank n and $\psi_{\nu B}^*: \mathfrak{K}(Q) \to \mathfrak{K}(R_{\nu B})$ is defined and injective. Let $i: Q \to H$ be the inclusion map. Then $i \circ \psi_{\nu B}: R_{\nu B} \to H$

has strict rank n and $(i \circ \psi_{\nu B})^*: \mathfrak{K}(H) \to \mathfrak{K}(R_{\nu B})$ is defined and injective. Clearly $i^*: \mathfrak{K}(H) \to \mathfrak{K}(Q)$ is defined and injective. Now,

$$\begin{array}{ccc} R_{\nu B} & \xrightarrow{\quad j \quad} & G \\ \Big\downarrow{\scriptstyle \psi_{\nu B}} & 0 & \Big\downarrow{\scriptstyle \hat{\varphi}_0} \\ Q & \xrightarrow[\quad i \quad]{} & H \end{array}$$

is a commutative diagram where j is the inclusion map. Here B is a branch of $U_{\nu 0}$ which is open in X_0. Also G is open in X_0. Therefore, $B \cap G$ is open in B and a union of branches of $U_{\nu 0} \cap B$. Now, $R_{\nu B}$ is open in B and contained in $B \cap G$. Hence, $R_{\nu B}$ is open in a union of branches of an open subset of G. Therefore, $j^*: \mathfrak{K}(G) \to \mathfrak{K}(R_{\nu B})$ is well defined and injective. The map $\hat{\varphi}_0$ has strict rank n. Hence $\hat{\varphi}_0^*: \mathfrak{K}(H) \to \mathfrak{K}(G)$ is well defined and injective. By Lemma 3.4.5, the diagram

$$\begin{array}{ccc} \mathfrak{K}(R_{\nu B}) & \xleftarrow{\quad j^* \quad} & \mathfrak{K}(G) \\ \Big\downarrow{\scriptstyle \psi_{\nu B}^*} & & \Big\downarrow{\scriptstyle \hat{\varphi}_0^*} \\ \mathfrak{K}(Q) & \xleftarrow[\quad i^* \quad]{} & \mathfrak{K}(H) \end{array}$$

is well defined and commutative. Hence,

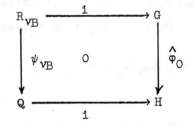

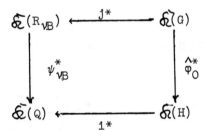

$$\psi^*_{\nu B}(h|Q) = \psi^*_{\nu B}(i^*(h)) = j^*\hat{\varphi}^*_0(h) = j^*(f|G) = f|R_{\nu B} = (f|B)|R_{\nu B}.$$

The map $\varphi_{\nu B} = \varphi_\nu : B \to N_\mu$ is surjective with $\varphi_{\nu B} = \varphi_{\nu 0} : B \to N_\mu$.

The assumptions of Proposition 9.8 are satisfied. Then one and only one meromorphic function $k_{\nu B} \in \mathfrak{K}(N_\mu)$ exists such that $f|B = \varphi^*_{\nu B}(k_{\nu B})$. Moreover, $k_{\nu B}|Q_\mu = h|Q_\mu$.

Define $\Lambda = \{(\nu,B) \mid \nu \in I \text{ and } B \in \mathfrak{B}_\nu\}$ and $\Delta = \{\rho \in \mathbb{N} \mid 1 \leqq \rho \leqq p\}$. A map $\mu : \Lambda \to \Delta$ is defined by $\varphi_\nu(B) = N_\mu$. Now, it will be shown that this map is surjective. Take $\rho \in \Delta$. Take a point $y \in N_\rho - S_Y$. Because $\varphi_0 : X_0 \to Y$ is surjective, a point $x \in X_0$ with $\varphi(x) = \varphi_0(x) = y$ exists. Then $x \in M = \varphi_0^{-1}(N)$. Therefore, $\nu \in I$ exists with $x \in U_{\nu 0}$. A branch $B \in \mathfrak{B}_\nu$ of $U_{\nu 0}$ exists with $x \in B$. Hence, $(\nu,B) \in I$ and $y = \varphi(x) = \varphi_\nu(x) \in N_\mu$ where $\mu = \mu(\nu,B)$. Because y is a simple point of Y, the point y is also a simple point of N and is contained in at most one branch of N. Hence $\rho = \mu(\nu,B)$. The map μ is surjective and $\Lambda_\rho = \mu^{-1}(\rho)$ is not empty for each $\rho \in \Delta$.

If $(\nu,B) \in \Lambda_\rho$ and $(\lambda,C) \in \Lambda_\rho$ then

$$k_{\nu B}|Q_\rho = h|Q_\rho = k_{\lambda,C}|Q_\rho$$

where $Q_\rho \neq \emptyset$ is open in the irreducible space N_ρ. Therefore, one and only one meromorphic function $k_\rho \in \mathfrak{K}(N_\rho)$ exists such that

$k_\rho = k_{\nu B}$ for all $(\nu, B) \in \Lambda_\rho$. Moreover, $k_\rho | Q_\rho = h | Q_\rho$ and $\varphi_{\nu B}^*(k_\rho) = f | B$ if $(\nu, B) \in \Lambda_\rho$.

This is true for each $\rho \in \Delta$. By Lemma 3.6, one and only one meromorphic function $k \in \mathfrak{K}(N)$ exists such that $k | N = k_\rho$. Then $k | Q \cap N_\rho = k_\rho | Q_\rho = h | Q_\rho = h | Q \cap N_\rho$ implies $k | Q = h | Q$. If $(\nu, B) \in \Lambda$, then $\psi_{\nu B} = \varphi_N \colon R_{\nu B} \to Q$. Hence,

$$\varphi_N^*(k) | R_{\nu B} = \psi_{\nu B}^*(k | Q) = \psi_{\nu B}^*(h | Q) = f | R_{\nu B}.$$

Hence, $\varphi_N^*(k) | B = f | B$ for each $B \in \mathfrak{B}_\nu$ which implies $\varphi_N^*(k) | U_{\nu 0} = f | U_{\nu 0}$ for each $\nu \in I$. Hence, $\varphi_N^*(k) = f | M$. Therefore, $b \in H$, by definition of the set H.

Hence, $S_Y \cup (Y - S_Y) \subseteq H$. Therefore, $Y = H$; q.e.d.

If $X = X_0$ is taken in Theorem 11.16, the following result is obtained.

Theorem 11.17. Let X be a complex space of pure dimension m. Let Y be an irreducible complex space of dimension n with $q = m - n \geqq 0$. Let $\varphi \colon X \to Y$ be a full, holomorphic map of strict rank n. Let $W \neq \emptyset$ be an open subset of Y. Define $U = \varphi^{-1}(W) \neq \emptyset$ and $\psi = \varphi \colon U \to W$. Let $f \in \mathfrak{K}(X)$ and $g \in \mathfrak{K}(W)$ be meromorphic functions such that $f | U = \psi^*(g)$. Then g is unique. One and only one meromorphic function $h \in \mathfrak{K}(Y)$ exists such that $f = \varphi^*(h)$. Moreover, $h | W = g$.

Let X and Y be complex spaces. Let $\varphi \colon X \to Y$ be a holomorphic map. Let p be a positive integer. Recall the notations introduced before and in Lemma 1.28. The complex space

$$X^p_\varphi = \{(x_1,\ldots,x_p) \in X^p \mid \varphi(x_1) = \ldots = \varphi(x_p)\}$$

is defined. Let $\pi_\nu\colon X^p_\varphi \to X$ be the projection $\pi_\nu(x_1,\ldots,x_p) = x_\nu$. Then $\delta_\varphi\colon X^p_\varphi \to Y$ is defined by $\delta_\varphi = \varphi\circ\pi_\nu$ and is independent of ν. If $\Delta_X = \Delta_X(p)$ is the diagonal of X^p, then $\Delta_X \subseteq X^p_\varphi$. If Id: $X \to X$ is the identity, then $\Delta_X = X^p_{Id}$ and $\delta_X = \delta_{Id}\colon \Delta_X \to X$ is biholomorphic.

Lemma 11.18. Let X and Y be irreducible complex spaces of dimension m and n respectively with $m - n = q \geqq 0$. Let $\varphi\colon X \to Y$ be a holomorphic map of rank n. Let p be a positive integer. Define $r = n + pq$. Define $\delta_\varphi\colon X^p_\varphi \to Y$ and

$$D = \{x \in X^p_\varphi \mid \operatorname{rank}_x \delta_\varphi < n\} .$$

Let $\mathfrak{B}$ be the set of branches of X^p_φ and define $\mathfrak{B}_0 = \{B \in \mathfrak{B} \mid B \cap D \neq B\}$. Then $\delta_\varphi|B$ has rank n, if and only if $B \in \mathfrak{B}_0$. Moreover, if $B \in \mathfrak{B}_0$ define $B' = B - D$. Then B' is open and dense in B and $\mathfrak{B}'_0 = \{B' \mid B \in \mathfrak{B}_0\}$ is the set of branches of $X^p_\varphi - D$. The open subset $X^p_\varphi - D$ of X^p_φ is not empty and pure r-dimensional. If $B \in \mathfrak{B}_0$, then $\dim B = r$ and $\operatorname{rank} \pi_\nu|B = m$. Obviously, $\mathfrak{B}_0 \neq \emptyset$.

Proof. If $1 \leqq \nu \leqq p$, then $\pi_\nu(X^p_\varphi) \supseteq \pi_\nu(\Delta_X) = X$. Hence, $\pi_\nu\colon X^p_\varphi \to X$ is surjective. Therefore,

$$\delta_\varphi(X^p_\varphi) = \varphi(\pi_\nu(X^p_\varphi)) = \varphi(X).$$

If $D = X^p_\varphi$, then rank $\delta_\varphi \leqq n - 1$, and $\delta_\varphi(X^p_\varphi) = \varphi(X)$ is almost thin of dimension $n - 1$ by Proposition 1.23. By Lemma 1.25, $\varphi(X)$ is not contained in any almost thin subset of Y. Hence, $X^p_\varphi - D \neq \emptyset$.

If $B \in \mathfrak{B}_0$, take a simple point x of X^p_φ with $x \in B - D$. Then $\operatorname{rank}_x \delta_\varphi | B = \operatorname{rank}_x \delta_\varphi = n$. Therefore, $\delta_\varphi | B$ has rank n. If $B \in \mathfrak{B}$ and if $\delta_\varphi | B$ has rank n, a simple point x of X^p_φ exists with $x \in B$ and $n = \operatorname{rank}_x \varphi | B = \operatorname{rank}_x \varphi$. Therefore, $x \in B - D$ and $B \in \mathfrak{B}_0$. Because $B \in \mathfrak{B}_0$ is a branch of X^p_φ; also $B' = B - D$ is a branch of $X^p_\varphi - D$. If $x \in X^p_\varphi - D$, then $x \in B \in \mathfrak{B}$ and $x \in B' = B \quad D \neq \emptyset$. Hence, $B \in \mathfrak{B}_0$. Therefore,

$$X^p_\varphi - D = \bigcup_{B \in \mathfrak{B}_0} B'$$

and $\mathfrak{B}_0' = \{B' | B \in \mathfrak{B}_0\}$ is the set of branches of $X^p_\varphi - D \neq \emptyset$. Hence, $\mathfrak{B}_0 \neq \emptyset$.

The analytic set $E = \{x \in X | \operatorname{rank}_x \varphi < n\}$ is thin in X, because φ has rank n and because X is irreducible. By Lemma 1.20, the image $E' = \varphi(E)$ is almost thin of dimension $n - 2$. Take $B \in \mathfrak{B}_0$. Let $Z \neq \emptyset$ be an open subset of $B - D$ which consists of simple points of X^p_φ only. Then Z is open in $X^p_\varphi - D$. Because $\delta_\varphi | B$ has rank n, the set $\delta_\varphi(Z)$ is not contained in E' by Lemma 1.25. Take $y \in \delta_\varphi(Z) - E'$. Then $y \in \delta_\varphi(X^p_\varphi) = \varphi(X)$. Hence, $\emptyset \neq \varphi^{-1}(y) \subseteq X - E$. If $x \in \varphi^{-1}(y)$, then

$$\dim_x \varphi^{-1}(y) = m - \widetilde{\mathrm{rank}}_x \varphi = m - \mathrm{rank}_x \varphi = m - n = q.$$

Therefore, $\varphi^{-1}(y)$ has pure dimension q. By (1.6), $\delta_\varphi^{-1}(y) = (\varphi^{-1}(y))^p$ has pure dimension $p \cdot q$. Because $y \in \delta_\varphi(Z)$, a point $x \in \delta_\varphi^{-1}(y) \cap Z$ exists which is a simple point of $X_\varphi^p - D$ and $B - D$.

$$\begin{aligned}\dim B = \dim_x B &= \mathrm{rank}_x \delta_\varphi | B + \dim_x (\delta_\varphi^{-1}(y) \cap B)\\ &= \mathrm{rank}_x \delta_\varphi + \dim_x \delta_\varphi^{-1}(y) \cap Z\\ &= \mathrm{rank}_x \delta_\varphi + \dim_x \delta_\varphi^{-1}(y) = n + pq = r\end{aligned}$$

for every $B \in \mathcal{L}_0$. Also $\dim B' = r$ if $B \in \mathcal{L}_0$. Therefore, $X_\varphi^p - D$ has pure dimension r.

Take $B \in \mathcal{L}_0$ and take Z and $y \in \delta_\varphi(Z) - E'$ as before. Take $x \in \delta_\varphi^{-1}(y) \cap Z$. Define $x_\nu = \pi_\nu(x)$. By (1.7)

$$\pi_\nu^{-1}(x_\nu) = (\varphi^{-1}(y))^{\nu-1} \times \{x_\nu\} \times (\varphi^{-1}(y))^{p-\nu+1}$$

has pure dimension $(p-1) \cdot q$. Because B is a neighborhood of x in X_φ^p this implies

$$\begin{aligned}\mathrm{rank}_x \pi_\nu | B &= r - \dim_x \pi_\nu^{-1}(x_\nu) \cap B\\ &= r - \dim_x \pi_\nu^{-1}(x_\nu)\\ &= n + pq - (p-1)q = n + q = m.\end{aligned}$$

Therefore, rank $\pi_\nu | B = m$; q.e.d.

Theorem 11.19. Let X and Y be irreducible complex spaces of dimension m and n respectively with $m - n = q \geqq 0$. Let $\varphi: X \to Y$ be a full holomorphic map of rank n. Let p be a positive integer. Let B be a branch of X_φ^p with rank $\delta_\varphi | B = n$. Then $\chi = \delta_\varphi | B: B \to Y$ is surjective. Let $W \neq \emptyset$ be an open subset of Y. Define $U = \chi^{-1}(W)$ and $\psi = \chi: U \to W$. Let $g \in \mathfrak{K}(W)$ and $f \in \mathfrak{K}(B)$ be meromorphic functions such that $\psi^*(g) = f | U$. Then g is unique. One and only one meromorphic function $h \in \mathfrak{K}(Y)$ exists such that $f = \chi^*(h)$. Moreover, $h | W = g$.

Proof. By Lemma 11.7, $\delta_\varphi: X_\varphi^p \to Y$ is a full holomorphic map. Define $D = \{x \in X_\varphi^p | \text{rank } \delta_\varphi < n\}$. Then $X_\varphi^p - D \neq \emptyset$ is open in X_φ^p and has pure dimension $r = n + pq$ by Lemma 11.18. Because χ has rank n, the intersection $B \cap D$ is thin in B and B has dimension $r = n + pq$ by Lemma 1.18. By Theorem 11.16, $\chi = \delta_\varphi: B \to Y$ is surjective. Also by Theorem 11.16, g is unique and one and only one meromorphic function $h \in \mathfrak{K}(Y)$ exists such that $f = \chi^*(h)$. Moreover, $h | W = g$; q.e.d.

§12. Globalization

Let X and Y be irreducible complex spaces of dimension m and n respectively with $m - n = q \geqq 0$. Let $\varphi: X \to Y$ be a full, holomorphic map of rank n. Let H be an open, irreducible subset of Y such that $\tilde{H} = \varphi^{-1}(H) \neq \emptyset$. Define $\varphi_0 = \varphi: \tilde{H} \to H$. Let $f_1,\ldots,f_k$ be meromorphic funcitons on X such that $f_1|\tilde{H},\ldots,f_k|\tilde{H}$ are algebraically dependent over $\varphi_0^* \mathfrak{K}(H)$. Then it will be shown that $f_1,\ldots,f_k$ are algebraically dependent over $\varphi^* \mathfrak{K}(Y)$. This globalization of algebraic dependence will be especially important in the pseudoconcave case. The globalization may fail if φ is not full as easy examples show.

Lemma 12.1. Let X and Y be irreducible complex spaces of dimension m and n respectively with $m - n = q \geqq 0$. Let $\varphi: X \to Y$ be a surjective, holomorphic map. For every open subset H of Y, define $\tilde{H} = \varphi^{-1}(H)$ and $\varphi_H = \varphi: \tilde{H} \to H$. Let $G \neq \emptyset$ be an open, irreducible subset of Y. Let $f_1,\ldots,f_p$ be meromorphic functions on X such that $f_1|\tilde{H},\ldots,f_p|\tilde{H}$ in $\mathfrak{K}(\tilde{H})$ are linearly independent over the field $\varphi_H^* \mathfrak{K}(H)$ for every open, irreducible subset $H \neq \emptyset$ of G. Let T be a thin analytic subset of X. Let S be an almost thin subset of Y. Let $W \neq \emptyset$ be any open subset of G.

Then a point $y_0 \in W - S$ and points $\xi_\nu \in \varphi^{-1}(y_0) - T$ for $\nu = 1,\ldots,p$ exist such that the following properties hold.

1. The point y_0 is a simple point of Y.
2. The points ξ_ν are simple points of X for $\nu = 1,\ldots,p$.
3. The map φ is regular at each point ξ_ν for $\nu = 1,\ldots,p$.

4. The functions $f_1,\dots,f_p$ are holomorphic at ξ_ν.

5. For $\mu \neq \nu$ in $\xi_\mu \neq \xi_\nu$ and

$$\begin{vmatrix} f_1(\xi_1), & \dots\dots, & f_1(\xi_p) \\ \vdots & & \vdots \\ f_p(\xi_1), & \dots\dots, & f_p(\xi_p) \end{vmatrix} \neq 0 .$$

<u>Proof.</u> Let S_X and S_Y be the sets of non-simple points of X and Y respectively. By Lemma 1.8 and Proposition 1.23, the surjective holomorphic map φ has rank n. The analytic set

$$E = \{x \in X \mid \operatorname{rank}_x \varphi < n\}$$

is thin in X. By Proposition 1.24, $E' = \varphi(E)$ is almost thin of dimension n - 2 in Y. Because φ is surjective, and because X is irreducible, the analytic set $\tilde{S}_Y = \varphi^{-1}(S_Y)$ is thin in X. The set R of non-regular points of the map $\varphi\colon X = (\tilde{S}_Y \cup S_X) \to Y - S_Y$ is analytic, and $\varphi(R)$ has measure zero on $Y - S_Y$. The map

$$\varphi\colon X - (\tilde{S}_Y \cup S_X \cup E) \to Y - S_Y$$

has pure rank n and is open by Theorem 1.22. Hence, $R \neq X - (\tilde{S}_Y \cup S_X)$. Because $X - (\tilde{S}_Y \cup S_X)$ is a connected complex manifold, R is a thin analytic subset of $X - (\tilde{S}_Y \cup S_X)$.

Now,

$$X_0 = X - (\tilde{S}_Y \cup S_X \cup R \cup T)$$

is an open, connected, dense subset of X and a complex manifold of dimension M. Also $Y_0 = Y - S_Y$ is an open, dense, connected subset of Y and a complex manifold. The map $\varphi_0 = \varphi: X_0 \to Y_0$ is holomorphic and regular. By Lemma 1.25, $\tilde{S} = \varphi^{-1}(S)$ is nowhere dense in X.

The proof proceeds by induction for p. At first consider the case $f = f_1$. Then $f \not\equiv 0$ on X. Because X is irreducible, the zero set $N = N_f(0)$ is a thin analytic subset of X. Let P_f be the pole set of f. Then $X - (N \cup P_f)$ is open and dense in X and f is holomorphic on $X - (N \cup P_f)$ with $f(x) \neq 0$ if $x \in X - (N \cup P_f)$. Because φ is surjective, $\tilde{W} = \varphi^{-1}(W) \neq \emptyset$. Take $\xi \in \tilde{W} \cap X_0 - (N \cup P_f \cup \tilde{S})$. Then $y_0 = \varphi(\xi) \in W - S$ is a simple point of Y. Also ξ is a simple point of X and φ is regular at ξ. Moreover, $\xi \in \varphi^{-1}(y) - T$. The function f is holomorphic at ξ with $f(\xi) \neq 0$. The case $p = 1$ is proved.

Now, assume that the Lemma is already proved for $p - 1$. Then it shall be proved for $p > 1$. Let $f_1, \ldots, f_p$ be meromorphic functions on X such that $f_1|\tilde{H}_1, \ldots, f_p|\tilde{H}_p$ are linearly independent over $\varphi_H^* \mathfrak{K}(H)$ for every open, irreducible subset $H \neq \emptyset$ of G. Let $W \neq \emptyset$ be an open subset of G. The functions $f_1|H_1, \ldots, f_{p-1}|H_{p-1}$ are linearly independent over $\varphi_H^* \mathfrak{K}(H)$ for every open, irreducible subset $H \neq \emptyset$ of G. By induction, a point $b \in W - (S \cup S_Y)$ and points $a_\nu \in \varphi^{-1}(b) \cap X_0$ exist for $\nu = 1, \ldots, p - 1$ such that the functions $f_1, \ldots, f_{p-1}$ are holomorphic at each a_ν and such that

$$\begin{vmatrix} f_1(a_1), & \dots\dots, & f_1(a_{p-1}) \\ \cdot & & \cdot \\ \cdot & & \cdot \\ \cdot & & \cdot \\ \cdot & & \cdot \\ f_{p-1}(a_1), & \dots\dots, & f_{p-1}(a_{p-1}) \end{vmatrix} \neq 0 .$$

For each $\nu = 1,\dots,p - 1$, a smooth product representation $(\alpha_\nu, \beta_\nu, \pi)$ of φ (and φ_0) exists

$$\begin{array}{ccc} a_\nu \in U_{\alpha_\nu} & \xrightarrow{\alpha_\nu} & U''_{\alpha_\nu} \times U'_{\beta_\nu} \subseteq \mathbb{C}^m \\ \downarrow \varphi & & \downarrow \pi_\nu \\ b \in U_{\beta_\nu} & \xrightarrow{\beta_\nu} & U'_{\beta_\nu} \qquad \subseteq \mathbb{C}^n \end{array}$$

with $\alpha_\nu(a_\nu) = 0 \in \mathbb{C}^m$ and $\beta_\nu(b) = 0 \in \mathbb{C}^n$. Here α_ν and β_ν are biholomorphic and π_ν is the projection. Clearly, $U_{\beta_\nu} \subseteq Y - S_Y = Y_0$. Also, the product representation can be taken such that U_{α_ν} is connected with $U_{\alpha_\nu} \subseteq X_0$, such that $f_1,\dots,f_{p-1}$ are holomorphic on U_{α_ν} and such that

$$\Delta(x_1,\dots,x_p) = \begin{vmatrix} f_1(x_1), & \dots\dots, & f_1(x_{p-1}) \\ \cdot & & \cdot \\ \cdot & & \cdot \\ \cdot & & \cdot \\ \cdot & & \cdot \\ f_{p-1}(x_1), & \dots\dots, & f_{p-1}(x_{p-1}) \end{vmatrix} \neq 0$$

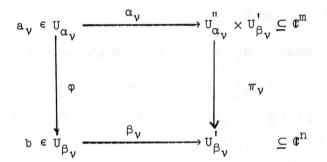

if $(x_1,\ldots,x_{p-1}) \in U_{\alpha_1} \times \ldots \times U_{\alpha_{p-1}}$.

Let V be an open connected neighborhood of b with $V \subseteq \bigcap_{\mu=1}^{p-1} U_{\beta_\mu}$. Define $V_\nu' = \beta_\nu(V)$ and $U_\nu = \alpha_\nu^{-1}(U_{\alpha_\nu}'' \times V')$. Then

$$\gamma_\nu = (\mathrm{Id} \times \beta_\nu^{-1})\circ\alpha_\nu\colon U_\nu \to U_{\alpha_\nu}'' \times V$$

is a biholomorphic map with $\gamma_\nu(a_\nu) = (0,b)$ and where U_{α_ν}'' is open and connected in $\mathbb{C}^q$, because U_{α_ν} is connected. Abbreviate $U_{\alpha_\nu}'' = U_\nu''$. Because V is also connected, U_ν is connected. Let $\psi_\nu = U_\nu'' \times V \to V$ Then $\psi_\nu\circ\gamma_\nu = \varphi|U_\nu = \varphi_0|U_\nu$ for $\nu = 1,\ldots,p$. Moreover, $f_1,\ldots,f_{p-1}$ are holomorphic on U_μ for $\mu = 1,\ldots,p-1$ and $\Delta(x_1,\ldots,x_{p-1}) \neq 0$ if $x_\mu \in U_\mu$ for $\mu = 1,\ldots,p-1$.

Let $P = P_{f_p}$ be the set of poles of $f_p \in \mathfrak{K}(X)$. Define $P_\nu = \gamma_\nu(P_\nu \cap U_\nu)$. Then P_ν is either empty or a pure (m-1)-dimensional analytic subset of $U_\nu'' \times V$. Define

$$Q_\nu = \{(z,y) \in P_\nu \mid \mathrm{rank}_{(y,z)}\psi_\nu|P_\nu \leqq n-1\}$$

$$Q_\nu' = \{y \in V \mid P_\nu \cap \psi_\nu^{-1}(y) = U_\nu'' \times \{y\}\}.$$

Now, $Q_\nu = U_\nu'' \times Q_\nu'$ is claimed. If $(z,y) \in Q_\nu$, then

$$q \geqq \dim_{(z,y)} P_\nu \cap (U_\nu'' \times \{y\}) = \dim_{(z,y)} P_\nu \cap \psi_\nu^{-1}(y)$$

$$= \dim_{(z,y)} P_\nu - \operatorname{rank}_{(z,y)} \psi_\nu | P_\nu \geqq m - 1 - (n-1) = q.$$

Because U_ν'' is connected and open in $\mathbb{C}^q$, this implies $P_\nu \cap \psi_\nu^{-1}(y) = U_\nu'' \times \{y\}$. Hence, $(z,y) \in U_\nu'' \times Q_\nu'$. If $(z,y) \in U_\nu'' \times Q_\nu'$ then $U_\nu'' \times \{y\} = P_\nu \cap \psi_\nu^{-1}(y)$ and

$$\operatorname{rank}_{(z,y)} \psi_\nu | P_\nu = \dim P_\nu - \dim_{(z,y)} P_\nu \cap \psi_\nu^{-1}(y)$$

$$= m - 1 - q = n - 1.$$

Therefore, $(z,y) \in Q_\nu$. Hence, $Q_\nu = U_\nu'' \times Q_\nu'$. Especially, Q_ν' is a thin analytic subset of V. Then $Q' = Q_1' \cup \ldots \cup Q_{p-1}'$ is a thin analytic subset of V. The set $\hat{V} = V - Q'$ is open, connected and dense in V. Define $\hat{U}_\nu = \varphi^{-1}(\hat{V}) \cap U_\nu = \gamma_\nu^{-1}(U_\nu'' \times \hat{V})$.

Pick $c \in \hat{V}$, then $d_\nu \in U_\nu''$ exists such that $(d_\nu, c) \notin P_\nu$. Define $c_\nu = \gamma_\nu^{-1}(d_\nu, c)$. Then $\varphi(c_\nu) = c$ for $\nu = 1,\ldots,p-1$. Open connected neighborhoods V^* of c with $V^* \subseteq \hat{V}$ and Z_ν of d_ν with $Z_\nu \subseteq U_\nu''$ exist for $\nu = 1,\ldots,p-1$ such that $P_\nu \cap (Z_\nu \times V^*) = \emptyset$. Then $U_\nu^* = \gamma_\nu^{-1}(Z_\nu \times V^*)$ is an open, connected neighborhood of c_ν with $U_\nu^* \subseteq \hat{U}_\nu$ and $\varphi(U_\nu^*) = V^*$. The functions $f_1,\ldots,f_p$ are holomorphic on each U_ν^* and $\Delta(x_1,\ldots,x_{p-1}) \neq 0$ if $(x_1,\ldots,x_{p-1}) \in U_1^* \times \ldots \times U_{p-1}^*$. Moreover, $\hat{U}_\nu \subseteq X_0$ and $V^* \subseteq W - S_Y$.

An injective, holomorphic map $\eta_\nu\colon V^* \to U_\nu^*$ is defined by

$\eta_\nu(y) = \gamma_\nu^{-1}(d_\nu, y)$ if $y \in V^*$. Here $\varphi \circ \eta_\nu$ is the identity on V^*. The functions $g_{\nu\mu} = f_\mu \circ \eta_\nu$ are holomorphic on V^* for $\mu = 1,\ldots,p$ and $\nu = 1,\ldots,p-1$. Define

$$D = \begin{vmatrix} g_{11}\circ\varphi & ,\ldots\ldots, & g_{1p}\circ\varphi \\ \vdots & & \vdots \\ g_{p-1,1}\circ\varphi & ,\ldots\ldots, & g_{p-1,p}\circ\varphi \\ f_1 & ,\ldots\ldots, & f_p \end{vmatrix}$$

The function D is meromorphic on $\varphi^{-1}(V^*) = \tilde{V}^*$. Then

$$D_\mu = (-1)^{p+\mu} \cdot \begin{vmatrix} g_{11}, \ldots, g_{1,\mu-1}, g_{1,\mu+1}, \ldots, g_{1p} \\ \vdots \qquad \vdots \qquad \vdots \qquad \vdots \\ g_{p-1,1}, \ldots, g_{p-1,\mu-1}, g_{p-1,\mu+1}, \ldots, g_{p-1,p} \end{vmatrix}$$

is holomorphic on V^* for $\mu = 1,\ldots,p$. If $y \in V^*$, then

$$D_p(y) = \Delta(\eta_1(y),\ldots,\eta_{p-1}(y)) \neq 0.$$

Moreover,

$$D = \sum_{\mu=1}^{p} (D_\mu \circ \varphi) \cdot f_\mu | \tilde{V}^*$$

on $\tilde{V}^*$. By assumption, $f_1|\tilde{V}^*,\dots,f_p|\tilde{V}^*$ are linearly independent over $\varphi^*_{V^*}\mathfrak{K}(V^*)$. Here $\varphi^*_{V^*}(D_p) = D_p\circ\varphi \not\equiv 0$ on $\tilde{V}^*$. Therefore, $D \not\equiv 0$ on $\tilde{V}^*$.

A point $e \in \tilde{V}^* \cap X_0$ exists such that $f_1,\dots,f_p$ are holomorphic at e and such that $D(e) \neq 0$. Then e and $\varphi(e)$ are simple points of X and Y respectively and φ is regular at e. A smooth product representation (α_p,β_p,π_p) of φ at e exists such that U_{α_p} is connected and contained in $\tilde{V}^* \cap X_0$, such that $f_1,\dots,f_p$ are holomorphic on U_{α_p} and such that $D(x) \neq 0$ if $x \in U_{\alpha_p}$. Then $U_{\beta_p} = \varphi(U_{\alpha_p}) \subseteq \varphi(\tilde{V}^*) = V^*$.

Pick $\xi_p \in U_{\alpha_p} - \tilde{S} \subseteq X_p$. Then $y_0 = \varphi(\xi_p) \in U_{\beta_p} - S$ and $y_0 \in V^*$. Define $\xi_\nu = \eta_\nu(y_0)$ for $\nu = 1,\dots,p-1$. Then $y_0 \in W - S$ is a simple point of Y. For each $\nu = 1,\dots,p$, the point $\xi_\nu \in \varphi^{-1}(y_0) - T$ simple point of X. The map φ is regular at $\xi_\nu \in X_0$ and the functions $f_1,\dots,f_p$ are holomorphic at ξ_ν, because $\xi_p \in U_{\alpha_p}$ and $\xi_\nu \in U^*_\nu$ for $\nu = 1,\dots,p$. Moreover, $D(\xi_p) \neq 0$. If $1 \leqq \nu \leqq p-1$ and if $1 \leqq \mu \leqq p$, then

$$g_{\nu\mu}(\xi_p) = f_\mu(\eta_\nu(\varphi(\xi_p))) = f_\mu(\eta_\nu(y_0)) = f_\mu(\xi_\nu).$$

Therefore,

$$\begin{vmatrix} f_1(\xi_1) & , \dots\dots , & f_p(\xi_1) \\ \cdot & & \cdot \\ \cdot & & \cdot \\ \cdot & & \cdot \\ \cdot & & \\ f_1(\xi_{p-1}) & , \dots\dots , & f_p(\xi_{p-1}) \\ f_1(\xi_p) & , \dots\dots , & f_p(\xi_p) \end{vmatrix} = D(\xi_p) \neq 0.$$

Especially, $\xi_\mu \neq \xi_\nu$ if $\mu \neq \nu$; q.e.d.

Let X and Y be complex spaces of pure dimension m and n respectively with $m - n = q \geqq 0$. Let $\varphi: X \to Y$ be a surjective holomorphic map of strict rank n. Then $\varphi^*: \mathfrak{K}(Y) \to \mathfrak{K}(X)$ is an injective homomorphism. If $\emptyset \neq H \subseteq Y$, define $\tilde{H} = \varphi^{-1}(H) \neq \emptyset$ and $\varphi_H = \varphi: \tilde{H} \to H$. If $H \neq \emptyset$ is open and irreducible in Y, then $\mathfrak{K}(H)$ is a field and $\varphi_H: \tilde{H} \to H$ is a surjective holomorphic map of strict rank n. Hence, $\varphi_H^*: \mathfrak{K}(H) \to \mathfrak{K}(\tilde{H})$ is an injective homomorphism and $\varphi_H^* \mathfrak{K}(H)$ is a subfield of the ring $\mathfrak{K}(\tilde{H})$. Especially, $\mathfrak{K}(\tilde{H})$ is a vector space over $\varphi_H^* \mathfrak{K}(H)$.

<u>Theorem 12.2.</u> Let X and Y be irreducible complex spaces of dimension m and n respectively with $m - n = q \geqq 0$. Let $\varphi: X \to Y$ be a full, holomorphic map of rank n. Then φ is surjective. Let $F \neq \emptyset$ be a set of meromorphic functions on X. If $U \neq \emptyset$ is open in X, define $F|U = \{f|U \mid f \in F\}$. Let $\mathfrak{H}$ be the set of all open, irreducible, non-empty subsets H of Y. If $H \in \mathfrak{H}$, regard $\mathfrak{K}(\tilde{H})$ as a vector space over $\varphi_H^* \mathfrak{K}(H)$. Let K(H) be the linear subspace of $\mathfrak{K}(\tilde{H})$, which is generated by $F|\tilde{H}$ over $\varphi_H^* \mathfrak{K}(H)$. Then the dimension d(H) of K(H) over $\varphi_H^* \mathfrak{K}(H)$ is independent of $H \in \mathfrak{H}$.

Proof. By Corollary 11.11, φ is surjective. Take $H_1 \in \mathfrak{H}$ with $\emptyset \neq H_1 \subseteq H_2$. Suppose that a subset $F' \neq \emptyset$ of F is given, such that $F'|\tilde{H}_2$ generates $K(H_2)$ over $\varphi^*_{H_2} \mathfrak{K}(H_2)$. Then it is claimed that $F'|\tilde{H}_1$ generates $K(H_1)$ over $\varphi^*_{H_1} \mathfrak{K}(H_1)$. Let $K'(H)$ be the vector space in $\mathfrak{K}(\tilde{H})$ which is generated by $F'|H$ over $\varphi^*_H \mathfrak{K}(H)$ if $H \in \mathfrak{H}$. Then $K'(H) \subseteq K(H)$. If $f \in F$, meromorphic functions $a_\mu \in \mathfrak{K}(H_2)$ and $g_\mu \in F'$ exist such that

$$f|H_2 = \sum_{\mu=1}^{r} \varphi^*_{H_2}(a_\mu) g_\mu|\tilde{H}_2$$

which implies

$$f|H_1 = \sum_{\mu=1}^{r} \varphi^*_{H_1}(a_\mu|H_1) g_\mu|\tilde{H}_1.$$

Hence, $f|H_1 \in K'(H_1)$ which implies $F|H_1 \subseteq K'(H_1)$. Hence, $K(H_1) \subseteq K'(H_1)$ which implies $K(H_1) = K'(H_1)$. Therefore, $F'|H_1$ generates $K(H_1)$ over $\varphi^*_{H_1} \mathfrak{K}(H_1)$.

A subset F' of F exists such that $F'|H_2$ is a base of $K(H_2)$. If $F' \neq \emptyset$, then $F'|H_1$ generates $K(H_1)$ over $\varphi^*_{H_1} \mathfrak{K}(H_1)$. Hence, $d(H_1) \leqq \#F'|H_1 = \#F'|H_2 = d(H_2)$. If $F' = \emptyset$, then F consists of the zero function only and $K(H_1) = \{0\}$. Hence, $d(H_1) = d(H_2) = 0$. Therefore, $d(H_1) \leqq d(H_2)$ if $H_1 \in \mathfrak{H}$ with $H_1 \subseteq H_2$.

Define $p = \mathrm{Min}\{d(H) \mid H \in \mathfrak{H}\}$. If $p = \infty$, then $d(H) = \infty$ for all $H \in \mathfrak{H}$ and the theorem is proved. Hence $p < \infty$ can be assumed. An open, irreducible subset $G \in \mathfrak{G}$ of Y exists such that $p = d(G)$. Meromorphic functions $f_1, \ldots, f_p$ in F exist such that $f_1|\tilde{G}, \ldots, f_p|\tilde{G}$ form a base of K(G) over $\varphi_G^* \mathfrak{K}(G)$. If $H \in \mathfrak{G}$ with $H \subseteq G$, then $p \leqq d(H) \leqq d(G) = p$. Hence, $d(H) = d(G)$. Moreover, $f_1|\tilde{H}, \ldots, f_p|\tilde{H}$ generate K(H) over $\varphi_H^* \mathfrak{K}(H)$. Because $p = d(H)$, these functions $f_1|\tilde{H}, \ldots, f_p|\tilde{H}$ form a base of K(H) over $\varphi_H^* \mathfrak{K}(H)$. Especially, $f_1|\tilde{H}, \ldots, f_p|\tilde{H}$ are linearly independent over $\varphi_H^* \mathfrak{K}(H)$ for each $H \in \mathfrak{G}$ with $H \subseteq G$.

Now, it is claimed, that $f_1, \ldots, f_p$ generate K(Y) over $\varphi^* \mathfrak{K}(Y)$. Because F generated K(Y) over $\varphi^* \mathfrak{K}(Y)$, it suffices to show that each $f \in F$ is linearly dependent on $f_1, \ldots, f_p$ over $\varphi^* \mathfrak{K}(Y)$. Take $f \in F$. The pole set P_f of f is a thin analytic subset of X and f is holomorphic on $X - P_f$. Let $\mathfrak{B}$ be the set of branches of X_φ^p. Define

$$\mathfrak{B}_0 = \{B \in \mathfrak{B} \mid \operatorname{rank} \delta_\varphi | B = n\}$$

$$\hat{S}_B = \{z \in B \mid \operatorname{rank}_z \delta_\varphi | B < n\} \qquad \text{if } B \in \mathfrak{B}.$$

If $B \in \mathfrak{B} - \mathfrak{B}_0$, then $\hat{S}_B = B$ and $S_B = \delta_\varphi(\hat{S}_B) = \delta_\varphi(B)$ is almost thin of dimension n - 1 by Proposition 1.23. If $B \in \mathfrak{B}_0$, then $\hat{S}_B$ is thin in B. By Lemma 1.20, $S_B = \delta_\varphi(\hat{S}_B)$ is almost thin of dimension n - 2 in Y. Hence, $S = \bigcup_{B \in \mathfrak{B}} S_B$ is almost thin in Y. Now, Lemma 12.1 can be applied with $T = P_f$ and $W = G$. Hence, a simple point $y_0 \in Y$ with $y_0 \in G - S$ and simple points ξ_ν of X with $\xi_\nu \in (\varphi^{-1}(y_0) - P_f)$ for $\nu = 1, \ldots, p$ exist such that φ is regular at each ξ_ν and such that $f_1,\dots,f_p$ are holomorphic at each ξ_ν with

$$\begin{vmatrix} f_1(\xi_1), & \dots, & f_1(\xi_p) \\ \vdots & & \vdots \\ f_p(\xi_1), & \dots, & f_p(\xi_p) \end{vmatrix} \neq 0.$$

Because $\xi_\nu \in X - P_f$, also f is holomorphic at each ξ_ν. Now, $\xi = (\xi_1,\dots,\xi_p) \in X_\varphi^p$. A branch B of X_φ^p with $\xi \in B$ and $\dim_\xi X_\varphi^p = \dim B$ exists. Because $\delta_\varphi(\xi) = y_0 = Y - S$, the point ξ does not belong to $\hat{S}_B$. Especially, $\operatorname{rank}_\xi \delta_\varphi | B = n$ and $B \in \mathfrak{L}_0$. By Lemma 11.18, $\dim B = n + pq$ and $\operatorname{rank} \pi_\nu | B = m$ where $\pi_\nu\colon X_\varphi^p \to X$ is the νth projection. By Remark 3.5.3, $\pi_\nu^*\colon \mathfrak{K}(X) \to \mathfrak{K}(B)$ is an injective homomorphism. Define $g_{\mu\nu} = \pi_\nu^*(f_\mu)$ for $\mu = 1,\dots,p$ and $\nu = 1,\dots,p$. Define $g_\nu = \pi_\nu^*(f)$ for $\nu = 1,\dots,p$. Then

$$\Delta = \begin{vmatrix} g_{11}, & \dots, & g_{1p} \\ \vdots & & \vdots \\ g_{p1}, & \dots, & g_{pp} \end{vmatrix} \in \mathfrak{K}(B).$$

All functions $f_1,\dots,f_p,f$ are holomorphic at each ξ_ν. Hence, $g_{\mu\nu}$ and g_ν are holomorphic at ξ with $g_{\mu\nu}(\xi) = f_\mu(\xi_\nu)$ and $g_\nu(\xi) = f(\xi_\nu)$.

Therefore, Δ is holomorphic at ξ with $\Delta(\xi) \neq 0$. Hence, Δ is not the zero element of the field $\mathfrak{K}(B)$. Consequently, meromorphic functions $A_\mu \in \mathfrak{K}(B)$ uniquely exist such that

(12.1) $$\sum_{\mu=1}^{p} A_\mu g_{\mu\nu} = g_\nu$$

for $\nu = 1,\ldots,p$.

Recall, that $f_1|\tilde{G},\ldots,f_p|\tilde{G}$ form a base of $K(G)$ over $\varphi_G^* \mathfrak{K}(G)$. Hence, meromorphic functions $a_\mu \in \mathfrak{K}(G)$ uniquely exist such that

(12.2) $$\sum_{\mu=1}^{p} \varphi_G^*(a_\mu) \cdot f_\mu|\tilde{G} = f|\tilde{G}.$$

Define $\hat{G} = \delta_\varphi^{-1}(G) \cap B = \pi_\nu^{-1}(\varphi^{-1}(G)) \cap B = \pi_\nu^{-1}(G) \cap B$. Because $\delta_\varphi|B$ and φ have rank n, the restrictions $\delta_\varphi|\hat{G}: \hat{G} \to G$ and $\varphi_G = \varphi: \tilde{G} \to G$ have strict rank n. The restriction $(\pi_\nu|\hat{G}): \hat{G} \to \tilde{G}$ has strict rank m, because $\pi_\nu|B: B \to X$ has rank m. Moreover, $\delta_\varphi|\hat{G} = \varphi_G \circ \pi_\nu|\hat{G}$. By Lemma 3.4, $(\delta_\varphi|\hat{G})^*: \mathfrak{K}(G) \to \mathfrak{K}(\hat{G})$ and $(\pi_\nu|\hat{G})^*: \mathfrak{K}(\tilde{G}) \to \mathfrak{K}(\hat{G})$ and $\varphi_G^*: \mathfrak{K}(G) \to \mathfrak{K}(\hat{G})$ are injective homomorphisms with $(\delta_\varphi|\hat{G})^* = (\pi_\nu|\hat{G})^* \circ \varphi_G^*$. Hence, (12.2) implies

(12.3) $$\sum_{\mu=1}^{p} (\delta_\varphi|\hat{G})(a_\mu) \cdot g_{\mu\nu}|\hat{G} = g_\nu|\hat{G}$$

for $\nu = 1,\ldots,p$. The zero set of Δ is thin on B. Hence, $\Delta|C \not\equiv 0$ if C is a branch of $\hat{G}$. Therefore, the equations (12.1) and (12.3) imply

(12.4) $$(\delta_\varphi|\hat{G})(a_\mu) = A_\mu|\hat{G}$$

for $\mu = 1,\dots,p$. By Theorem 11.19, one and only one meromorphic function b_μ on Y exists such that $(\delta_\varphi|B)^*(b_\mu) = A_\mu$. Moreover, $b_\mu|G = a_\mu$. Hence, $\varphi_G^*(a_\mu) = \varphi^*(b_\mu)|\tilde{G}$. Because $\tilde{G} \neq \emptyset$ is open on the irreducible complex space X, (12.2) implies

(12.5) $$\sum_{\mu=1}^{p} \varphi^*(b_\mu)f_\mu = f.$$

on X. Hence, f is linearly dependent on $f_1,\dots,f_p$ over $\varphi^*\mathfrak{K}(Y)$. Therefore, $f_1,\dots,f_p$ generate K(Y) over $\varphi^*\mathfrak{K}(Y)$. Because $d(Y) \geqq p$, the functions $f_1,\dots,f_p$ form a base of K(Y) over $\varphi^*\mathfrak{K}(Y)$ and $p = d(Y)$.

If $H \in \mathfrak{H}$, then $p \leqq d(H) \subseteq d(Y) = p$. Hence, $d(H) = p$ for all $H \in \mathfrak{H}$; q.e.d.

<u>Corollary 12.3.</u> Let X and Y be irreducible complex spaces of dimension m and n respectively with $m - n = q \geqq 0$. Let $\varphi: X \to Y$ be a full, holomorphic map of rank n. Then φ is surjective. Let $G \neq \emptyset$ and $H \neq \emptyset$ be open, irreducible subsets of Y. Let $f_1,\dots,f_p$ be meromorphic functions on X. Then $f_1|\tilde{G},\dots,f_p|\tilde{G}$ are linearly independent (resp. dependent) over $\varphi_G^*\mathfrak{K}(G)$ if and only if $f_1|\tilde{H},\dots,f_p|\tilde{H}$ are linearly independent (resp. dependent) over $\varphi_H^*\mathfrak{K}(H)$.

<u>Proof.</u> Define $F = \{f_1,\dots,f_k\}$. Let $\mathfrak{H}$ be the set of all

open, irreducible subsets $W \neq \emptyset$ of Y. Let K(W) be the linear subspace of $\mathfrak{K}(\tilde{W})$ which is generated by F over $\varphi_W^* \mathfrak{K}(W)$. Let d(W) be the dimension of K(W) over $\varphi_W^* \mathfrak{K}(W)$. Then $d(W) = p$ (resp. $d(W) < p$) if and only if $f_1|\tilde{W}, \ldots, f_p|\tilde{W}$ are linearly independent (resp., dependent) over $\varphi_W^* \mathfrak{K}(W)$. By Theorem 12.2, $d(H) = d(G)$, which proves the Corollary; q.e.d.

Theorem 12.4. Let X and Y be irreducible complex spaces of dimension m and n respectively with $m - n = q > 0$. Let $\varphi: X \to Y$ be a full, holomorphic map of rank n. Then φ is surjective. Let $\mathfrak{H}$ be the set of all open, irreducible non-empty subsets H of Y. Let $G \in \mathfrak{H}$. Let $f_1, \ldots, f_k$ be meromorphic functions on X such that $f_1|\tilde{G}, \ldots, f_k|\tilde{G}$ are algebraically dependent over $\varphi_G^* \mathfrak{K}(G)$. Then $f_1|\tilde{H}, \ldots, f_k|\tilde{H}$ are algebraically dependent over $\varphi_H^* \mathfrak{K}(H)$ for each $H \in \mathfrak{H}$. If

$$r(H) = [(f_1|H, \ldots, f_p|H) : \varphi_H^* \mathfrak{K}(H)]$$

then the degree r(H) does not depend on $H \in \mathfrak{H}$. Especially, $H = Y$ can be taken.

Proof. By Corollary 11.11, φ is surjective. A polynomial

$$P = \sum_{0 \leq \mu_1 + \ldots + \mu_k \leq r} a_{\mu_1, \ldots, \mu_k} \xi_1^{\mu_1}, \ldots, \xi_k^{\mu_k} \in \mathfrak{K}(G)[\xi_1, \ldots, \xi_k]$$

exists with $a_{\mu_1, \ldots, \mu_k} \in \mathfrak{K}(G)$ such that $a_{\nu_1, \ldots, \nu_r} \not\equiv 0$ for some index set $\nu_1, \ldots, \nu_k$ with $\nu_1 + \ldots + \nu_k = r = r(G)$ and such that

$$0 = \sum_{0 \leqq \mu_1 + \ldots + \mu_k \leqq r} \varphi_G^*(a_{\mu_1,\ldots,\mu_k})(f_1|\tilde{G})^{\mu_1} \ldots (f_p|\tilde{G})^{\mu_k}.$$

Hence, $f_1^{\mu_1} \ldots f_k^{\mu_k}$ with $0 \leqq \mu_1 + \ldots + \mu_k \leqq r$ is a finite set of meromorphic functions on X such that $(f_1^{\mu_1} \ldots f_k^{\mu_k})|\tilde{G}$ are linearly dependent over $\varphi_G^* \mathfrak{K}(G)$. Hence, $(f_1^{\mu_1} \ldots f_k^{\mu_k})|\tilde{H}$ are linearly dependent over $\varphi_H^* \mathfrak{K}(H)$ if $H \in \mathfrak{g}$ by Corollary 12.3. Meromorphic functions $b_{\mu_1,\ldots,\mu_k} \in \mathfrak{K}(H)$ exist such that

$$(12.6) \qquad 0 = \sum_{0 \leqq \mu_1 + \ldots + \mu_k \leqq r} \varphi_H^*(b_{\mu_1,\ldots,\mu_k})(f_1|\tilde{H})^{\mu_1} \ldots (f_p|\tilde{H})^{\mu_k}$$

and such that $b_{\nu_1,\ldots,\nu_k} \not\equiv 0$ for at least one index set $\nu_1,\ldots,\nu_k$ with $0 \leqq \nu_1 + \ldots + \nu_k \leqq r$. Because $r = r(G)$, the set

$$\{f_1^{\mu_1}|\tilde{G} \ldots f_k^{\mu_k}|\tilde{G} | 0 \leqq \mu_1 + \ldots + \mu_k \leqq r-1\}$$

is a finite set of meromorphic functions on $\tilde{G}$ which is linearly independent over $\varphi_G^* \mathfrak{K}(G)$. Hence, the set

$$\{f_1^{\mu_1}|\tilde{H} \ldots f_k^{\mu_k}|\tilde{H} | 0 \leqq \mu_1 + \ldots + \mu_k \leqq r-1\}$$

is a finite set of meromorphic functions on $\tilde{H}$ which is linearly independent over $\varphi_H^* \mathfrak{K}(H)$ by Corollary 12.3. Hence, $r(H) \geqq r$. Also $b_{\nu_1,\ldots,\nu_k} \not\equiv 0$ for at least one index set $\nu_1 + \ldots + \nu_k = r$. Now, equation (12.6) implies that $f_1|\tilde{H},\ldots,f_k|\tilde{H}$ are algebraically dependent over $\varphi_H^* \mathfrak{K}(H)$ with $r(H) \leqq r$; hence, $r(H) = r = r(G)$; q.e.d.

Theorem 12.5. Let X and Y be irreducible complex spaces of dimension m and n respectively with $m - n = q \geqq 0$. Let $\varphi: X \to Y$ be a full holomorphic map of rank n. Then φ is surjective. Let $\mathfrak{g}$ be the set of all open, irreducible non-empty subsets of Y. Let $G \in \mathfrak{g}$. Let $f_1,\ldots,f_k$ be meromorphic functions on X such that $f_1|\tilde{G},\ldots,f_k|\tilde{G}$ are algebraically independent over $\varphi_G^* \mathfrak{K}(G)$. Then $f_1|\tilde{H},\ldots,f_k|\tilde{H}$ are algebraically independent over $\varphi_H^* \mathfrak{K}(H)$ for each $H \in \mathfrak{g}$.

Proof. Suppose that $f_1|\tilde{H},\ldots,f_k|\tilde{H}$ are algebraically dependent over $\varphi_H^* \mathfrak{K}(H)$ for some $H \in \mathfrak{g}$. By Theorem 12.4, $f_1|\tilde{G},\ldots,f_k|\tilde{G}$ are algebraically dependent over $\varphi_G^* \mathfrak{K}(G)$ which is wrong. Hence, $f_1|\tilde{H},\ldots,f_k|\tilde{H}$ are algebraically independent over $\varphi_H^* \mathfrak{K}(H)$ for each $H \in \mathfrak{g}$; q.e.d.

Theorem 12.6. Let X and Y be irreducible complex spaces of dimension m and n respectively with $m - n = q \geqq 0$. Let $\varphi: X \to Y$ be a full holomorphic map of rank n. Let $G \neq \emptyset$ be an open irreducible subset of Y. Let f be a meromorphic function on X such that $f|\tilde{G}$ is algebraic over $\varphi_G^* \mathfrak{K}(G)$. Let $H \neq \emptyset$ be an open irreducible subset of Y. Then $f|\tilde{H}$ is algebraic over $\varphi_H^* \mathfrak{K}(H)$ with

$$[f|\tilde{H}: \varphi_H^* \mathfrak{K}(H)] = [f|\tilde{G}: \varphi_G^* \mathfrak{K}(G)].$$

Proof. Apply Theorem 12.4 with $k = 1$; q.e.d.

This theorem solves the globalization problem for a single function if $H = Y$ is taken. The general situation is given by:

<u>Theorem 12.7.</u> Let X and Y be irreducible complex spaces of dimension m and n respectively with $m - n = q \geqq 0$. Let $\varphi: X \to Y$ be a full, holomorphic map of rank n. Let $G \neq \emptyset$ be an open, irreducible subset of Y. Let $f_1, \ldots, f_k, f$ be meromorphic functions on X such that $f_1|\tilde{G}, \ldots, f_k|\tilde{G}$, are algebraically independent over $\varphi_G^* \mathfrak{K}(G)$ and such that $f_1|\tilde{G}, \ldots, f_k|\tilde{G}$, $f|\tilde{G}$ are algebraically dependent over $\varphi_G^* \mathfrak{K}(G)$. Then $f|\tilde{G}$ is algebraic over $\varphi_G^* \mathfrak{K}(G)(f_1|\tilde{G}, \ldots, f_k|\tilde{G})$ with

$$(12.7) \qquad r = [f|\tilde{G}: \varphi_G^* \mathfrak{K}(G)(f_1|\tilde{G}, \ldots, f_k|\tilde{G})].$$

Let $H \neq 0$ be an open, irreducible subset of Y. Then $f_1|\tilde{H}, \ldots, f_k|\tilde{H}$, are algebraically independent over $\varphi_H^* \mathfrak{K}(H)$ and $f_1|\tilde{H}, \ldots, f_k|\tilde{H}$, $f|\tilde{H}$ are algebraically dependent over $\varphi_H^* \mathfrak{K}(H)$ with

$$(12.8) \qquad r = [f|\tilde{H}: \varphi_H^* \ (H)(f_1|\tilde{H}, \ldots, f_k|\tilde{H})].$$

Especially, $H = Y$ with $\tilde{H} = X$ is a possible choice.

Proof. By Theorem 12.5, $f_1|\tilde{H},\ldots,f_k|\tilde{H}$ are algebraically independent over $\varphi_H^*\mathfrak{K}(H)$. On G, meromorphic functions $a_{\mu\nu_1\ldots\nu_k}$ exist such that

$$\sum_{\mu=0}^{r}\sum_{\nu_1\ldots\nu_k=0}^{s}\varphi_G^*(a_{\mu\nu_1\ldots\nu_k})f_1^{\nu_1}|\tilde{G}\ \ldots\ f_k^{\nu_k}|\tilde{G}\ f^{\mu}|\tilde{G} = 0$$

on $\tilde{G}$ where $a_{r\nu_1\ldots\nu_k} \not\equiv 0$ on G for at least one index set $(\nu_1,\ldots,\nu_k)$ and where r is given by (12.8). Hence, the functions $f_1^{\nu_1}|\tilde{G}\ldots f_k^{\nu_k}|\tilde{G}$ $f^{\mu}|\tilde{G}$ for $0 \leqq \mu \leqq r$ and $0 \leqq \nu_\lambda \leqq s$ if $\lambda = 1,\ldots,k$ are linearly dependent over $\varphi_G^*\mathfrak{K}(G)$. By Corollary 12.3, meromorphic functions $b_{\mu\nu_1\ldots\nu_k}$ on H exist such that

$$\sum_{\mu=0}^{r}\sum_{\nu_1,\ldots,\nu_k=0}^{s}\varphi_H^*(b_{\mu\nu_1\ldots\nu_k})f_1^{\nu_1}|\tilde{H}\ \ldots\ f_k^{\nu_k}|\tilde{H}\ f^{\mu}|\tilde{H} = 0$$

where $b_{\mu\nu_1,\ldots,\nu_k} \not\equiv 0$ for some index set $(\mu,\nu_1,\ldots,\nu_k)$. Hence, $f_1|\tilde{H},\ldots,f_k|\tilde{H}$, $f|\tilde{H}$ are algebraically dependent and $f_1|\tilde{H},\ldots,f_k|\tilde{H}$ are algebraically independent over $\varphi_H^*\mathfrak{K}(H)$. Especially, $b_{r\nu_1,\ldots,\nu_k} \not\equiv 0$ for some index set $(\nu_1,\ldots,\nu_k)$. Hence, $f|\tilde{H}$ is algebraic over $\varphi_H^*\mathfrak{K}(H)(f_1|\tilde{H},\ldots,f_k|\tilde{H})$ with

$$r \geqq [f|\tilde{H}\colon \varphi_H^*\mathfrak{K}(H)(f_1|\tilde{H},\ldots,f_k|\tilde{H})] = r'.$$

Now, by symmetry, $r' \geqq r$; hence, $r' = r$; q.e.d.

In general, not more can be said, because $\tilde{G}$ may not be irreducible, such that $\mathfrak{K}(\tilde{G})$ is not a field. However, if $\tilde{G}$ is irreducible, then the following theorem holds:

<u>Theorem 12.8.</u> Let X and Y be irreducible complex spaces of dimension m and n with $m - n = q \geqq 0$. Let $\varphi: X \to Y$ be a full holomorphic map of rank n. Let $G \neq \emptyset$ be an open, irreducible subset of Y. Suppose that $\tilde{G} = \varphi^{-1}(G) \neq \emptyset$ is irreducible. Define $\varphi_G = \varphi: \tilde{G} \to G$. Suppose that $\mathfrak{K}(\tilde{G})$ is an algebraic function field over $\varphi_G^* \mathfrak{K}(G)$. Then $\mathfrak{K}(X)$ is an algebraic function field over $\varphi^* \mathfrak{K}(Y)$.

<u>Proof.</u> Let t be the transcendence degree of $\mathfrak{K}(\tilde{G})$ over $\varphi_G^* \mathfrak{K}(G)$. Suppose that $f_1, \ldots, f_k$ are meromorphic functions on X which are algebraically independent over $\varphi^* \mathfrak{K}(Y)$. By Theorem 12.5, $f_1|\tilde{G}, \ldots, f_k|\tilde{G}$ are algebraically independent over $\varphi_G^* \mathfrak{K}(G)$. Therefore, $k \leqq t$. Hence, $\mathfrak{K}(X)$ has a finite transcendency degree s over $\varphi^* \mathfrak{K}(Y)$. Let $f_1, \ldots, f_s$ be meromorphic functions on X, such that $f_1, \ldots, f_s$ are algebraically independent over $\varphi^* \mathfrak{K}(Y)$. Take any $f \in \mathfrak{K}(X)$. Then $f_1, \ldots, f_s, f$ are algebraically dependent over $\varphi^* \mathfrak{K}(Y)$. Then $f_1|\tilde{G}, \ldots, f_s|\tilde{G}$, $f|\tilde{G}$ are algebraically dependent over $\varphi_G^* \mathfrak{K}(G)$. Extend $f_1|\tilde{G}, \ldots, f_s|\tilde{G}$ to a transcendence base $f_1|\tilde{G}_1, \ldots, f_1|\tilde{G}$, $u_1, \ldots, u_p$ of $\mathfrak{K}(\tilde{G})$ over $\varphi_G^* \mathfrak{K}(G)$. Define

$$S = \varphi_G^* \mathfrak{K}(G)[f_1|\tilde{G},\ldots,f_s|\tilde{G}]$$

$$\hat{S} = \varphi_G^* \mathfrak{K}(G)(f_1|\tilde{G},\ldots,f_s|\tilde{G})$$

$$T = \varphi_G^* \mathfrak{K}(G)[f_1|\tilde{G},\ldots,f_1|\tilde{G},u_1,\ldots,u_p]$$

$$\hat{T} = \varphi_G^* \mathfrak{K}(G)(f_1|\tilde{G},\ldots,f_s|\tilde{G},u_1,\ldots,u_p).$$

Then $\mathfrak{K}(\tilde{G})$ is finite algebraic over $\hat{T}$ with $[\mathfrak{K}(\tilde{G}): \hat{T}] = r$. Also, $f|\tilde{G}$ is algebraic over $\hat{S}$. By Proposition 4.22 a, $[f|\tilde{G} : \hat{S}] \leqq r$ By Theorem 12.7, f is algebraic over $\varphi^* \mathfrak{K}(Y)(f_1,\ldots,f_s)$ with $[f: \varphi^* \mathfrak{K}(Y)(f_1,\ldots,f_s)] \leqq r$. By Lemma 4.22, $\mathfrak{K}(X)$ is finite algebraic over $\varphi^* \mathfrak{K}(Y)(f_1,\ldots,f_s)$ with $[\mathfrak{K}(X): \varphi^* \mathfrak{K}(Y)(f_1,\ldots,f_s)] \leqq r$. By construction $\varphi^* \mathfrak{K}(Y)(f_1,\ldots,f_s)$ is a pure transcendental extension of $\varphi^* \mathfrak{K}(Y)$. Therefore, $\mathfrak{K}(X)$ is an algebraic function field over $\mathfrak{K}(Y)$.

The results of Theorem 12.4 to Theorem 12.8 are the second main results of this investigation. They will be of importance for the study of pseudoconcave maps where the method of Siegel will be used.

The following Lemma may be of help in interpreting Theorem 12.8.

<u>Lemma 12.9.</u> Let X and Y be irreducible complex spaces of dimension m and n respectively with $m - n = q \geqq 0$. Let $\varphi: X \to Y$ be a holomorphic map of rank n. Let $H \neq \emptyset$ be an open, irreducible subset of Y with $\tilde{H} = \varphi^{-1}(H) \neq \emptyset$. Define $\varphi_H = \varphi: \tilde{H} \to H$. Let $f_1,\ldots,f_k$, φ-independent meromorphic functions on X. Then

1. The holomorphic map φ_H has strict rank n.

2. The subring

$$R = \varphi_H^* \mathfrak{K}(H)[f_1|\tilde{H},\ldots,f_k|\tilde{H}]$$

of $\mathfrak{K}(\tilde{H})$ is an integral domain.

3. The ring of quotients

$$K = \varphi_H^* \mathfrak{K}(H)(f_1|\tilde{H},\ldots,f_k|\tilde{H})$$

of R is a subfield of $\mathfrak{K}(\tilde{H})$.

4. If B is a branch of $\tilde{H}$, and if $\psi_B = \varphi: B \to H$, then $f_1|B,\ldots,f_k|B$ are ψ_B-independent and algebraically independent over $\psi_B^* \mathfrak{K}(B)$.

<u>Proof.</u> By Lemma 4.14, φ_H has strict rank n and $f_1|\tilde{H},\ldots,f_k|\tilde{H}$ are φ_H-independent. Let $\mathfrak{L}$ be the set of branches of $\tilde{H}$. If $B \in \mathfrak{L}$, then $f_1|B,\ldots,f_k|B$ are ψ_B-independent by Lemma 4.4. By Proposition 4.23, $f_1|B,\ldots,f_k|B$ are algebraically independent over $\psi_B^* \mathfrak{K}(H)$. Also ψ_B has rank n, (Lemma 1.16). Hence, $\psi_B^*: \mathfrak{K}(H) \to \psi_B^* \mathfrak{K}(H)$ is an isomorphism. Especially, $\psi_B^* \mathfrak{K}(H)$ is a subfield of $\mathfrak{K}(B)$. Let $\xi_1,\ldots,\xi_k$ be indeterminants over $\mathfrak{K}(H)$ and $\eta_1,\ldots,\eta_k$ be indeterminants over $\psi_B^* \mathfrak{K}(H)$. Then ψ_B^* extends to one and only one ring isomorphism

$$\tilde{\psi}_B^*: \mathfrak{K}(H)[\xi_1,\ldots,\xi_k] \to \psi_B^* \mathfrak{K}(H)[\eta_1,\ldots,\eta_k]$$

by requiring that $\tilde{\psi}_B^*(\xi_\mu) = \eta_\mu$ for $\mu = 1,\ldots,k$. A substitution ring-epimorphism

$$\sigma_B\colon \psi_B^*\mathfrak{K}(B)[\eta_1,\ldots,\eta_k] \to \psi_B^*\mathfrak{K}(B)[f_1|B,\ldots,f_k|B]$$

uniquely defined by requiring that σ_H is the identity on $\psi_B^*\mathfrak{K}(B)$ and such that $\sigma_H(\eta_\mu) = f_\mu|B$ for $\mu = 1,\ldots,k$. Because $f_1|B,\ldots,f_k|B$ are algebraically independent over $\psi_B^*\mathfrak{K}(B)$, the epimorphism σ_B is an isomorphism. Then $\rho_B = \sigma_B\circ\tilde{\psi}_B^*$ is an isomorphism.

Let $j_B\colon B \to \tilde{H}$ be the injection. Then $j_B^*\colon \mathfrak{K}(\tilde{H}) \to \mathfrak{K}(B)$ is a ring epimorphism. The map $\varphi_H^*\colon \mathfrak{K}(H) \to \varphi_H^*\mathfrak{K}(H)$ is a field isomorphism onto a subfield of $\mathfrak{K}(\tilde{H})$. Now, $\psi_B = \varphi_H\circ j_B$ implies $j_B^*\circ\varphi_H^* = \psi_B^*$. Let $\zeta_1,\ldots,\zeta_k$ be indeterminants over $\varphi_H^*\mathfrak{K}(H)$. Then φ_H^* extends to a ring isomorphism

$$\tilde{\varphi}_H^*\colon \mathfrak{K}(H)[\xi_1,\ldots,\xi_k] \to \varphi_H^*\mathfrak{K}(H)[\zeta_1,\ldots,\zeta_k]$$

which is uniquely extended by the requirement that $\varphi_H^*(\xi_\mu) = \zeta_\mu$ for $\mu = 1,\ldots,k$. Also, j_B^* extends uniquely to an epimorphism

$$\tilde{j}_B^*\colon \varphi_H^*\mathfrak{K}(H)[\zeta_1,\ldots,\zeta_k] \to \psi_B^*\mathfrak{K}(H)[\eta_1,\ldots,\eta_k]$$

by $j_B^*(\zeta_\mu) = \eta_\mu$ for $\mu = 1,\ldots,k$. Then $\tilde{j}_B^*\circ\tilde{\varphi}_H^*(\xi_\mu) = \eta_\mu$ for $\mu = 1,\ldots,k$. Hence, $\tilde{j}_B^*\circ\tilde{\varphi}_H^* = \tilde{\psi}_B^*$. A substitution ring-homomorphism

$$\sigma_H\colon \varphi_H^*\mathfrak{K}(H)[\zeta_1,\ldots,\zeta_k] \to \varphi_H^*\mathfrak{K}(H)[f_1|\tilde{H},\ldots,f_k|\tilde{H}] = R$$

is defined by $\sigma_H(\zeta_\mu) = f_\mu|\tilde{H}$ such that $\sigma_\mu|\varphi_H^*\mathfrak{K}(H)$ is the identity. Obviously, σ_H is surjective. Observe that R is a subring of $\mathfrak{K}(\tilde{H})$. Therefore, j_B: $R \to \mathfrak{K}(B)$ is defined. Obviously $j_B^*(R) \subseteq \psi_B^*\mathfrak{K}(H)[f_1|B,\ldots,f_k|B]$ and $j_B^*\circ\sigma_H = \sigma_B\circ\tilde{j}_B^*$ since this is true on $\varphi_H^*\mathfrak{K}(H)$ and since $j_B^*\circ\sigma_H(\zeta_\mu) = f_\mu|B = \sigma_B(\eta_\mu) = \sigma_B\circ\tilde{j}_B^*(\zeta_\mu)$. Now: $\rho_H = \sigma_H\circ\tilde{\varphi}_H^*$ is an epimorphism with

$$j_B^*\circ\rho_H = j_B^*\circ\sigma_H\circ\tilde{\varphi}_H^* = \sigma_B\circ\tilde{j}_B^*\circ\tilde{\varphi}_H^* = \sigma_B\circ\tilde{\psi}_B^* = \rho_B.$$

Hence, $j_B^*\circ\rho_H$ is an isomorphism. Therefore, ρ_H is injective. Hence, $\rho_H = \sigma_H\circ\tilde{\varphi}_H^*$ is an isomorphism. Because $\tilde{\varphi}_H^*$ is an isomorphism, σ_H is an isomorphism. Hence, R is an integral domain and $f_1|\tilde{H},\ldots,f_k|\tilde{H}$ are algebraically independent over $\varphi_H^*\mathfrak{K}(H)$. Hence, the quotient ring

$$K = \varphi_H^*\mathfrak{K}(H)(f_1|\tilde{H},\ldots,f_k|\tilde{H}) \subseteq \mathfrak{K}(\tilde{H})$$

is a field and a pure transcendental extension of $\varphi_H^*\mathfrak{K}(H)$ with transcendency degree k; q.e.d.

§13. The Schwarz Lemma

Let X and Y be complex spaces. Let $\varphi: X \to Y$ be a holomorphic map. If φ is light, then the order $\tilde{\nu}_\varphi(a)$ and the multiplicity $\nu_\varphi(a)$ of φ at $a \in X$ have been defined in §2. If Y is locally irreducible and if X is pure dimensional, then $\nu_\varphi(a) = \tilde{\nu}_\varphi(a)$ by [21] Lemma 1.6.

Now, the <u>zero order</u> $\sigma_f(a)$ of a holomorphic function f on a complex space X at $a \in X$ shall be defined. If $f(a) \neq 0$, define $\sigma_f(a) = 0$. If the germ f_a of f at a is zero, define $\sigma_f(a) = \infty$. If $f(a) = 0$ but $f_a \neq 0$, let $\mathcal{L}_a(f)$ be the set of all subsets L of X satisfying the following conditions.

1. L is a pure one-dimensional analytic subset of an open neighborhood U_L of a.

2. The closure $\overline{L}$ is compact.

3. $\overline{L} \cap f^{-1}(0) = \{a\}$.

Then $\mathcal{L}_a(f) \neq \emptyset$. Because L is closed in U_L, condition 3 implies $L \cap f^{-1}(0) = \{a\}$ if $L \in \mathcal{L}_a(f)$. Moreover, $f|L: L \to \mathbb{C}$ is a light map. Define

$$\sigma_f(a) = \operatorname{Min}\{\nu_{f|L}(a) \mid L \in \mathcal{L}_a(\varphi)\}.$$

Obviously, if X has pure dimension one, if X is irreducible at a, if $f: X \to \mathbb{C}$ is light and if $f(a) = 0$, then $\sigma_f(a) = \nu_f(a) = \tilde{\nu}_f(a)$.

Again, let X be a complex space and f be a holomorphic function on X. Then $f^{-1}(0) = \{x \in X \mid \sigma_f(x) > 0\}$ and $\{x \in X \mid f_x = 0\} = \{z \in X \mid \sigma_f(x) = \infty\}$. If a is a simple point of X, if $f(a) = 0$ but $f_a \neq 0$, then $\sigma_f(a) = \nu_f(a)$ in the sense of [21] p. 48. Hence,

Lemma 5.1 [20] implies $p = \sigma_f(a)$ and $f_a \in \mathfrak{m}_a^p - \mathfrak{m}_a^{p+1}$ where $\mathfrak{m}_a$ is the maximal ideal in the local ring $\mathcal{O}_a$ of holomorphic germs at a. Especially

$$\sigma_{f+g}(a) \geqq \mathrm{Min}(\sigma_f(a), \sigma_g(a))$$
$$\sigma_{f\cdot g}(a) = \sigma_f(a) + \sigma_g(a)$$

if a is a simple point of X and if f and g are holomorphic on X. If $\sigma_g(x) < \infty$ for all $x \in X$, then f/g is meromorphic on X. If a is a simple point of X, then f/g is holomorphic at a, if and only if an open neighborhood U of a exists such that $\sigma_f(x) \geqq \sigma_g(x)$ for all $x \in U$.

Lemma 13.1. Let Y be an analytic subset of the complex space X. Let $f: X \to \mathbb{C}$ be a holomorphic function on X. Define $g = f|Y$. Then $\sigma_g(a) \geqq \sigma_f(a)$ if $a \in Y$.

Proof. If $f(a) \neq 0$, then $g(a) \neq 0$ and $\sigma_g(a) = 0 = \sigma_f(a)$. Assume $f(a) = 0$. Then $g(a) = 0$. If $g_a = 0$, then $\sigma_g(a) = \infty$. Hence, $\sigma_g(a) \geqq \sigma_f(a)$. Assume $g_a \neq 0$, then $f_a \neq 0$. Obviously $\mathcal{L}_a(g) \subseteq \mathcal{L}_a(f)$ which implies $\sigma_g(a) \geqq \sigma_f(a)$; q.e.d.

If $\emptyset \neq S \subseteq X$, and if $f: X \to \mathbb{C}$ is a function, define

$$\|f\|_S = \sup\{|f(x)| \mid x \in S\}.$$

Then $0 \leqq \|f\|_S \leqq \infty$. Here $\| \ \|_S$ is a seminorm on the vector space of all functions on X which are bounded on S. If X is a Hausdorff space, if S is compact and if f is continuous on S, then $\|f\|_S < \infty$. If S is a compact subset of a complex space X, if each branch of X contains an interior point of S, then $\| \ \|_S$ is a norm on the algebra $\mathfrak{H}(X)$ of holomorphic functions on X.

<u>Lemma 13.2.</u> Let f be a holomorphic function on the complex space X. Define $D = \{z \in \mathbb{C} \mid |z| < r\}$ with $0 < r \in \mathbb{R}$. Let $h: X \to D$ be a surjective, proper, holomorphic map. Define $h \cdot f = g$. Then $r\|f\|_X \leqq \|g\|_X$.

<u>Proof.</u> Take any $z \in X$. Then $|h(z)| < r$. Take any $\rho \in \mathbb{R}$ with $|h(z)| < \rho < r$. Then $H = \{x \in X \mid |h(x)| < \rho\}$ is an open neighborhood of z and $K = \{x \in X \mid |h(x)| \leqq \rho\}$ is compact. By the maximum principal $\|f\|_K = \|f\|_{K-H} = |f(a)|$ where $a \in K - H$. Then

$$\rho|f(a)| = |h(a)| \ |f(a)| = |g(a)| \leqq \|g\|_X$$

Hence,

$$\rho|f(z)| \leqq \rho\|f\|_K = \rho|f(a)| \leqq \|g\|_X.$$

Now, $\rho \to r$ implies $r|f(z)| \leqq \|g\|_X$. Therefore, $r\|f\|_X \leqq \|g\|_X$;

q.e.d.

A <u>gauge</u> p on a complex vector space V is a real valued, upper-semi-continuous function p such that $p(v) > 0$ if $0 \neq v \in V$ and $p(zv) = |z|p(v)$ if $z \in \mathbb{C}$ and $v \in V$. Then $p(0) = 0$. A subset G of V is said to <u>be circular</u>, if $v \in G$ and $z \in \mathbb{C}$ with $|z| \leqq 1$ implies $zv \in G$. If p is a gauge on V and if $0 < r \in \mathbb{R}$, then $\{v \mid p(v) < r\}$

is an open, bounded circular subset of G. If G is an open, bounded, circular subset of V, a gauge p exists on V such that $G = \{v \mid p(v) < 1\}$.

<u>Proposition 13.3.</u> (Schwarz) Let X be a complex space of pure dimension m. Let p be a gauge on the m-dimensional complex vector space V. Let $0 < r \in \mathbb{R}$ and define $G = \{v \in V \mid p(v) < r\}$. Suppose that a proper, light, holomorphic map $\varphi: X \to G$ with sheet number s is given. Let f be a bounded holomorphic function on X. Let h be a non-negative integer, such that $h \leqq \sigma_f(x)$ for all $x \in \varphi^{-1}(0)$. Then

$$(13.1) \qquad |f(x)| \leqq \left(\frac{p(\varphi(x))}{r}\right)^{\frac{h}{s}} \|f\|_X \qquad \text{if } x \in X.$$

<u>Proof.</u> By Lemma 2.2, φ is surjective and open. If $h = 0$ or $f(x) = 0$, the statement is trivial. Assume $h > 0$ and $f(x) \neq 0$. Define $b = \varphi(x)$. If $b = 0$, then $\sigma_f(x) \geqq h > 0$. Hence, $f(x) = 0$ which is wrong. Therefore, $b \neq 0$. Let $L = \{zb \mid z \in \mathbb{C}\}$ be the straight line through 0 and b. Define

$$E = L \cap G = \{zb \mid \ |z| < \frac{r}{p(b)}\}$$

$$D = \{z \in C \mid \ |z| < \frac{r}{p(b)}\} .$$

A biholomorphic map $u: E \to D$ is defined by $u(zb) = z$ with $u(b) = 1$.

The set $\tilde{L} = \varphi^{-1}(L) = \varphi^{-1}(E)$ is analytic in X with $x \in \tilde{L}$. The map $\varphi_0 = \varphi: \tilde{L} \to E$ is proper, light, surjective, holomorphic. If U is open in $\tilde{L}$, then $U = U' \cap \tilde{L}$ where U' is open in X. Then $\varphi(U')$

is open in V. Hence, $\varphi_0(U) = E \cap \varphi(U')$ is open in E. The map φ_0 is open. By Lemma 1.10, $\tilde{L}$ has pure dimension one. The sheet number s_0 of φ_0 is defined and $s_0 \leqq s$.

Let T be a branch of $\tilde{L}$ with $x \in T$. Then $\varphi_1 = \varphi_0 \colon T \to E$ is a proper, light, surjective, open, holomorphic map with a sheet number $t \leqq s_0 \leqq s$. Let $\rho\colon \hat{T} \to T$ be the normalization of T. Then $\hat{T}$ is a connected complex manifold of dimension 1 and $\psi = \varphi_1 \circ \rho = \varphi_0 \circ \rho = \varphi \circ \rho\colon \hat{T} \to E$ and $v = u \circ \psi\colon \hat{T} \to D$ are proper, light, surjective, open, holomorphic maps of sheet number t. The functions $f_0 = f|T$ and $f_1 = f_0 \circ \rho$ are holomorphic on T respectively $\hat{T}$. Take $\tilde{x} \in \rho^{-1}(x)$. Then $f_1(\tilde{x}) = f_0(x) = f(x) \neq 0$ and

$$v(\tilde{x}) = u(\varphi(\rho(\tilde{x}))) = u(\varphi(x)) = u(b) = 1.$$

Take $a \in v^{-1}(0)$. Then $0 < \sigma_v(a) = \nu_v(a) \leqq t \leqq s$. Define $w = \rho(a)$. Then $0 = v(a) = u(\varphi(w))$. Hence, $\varphi(w) = 0$ and $w \in \varphi^{-1}(0)$ which implies $\sigma_f(w) \geqq h > 0$. Therefore, $f_1(a) = f_0(w) = f(w) = 0$. The functions f_0 and f_1 are not constant and $f_0\colon T \to \mathbb{C}$ and $f_1\colon \hat{T} \to \mathbb{C}$ are open, light, holomorphic maps.

An open neighborhood U of w in X exists, such that

$$U \cap T \cap f^{-1}(0) = U \cap f_0^{-1}(0) = \{w\}.$$

Then $\hat{U} = \rho^{-1}(U)$ is an open neighborhood of a. An open, connected subset W of $\hat{T}$ exists such that $\overline{W}$ is compact and $a \in W \subseteq \overline{W} \subseteq \hat{U}$ and such that $L = \rho(W)$ is a branch of $U_L \cap T$ where U_L is an open neighborhood of w with $\overline{U}_L \subseteq U$. Then $\overline{L}$ is compact and $\overline{L} \cap f^{-1}(0) =$

$\{w\}$. Hence, $L \in \mathcal{L}_w(f)$. Moreover, W can be chosen, such that $L - \{w\}$ consists of simple points of T only. Then $\rho_0 = \rho\colon W \to L$ is bijective and holomorphic. Therefore, $\tilde{\nu}_{f_0|L}(w) = \tilde{\nu}_{f_1|W}(a)$ by Lemma 1.7. Because $f_0|L = f|L$ and $L \in \mathcal{L}_w(f)$, the definition of the zero order implies $\tilde{\nu}_{f_0|L}(w) \geqq \sigma_f(w)$. Because W is open in $\hat{T}$ and because a is a simple point of $\hat{T}$, $\tilde{\nu}_{f_1|W}(a) = \sigma_{f_1|W}(a) = \sigma_{f_1}(a)$. Together

$$\sigma_{f_1}(a) = \tilde{\nu}_{f_1|W}(a) = \tilde{\nu}_{f_0|L}(w) \geqq \sigma_f(w) \geqq h$$

if $a \in v^{-1}(0)$.

Define $G = f_1^s$ and $H = v^h$. Define $D_0 = \{z \in \mathbb{C} \mid |z| < (\frac{r}{p(b)})^h\}$. Then $H\colon \hat{T} \to D_0$ is surjective, proper and holomorphic. If $a \in H^{-1}(0)$ then

$$\sigma_G(a) = s\sigma_{f_1}(a) \geqq s\cdot h \geqq h\cdot\sigma_v(a) = \sigma_H(a).$$

Hence, $F = G/H$ is a holomorphic function on $\hat{T}$ with $HF = G$. By Lemma 13.2

$$\left(\frac{r}{p(b)}\right)^h \|F\|_{\hat{T}} \leqq \|G\|_{\hat{T}} .$$

Now,

$$|f(x)|^s = |(f_1(\tilde{x}))^s \cdot v(\tilde{x})^{-h}| = |G(\tilde{x})/H(\tilde{x})| = |F(\tilde{x})|$$

$$\leqq \left(\frac{p(b)}{r}\right)^h \|G\|_{\hat{T}} = \left(\frac{p(b)}{r}\right)^h \|f_1\|_{\hat{T}}^s \leqq \left(\frac{p(b)}{r}\right)^h \|f\|_{\hat{X}}^s$$

which implies (13.1); q.e.d.

Take $a = (a_1,\ldots,a_m) \in \mathbb{C}^m$ and $r = (r_1,\ldots,r_m) \in \mathbb{R}^m$ with $r_\mu > 0$ for $\mu = 1,\ldots,p$. Then

$$P = \prod_{\mu=1}^{m} \{z_\mu \mid |z_\mu - a_\mu| < r_\mu\}$$

is said to be a <u>poly-disc</u> with radius r and center a.

$$\partial P = \prod_{\mu=1}^{m} \{z_\mu \mid |z_\mu - a_\mu| = r_\mu\}$$

is called the <u>Bergman boundary</u> of P.

Let X be a complex space of dimension m. Then (P,α) is said to be a <u>poly-disc chart</u> at $a \in X$, if and only if $\alpha\colon U_\alpha \to U'_\alpha$ is a chart of X at a, if $\alpha(a) = 0$, and if a poly-disc P' with center 0 in $\mathbb{C}^m$ exists such that $\overline{P'} \subseteq U'_\alpha$ and $P = \alpha^{-1}(P')$. If (P,α) is a poly-disc chart at $a \in X$, then P is open, $\overline{P}$ is compact with $\overline{P} = \alpha^{-1}(\overline{P'}) \subseteq U_\alpha$ and $\alpha\colon P \to P'$ is a chart of X which has the same sheet number as the chart $\alpha\colon U_\alpha \to U'_\alpha$. Define $\partial_\alpha P = \alpha^{-1}(\partial P')$ as the Bergman boundary of (P,α). The poly-disc chart is said to be schlicht, if $\alpha\colon U_\alpha \to U'_\alpha$ is biholomorphic.

<u>Lemma 13.4.</u> Let X be a complex space of pure dimension m. Let (P,α) be a poly-disc chart on X. Let $f: \overline{P} \to \mathbb{C}$ be continuous and $f|P$ be holomorphic. Then $\|f\|_{\overline{P}} = \|f\|_{\partial_\alpha P}$.

<u>Proof.</u> At first assume, that f is holomorphic in an open neighborhood U of $\overline{P}$. If $m = 1$, the theorem is trivial by the maximum principle, since $\overline{P} - P = \partial_\alpha P$ in this case. Assume, the theorem is true for $1,2,\ldots,m - 1$. Then it shall be proved for m (with f holomorphic on $U \supset \overline{P}$). Let $r = (r_1,\ldots,r_m)$ be the radius of P'. Then $c \in \overline{P} - P$ exists such that $\|f\|_{\overline{P}} = |f(c)|$. Define $\alpha(c) = (c_1,\ldots,c_m)$. Then $|c_\mu| = r_\mu$ for some index μ. Without loss of generality $\mu = m$ can be assumed. Number $s_\nu > r_\nu$ exist such that

$$R' = \prod_{\mu=1}^{m}\{z_\mu \mid |z_\mu| < s_\mu\} \subseteq \overline{R}' \subset U_\alpha$$

$$R = \alpha^{-1}(R').$$

Then (R,α) is a poly-disc chart of X with $\overline{P} \subset R$. Define $L = R' \cap (\mathbb{C}^{m-1} \times \{c_m\})$ and $U'_\beta = \prod_{\mu=1}^{m-1}\{z_\mu \mid |z_\mu| < s_\mu\}$. Then the projection $\pi: L \to U'_\beta$ is biholomorphic and L is a pure (m-1)-dimensional complex submanifold of R'. The set $U_\beta = \alpha^{-1}(L)$ is analytic in R. The map $\beta = \pi\circ\alpha: U_\beta \to U'_\beta$ is proper, light, surjective, open and holomorphic. By Lemma 1.10, U_β has pure dimension $m - 1$. Define

$$Q' = \prod_{\mu=0}^{m-1}\{z_\mu \mid |z_\mu| < r_\mu\}.$$

Then $Q' \subseteq \overline{Q}' \subseteq U'_\beta$ and $Q = \beta^{-1}(Q') \subseteq \overline{Q} \subseteq \overline{P} \cap U_\beta \subseteq U$. Then (Q,β) is a poly-disc chart on U_β. By induction assumption $\|f\|_{\overline{Q}} \leqq \|f\|_{\partial_\beta Q}$. Hence, $d \in \partial_\beta Q$ exists with $\|f\|_{\overline{Q}} = |f(d)|$. Then $\alpha(d) = (d_1,\ldots,d_m)$ with $d_m = c_m$ and $|d_\mu| = r_\mu$ for $\mu = 1,\ldots,m-1$. Therefore, $|d_\mu| = r_\mu$ for $\mu = 1,\ldots,m$ and $d \in \partial_\alpha P$. Hence, $|f(d)| \leqq \|f\|_{\partial_\alpha P}$. Now, $c \in \overline{Q}$ implies

$$\|f\|_{\overline{P}} = |f(c)| \leqq \|f\|_{\overline{Q}} = |f(d)| \leqq \|f\|_{\partial_\alpha P} \leqq \|f\|_{\overline{P}}.$$

Hence, $\|f\|_{\overline{P}} = \|f\|_{\partial_\alpha P}$.

If $f\colon \overline{P} \to \mathbb{C}$ is continuous and $f|P$ is holomorphic, let $r = (r_1,\ldots,r_m)$ be the radius of $P' = \alpha(P)$. Define $r_\mu(n) = r_\mu(1 - \frac{1}{2n})$ for $n \in \mathbb{N}$. Then $0 < r_\mu(n) < r_\mu$ and $r_\mu(n) \to r_\mu$ for $n \to \infty$. Define

$$P'_n = \prod_{\mu=1}^{m}\{z_\mu \mid |z_\mu| < r_\mu(n)\} \subseteq \overline{P}'_n \subset P'$$

$$P_n = \alpha^{-1}(P'_n) \subseteq \overline{P}_n \subset P .$$

Then (P_n,α) is a poly disc chart. Take $x \in P$. Then $\alpha(x) = (x_1,\ldots,x_m)$ with $|x_\mu| < r_\mu$. A number q exists such that $|x_\mu| < r_\mu(n)$ for all $n \geqq q$ and $\mu = 1,\ldots,m$. Then $x \in P_n$ for $n \geqq q$. Hence, $|f(x)| \leqq \|f\|_{P_n} = \|f\|_{\partial_\alpha P_n}$. Hence, $y(n) \in \partial_\alpha P_n$ with $|f(x)| \leqq |f(y(n))|$ exists for each $n \geqq q$. Then $\alpha(y(n)) = (y_1(n),\ldots,y_m(n))$ with $|y_\mu(n)| = r_\mu(n)$. Because $\overline{P}$ is compact a subsequence of $\{y(n)\}_{n \geqq q}$ converges to a point $y \in \overline{P}$ with $\alpha(y) = (y_1,\ldots,y_m)$.

Then $|f(x)| \leqq |f(y)|$ and $|y_\mu| = r_\mu$. Hence, $(y_1,\ldots,y_m) \in \partial P'$ and $y \in \partial_\alpha P$. Therefore,

$$|f(x)| \leqq |f(y)| \leqq \|f\|_{\partial_\alpha P}$$

for all $x \in P$. By continuity $|f(x)| \leqq \|f\|_{\partial_\alpha P}$ for all $x \in \overline{P}$. Hence, $\|f\|_{\overline{P}} \leqq \|f\|_{\partial_\alpha P} \leqq \|f\|_{\overline{P}}$ which implies $\|f\|_{\overline{P}} = \|f\|_{\partial_\alpha P}$; q.e.d.

If $r = (r_1,\ldots,r_m) \in \mathbb{R}^m$ with $0 < r_\mu$ for $\mu = 1,\ldots,m$, define

$$p(z) = \operatorname{Max}\left(\frac{|z_1|}{r_1},\ldots,\frac{|z_m|}{r_m}\right) \tag{13.2}$$

for $z = (z_1,\ldots,z_m) \in \mathbb{C}^m$. Then p is a gauge on $\mathbb{C}^m$ and

$$\{z \in \mathbb{C}^m \mid p(z) < 1\} = \prod_{\mu=1}^{m}\{z_\mu \mid |z_\mu| < r_\mu\} \tag{13.3}$$

is the poly-disc with radius r and center 0.

Let X and Y be complex spaces of pure dimension m and n respectively with $m - n = q \geqq 0$. Let $\varphi: X \to Y$ be a holomorphic map. Then (P,Q,α,β) is said to be a <u>poly-disc product representation</u> of φ at $a \in X$ over $b = \varphi(a) \in Y$ if and only if

1. The pair (α,β) is a product representation of φ at a and over b with $\alpha(a) = 0 \in \mathbb{C}^m$ and $\beta(b) = 0 \in \mathbb{C}^n$.

2. (P,α) and (Q,β) are poly-disc charts on X and Y respectively.

3. If $P' = \alpha(P)$ and $Q' = \beta(Q)$ then $P' = N' \times Q'$.

Here $U'_\alpha = U''_\alpha \times U'_\beta$. The set N' is a poly-disc with center $0 \in \mathbb{C}^q$ with $\overline{N}' \subseteq U''_\alpha$. Moreover, $(\alpha|P,\beta|Q)$ is a product representation of φ at a and over b. By Lemma 2.5.5

$$S = S(P,Q,\alpha,\beta) = \alpha^{-1}(\{0\} \times Q')$$

is a strictly central section of φ in P over Q. It is called the <u>center section</u> of (P,Q,α,β).

<u>Lemma 13.5.</u> Let X and Y be complex spaces of pure dimension m and n respectively with $q = m - n \geqq 0$. Let $\varphi: X \to Y$ be a q-fibering, holomorphic map. Take $a \in X$ and $b = \varphi(a)$. Let U be an open neighborhood of a. Let $\delta: U_\delta \to U'_\delta$ be a chart centered at b with $\delta(b) = 0$. Then a poly-disc product representation (P,Q,α,β) of φ at a and over b exists such that $\alpha: U_\alpha \to U'_\alpha$ is a chart centered at a, such that $U_\alpha \subseteq U$ and $U_\beta \subseteq U_\delta$ with $\beta = \delta|U_\beta$.

<u>Proof.</u> By Proposition 2.4, a product representation (α,β) of φ centered at a and over b exists such that $U_\alpha \subseteq U$ and $U_\beta \subseteq U_\delta$ with $\beta = \delta|U_\beta$. Especially α is centered at a. Then $\alpha(a) = (a',\beta(b)) = (a',0)$. By using a parallel translation of the first q coordinates of $\mathbb{C}^m = \mathbb{C}^q \times \mathbb{C}^n$, it can be assumed that $a' = 0$. Hence, $\alpha(a) = 0$. A positive real number r exists such that

$$P' = \prod_{\mu=1}^{m}\{z_\mu \mid |z_\mu| < r\} \subseteq \overline{P}' \subseteq U''_\alpha \times U'_\beta$$

$$Q' = \prod_{\mu=q+1}^{m}\{z_\mu \mid |z_\mu| < r\} \subseteq \overline{Q}' \subset U'_\beta$$

$$N' = \prod_{\mu=1}^{q}\{z_\mu \mid |z_\mu| < r\} \subseteq \overline{N}' \subset U''_\alpha$$

Define $P = \alpha^{-1}(P')$ and $Q = \beta^{-1}(Q')$. Obviously (P,Q,α,β) is a poly-disc product representation at a and over b; q.e.d.

Let (P,Q,α,β) and (R,Q,α,β) be poly-disc product representations of φ. Then $\alpha(P) = P' = N' \times Q'$ and $\alpha(R) = R' = M' \times Q'$ where $Q' = \beta(Q)$. Then (R,Q,α,β) is called a <u>retraction</u> of (P,Q,α,β) if and only if $\overline{M}' \subseteq N'$. If so, let $r = (r_1,\ldots,r_q)$ be the radius of N', let $t = (t_1,\ldots,t_q)$ be the radius of M' and let s be the sheet number of α. Then

$$(13.4) \qquad 0 < \theta_0 = \operatorname{Max}({}^{t_1}/_{r_1},\ldots,{}^{t_q}/_{r_q}) < 1.$$

The number $\theta = \theta_0^{\frac{1}{s}}$ is called the <u>condensor</u> of the retraction (R,Q,α,β) of (P,Q,α,β). Observe, that (P,Q,α,β) and (R,Q,α,β) have the same center section

$$S = \alpha^{-1}(\{0\} \times Q') = S(P,Q,\alpha,\beta) = S(R,Q,\alpha,\beta).$$

<u>Proposition 13.6.</u> (Schwarz Lemma for maps). Let X and Y be complex spaces of pure dimension m and n respectively with $m - n = q \geqq 0$. Let $\varphi\colon X \to Y$ be a q-fibering, holomorphic map. Let (P,Q,α,β) and (R,Q,α,β) be a poly-disc product representations of φ such that (R,Q,α,β) is a retraction of (P,Q,α,β) with θ as the condensor. Let S be the common center section of (P,Q,α,β) and (R,Q,α,β).

Let $f\colon \overline{P} \to \mathbb{C}$ be a continuous function such that $f|P$ is holomorphic. Let $h \geqq 0$ be an integer, such that $\sigma_f(x) \geqq h$ for all $x \in S$. Then

(13.5) $$\|f\|_{\overline{R}} \leqq \theta^h \|f\|_{\overline{P}}.$$

<u>Proof.</u> Let $r = (r_1, \ldots, r_m)$ be the radius of $P' = \alpha(P)$. Then $(r_{q+1}, \ldots, r_m)$ is the radius of $Q' = \beta(Q)$. Let $t = (t_1, \ldots, t_m)$ be the radius of $R' = \alpha(R)$. Then $t_\mu = r_\mu$ for $\mu = q + 1, \ldots, m$. Now, $P' = N' \times Q'$ and $R' = M' \times Q'$ where $(r_1, \ldots, r_q)$ is the radius of N' and where $(t_1, \ldots, t_q)$ is the radius of M' with $t_\mu < r_\mu$ for $\mu = 1, \ldots, q$. Define θ_0 by (13.4). Let s be the sheet number of α. Then $\theta = \theta_0^{\frac{1}{s}}$.

Take $x \in R$. Then

(13.6) $$|f(x)| \leqq \theta^h \|f\|_{\overline{P}}$$

is claimed. If $f(x) = 0$, (13.6) is true. Assume $f(x) \neq 0$ and define $y = \varphi(x)$. Let B be a branch of $\varphi^{-1}(y) \cap U_\alpha$ with $x \in B$. Then B is an irreducible, q-dimensional analytic subset of U_α with $x \in B \cap R$. If $z \in B$, then $\varphi(z) = y$ and $\alpha(z) = (z_0, z_1)$ with $z_0 \in U''_\alpha$ and $z_1 \in U'_\beta$. Then $\beta(\varphi(z)) = \pi(\alpha(z)) = z_2$, where $\pi: U''_\alpha \times U'_\beta \to U'_\beta$ if the projection. Hence, $z_2 = \beta(y)$ and $B \subseteq \alpha^{-1}(U''_\alpha \times \{\beta(y)\})$. Let $\lambda: U''_\alpha \times \{\beta(y)\} \to U''_\alpha$ be the projection. Then λ is biholomorphic. The map $\gamma = \lambda \circ \alpha: B \to U''_\alpha$ is proper, light and holomorphic. By Lemma 2.2, γ is also open and surjective. The map γ has a sheet number $u \leqq s$. Then $\gamma: B \to U''_\alpha$ is a chart on B. Define $M = \gamma^{-1}(M') = R \cap B$ and $N = \gamma^{-1}(N') = P \cap B$. Then (M, γ) and (N, γ) are poly-disc charts on B.

Define $g = f|\overline{N}$. Then g is continuous on $\overline{N}$ and holomorphic on N. Especially g is bounded on N. If $z \in \gamma^{-1}(0)$, then $\sigma_g(z) \geqq \sigma_f(z)$ by Lemma 13.1. Moreover, $\lambda(\alpha(z)) = 0$ implies $\alpha(z) = (0,\beta(y))$. Hence, $z \in \alpha^{-1}(\{0\} \times U'_\alpha) = S$. Hence, $\sigma_f(z) \geqq h$, which implies $\sigma_g(z) \geqq h$ for all $z \in \gamma^{-1}(0)$.

For $z = (z_1,\ldots,z_q) \in \mathbb{C}^q$ define

$$p(z) = \operatorname{Max}\left(\frac{|z_1|}{r_1},\ldots,\frac{|z_q|}{r_q}\right).$$

Then p is a gauge on $\mathbb{C}^q$ with $N' = \{z \in \mathbb{C}^q | p(z) < 1\}$. Proposition 13.3 implies

$$|g(x)| \leqq (p(\gamma(x)))^{\frac{h}{u}} \|g\|_M$$

because $x \in R \cap B$. Now, $\alpha(x) = (x_1,\ldots,x_m)$ with $|x_\mu| < t_\mu$ for $\mu = 1,\ldots,m$ implies

$$p(\gamma(x)) = \operatorname{Max}\left(\frac{|x_1|}{r_1},\ldots,\frac{|x_q|}{r_q}\right) \leqq \operatorname{Max}\left(\frac{t_1}{r_1},\ldots,\frac{t_q}{r_q}\right) = \theta_0.$$

Hence,

$$|f(x)| \leqq \theta_0^{h/u}\|f\|_{\overline{P}} \leqq \theta_0^{h/s}\|f\|_{\overline{P}} = \theta^h\|f\|_{\overline{P}}$$

for all $x \in R$. Hence, $|f(x)| \leqq \theta^h\|f\|_{\overline{P}}$ for all $x \in \overline{R}$ by continuity, which proves (13.5); q.e.d.

Remark: Observe that the factor θ in (13.5) does not depend on the choice of the function f.

§14. Sections in meromorphic line bundles

Let X be a complex space. Let $\mathfrak{V} = \{V_i\}_{i \in I}$ be a family of subsets of X. Then $\mathfrak{V}$ is said to be open, closed, compact or Stein if and only if each V_i is open, closed, compact or Stein respectively. $\mathfrak{V}$ is said to be finite if I is finite, $\mathfrak{V}$ is said to be locally finite, if every point $x \in X$ is contained in a neighborhood U such that $\{i \in I \mid U \cap V_i \neq \emptyset\}$ is finite. Define

$$\overline{\mathfrak{V}} = \{\overline{V}_i\}_{i \in I} \qquad \text{(closure of } \mathfrak{V}\text{)}$$

$$\operatorname{supp} \mathfrak{V} = \bigcup_{i \in I} V_i \qquad \text{(support of } \mathfrak{V}\text{)}$$

$$\mathfrak{V} \cap M = \{V_i \cap M\}_{i \in I} \qquad \text{if } M \subseteq X$$

$$V_{i_0 \ldots i_p} = V_{i_0} \cap \ldots \cap V_{i_p} \qquad \text{if } i_\mu \in I \text{ for } \mu = 0,\ldots,p$$

$$I[p] = I[p,\mathfrak{V}] = \{(i_0,\ldots,i_p) \mid V_{i_0 \ldots i_p} \neq \emptyset\}.$$

$\mathfrak{V}$ is said to be a covering of X, if and only if $X = \operatorname{supp} \mathfrak{V}$. A family $\mathfrak{W} = \{W_i\}_{i \in I}$ is said to be contained in $\mathfrak{V}$ if and only if $W_i \subseteq V_i$ for each $i \in I$. The family $\mathfrak{W}$ is said to be a shrinkage of $\mathfrak{V}$ if and only if $\overline{\mathfrak{W}}$ is compact and contained in $\mathfrak{V}$. The pair $(\mathfrak{W},\tau)$ with $\mathfrak{W} = \{W_j\}_{j \in J}$ is said to be a refinement of $\mathfrak{V}$ if and only if $\tau\colon J \to I$ is a map with $W_j \subseteq V_{\tau(j)}$. Then a map $\tau\colon J[p] \to I[p]$ is defined by $\tau(j_0,\ldots,j_p) = (\tau(j_0),\ldots,\tau(j_p))$. If $(\mathfrak{W},\tau)$ is a refinement of $\mathfrak{V}$, and if $(\mathfrak{Z},\sigma)$ is a refinement of $\mathfrak{W}$, then $(\mathfrak{Z},\tau\circ\sigma)$ is a refinement of $\mathfrak{V}$.

Let $A = \{A(U), r_V^U\}$ be a presheaf (of sets, groups, rings, etc.) on **X**. Let $\mathfrak{V} = \{V_i\}_{i \in I}$ a family of subsets of X. Define

$$C^p(\mathfrak{W},A) = \prod_{(i_0,\ldots,i_p)\in I[p]} A(V_{i_0,\ldots,i_p})$$

as the set <u>p-dimensional cochains</u> on $\mathfrak{W}$ with coefficients in A. Each cochain $s \in C^p(\mathfrak{W},A)$ is written as a family $s = \{s_{i_0,\ldots,i_p}\}_{(i_0,\ldots,i_p)\in I[p]}$. If the category of A permits direct products, then $C^p(\mathfrak{W},A)$ belongs to the same category.

Let $(\mathfrak{M},\tau)$ be an open refinement of $\mathfrak{W}$. Then a map

$$\tau^*: C^p(\mathfrak{W},A) \to C^p(\mathfrak{M},A)$$

is defined by

$$\tau(s)_{j_0,\ldots,j_p} = r^V_W(s_{\tau(j_0,\ldots,j_p)})$$

with $V = V_{\tau(j_0,\ldots,j_p)}$ and $W = W_{j_0,\ldots,j_p} \subseteq V$. The map τ is a homomorphism within the appropriate category. If $(\mathfrak{Z},\sigma)$ is a refinement of $\mathfrak{M}$, and if $(\mathfrak{M},\tau)$ is a refinement of $\mathfrak{W}$, then $(\mathfrak{Z},\tau\circ\sigma)$ is a refinement of $\mathfrak{W}$ and $(\tau\circ\sigma)^* = \sigma^*\circ\tau^*$. If $\mathfrak{M} = \{W_i\}_{i\in I}$ is contained in $\mathfrak{W}$, then $(\mathfrak{M},\mathrm{Id})$ is a refinement of $\mathfrak{W}$. Define $s|\mathfrak{M} = \mathrm{Id}^*(I)$ for $s \in C^p(\mathfrak{W},A)$. If Z is open in X, write $s|Z = s|\mathfrak{W}\cap Z$.

Let $\mathfrak{A}$ be a sheaf on X. Let $\Gamma(\mathfrak{A})$ be the canonical presheaf. Define

$$C^p(\mathfrak{U},\mathfrak{A}) = C^p(\mathfrak{U},\Gamma(\mathfrak{A})).$$

If A is a complete presheaf and if $\mathfrak{A}$ is the associated sheaf of A, then the natural homomorphism $A \to \Gamma(\mathfrak{A})$ is an isomorphism. Hence, write also $C^p(\mathfrak{U},A) = C^p(\mathfrak{U},\mathfrak{A}) = C^p(\mathfrak{U},\Gamma(A))$ in this case.

Let $\mathfrak{O}$ be the sheaf of germs holomorphic functions on X. Its canonical presheaf $\Gamma(\mathfrak{O})$ is identified with the presheaf $\{\mathfrak{H}(U),r_V^U\}$ of holomorphic functions. Let $\mathfrak{M}$ be the sheaf of germs of meromorphic functions on X. Its canonical presheaf $\Gamma(\mathfrak{M})$ is the presheaf $\{\mathfrak{K}(U),r_V^U\}$ of meromorphic functions. Of course, $\mathfrak{O}$ is a subsheaf of $\mathfrak{M}$. Both $\mathfrak{O}$ and $\mathfrak{M}$ are sheaves of rings. Let $\mathfrak{K}^*(U)$, $\mathfrak{H}^*(U)$, $\mathfrak{O}_x^*$, $\mathfrak{M}_x^*$ be the set of units in $\mathfrak{K}(U)$, $\mathfrak{H}(U)$, $\mathfrak{O}_x$, $\mathfrak{M}_x$ respectively. Then $\{\mathfrak{K}^*(U),r_V^U\}$ and $\{\mathfrak{H}^*(U),v_V^U\}$ are complete presheaves of abelian groups defining the subsheaves $\mathfrak{M}^* = \bigcup_{x\in X}\mathfrak{M}_x^*$ of $\mathfrak{M}$ and $\mathfrak{O}^* = \bigcup_{x\in X}\mathfrak{O}_x^*$ of $\mathfrak{O}$ respectively. Then $\mathfrak{T}(U) = \mathfrak{K}^*(U)\cap\mathfrak{H}(U)$ is the set of non-zero divisors in $\mathfrak{H}(U)$. Let $\mathfrak{T}$ be the sheaf defined by the complete presheaf $\{\mathfrak{T}(U),r_V^U\}$. Then $\mathfrak{T}_x$ is the set of non-zero divisors in $\mathfrak{O}_x$. Now,

$$\begin{array}{ccccc} \mathfrak{O}^* & \subseteq & \mathfrak{T} & \subseteq & \mathfrak{O} \\ & & \cap| & & \cap| \\ & & \mathfrak{M}^* & \subseteq & \mathfrak{M} \end{array}$$

Let $\mathfrak{W} = \{V_i\}_{i\in I}$ be an open family on X with $V = \operatorname{supp}\mathfrak{W}$. If $0 \leqq p \in \mathbb{N}$, then

$$\begin{array}{ccccc} D^p(\mathfrak{W},\mathfrak{O}^*) & \subseteq & C^p(\mathfrak{W},\mathfrak{T}) & \subseteq & C^p(\mathfrak{W},\mathfrak{O}) \\ & & \cap| & & \cap| \\ & & C^p(\mathfrak{W},\mathfrak{M}^*) & \subseteq & C^p(\mathfrak{W},\mathfrak{M}). \end{array}$$

Here $C^p(\mathfrak{W}, \mathcal{O}^*)$ and $C^p(\mathfrak{W}, \mathfrak{M}^*)$ are multiplicative abelian groups, $C^p(\mathfrak{W}, \mathfrak{T})$ is a multiplicative abelian semigroup, $C^p(\mathfrak{W}, \mathfrak{M})$ is a ring and $\mathfrak{K}(V)$-module, $C^p(\mathfrak{W}, \mathcal{O})$ is a ring and a $\mathfrak{G}(V)$-module.

A cochain $g \in C^1(\mathfrak{W}, \mathfrak{M}^*)$ is called a meromorphic line bundle on $\mathfrak{W}$ if and only if $g_{ij}g_{jk} = g_{ik}$ on V_{ijk} if $(i,j,k) \in I[2]$. If g is a <u>meromorphic line bundle</u> on $\mathfrak{W}$, then $g_{ii}g_{ii} = g_{ii}$ hence, $g_{ii} \equiv 1$ on V_i if $i \in I[0]$. Moreover, $g_{ij}g_{ji} \equiv 1$ or $g_{ij} = \dfrac{1}{g_{ji}}$ if $(i,j) \in I[1]$. The set $Z^1(\mathfrak{W}, \mathfrak{M}^*)$ of all <u>meromorphic line bundles</u> on $\mathfrak{W}$ is a multiplicative subgroup of $C^1(\mathfrak{W}, \mathfrak{M}^*)$. If $(\mathfrak{V}, \tau)$ is a refinement of $\mathfrak{W}$, then $\tau^*: Z^1(\mathfrak{W}, \mathfrak{M}^*) \to Z^1(\mathfrak{V}, \mathfrak{M}^*)$.

Let $g \in Z^1(\mathfrak{W}, \mathfrak{M}^*)$ be a meromorphic line bundle on $\mathfrak{W}$. Then $s \in C^0(\mathfrak{W}, \mathfrak{M})$ is said to be a <u>meromorphic section</u> on $\mathfrak{W}$ if and only if $s_i = g_{ij}s_j$ on V_{ij} if $(i,j) \in I[1]$. The set $\Gamma(\mathfrak{W}, g, \mathfrak{M})$ of meromorphic sections is a $\mathfrak{K}(V)$-module. Define

$$\Gamma(\mathfrak{W}, g) = \Gamma(\mathfrak{W}, g, \mathcal{O}) = \Gamma(\mathfrak{W}, g, \mathfrak{M}) \cap C^0(\mathfrak{W}, \mathcal{O})$$

$$\Gamma(\mathfrak{W}, g, \mathfrak{M}^*) = \Gamma(\mathfrak{W}, g, \mathfrak{M}) \cap C^0(\mathfrak{W}, \mathfrak{M}^*)$$

$$\Gamma(\mathfrak{W}, g, \mathfrak{T}) = \Gamma(\mathfrak{W}, g, \mathfrak{M}) \cap C^0(\mathfrak{W}, \mathfrak{T})$$

$$\Gamma(\mathfrak{W}, g, \mathcal{O}^*) = \Gamma(\mathfrak{W}, g, \mathfrak{M}) \cap C^0(\mathfrak{W}, \mathcal{O}^*).$$

The elements of $\Gamma(\mathfrak{W}, g)$ are called <u>holomorphic sections</u> or simply sections of g on $\mathfrak{W}$. The elements of $\Gamma(\mathfrak{W}, g, \mathfrak{M}^*)$ and $\Gamma(\mathfrak{W}, g, \mathcal{O}^*)$ are called <u>invertable sections</u>. An element of $\Gamma(\mathfrak{W}, g, \mathfrak{T})$ is called a <u>holomorphic, non-zero divisor section</u>. Obviously

$$\begin{array}{ccccc} \Gamma(\mathfrak{W},g,\mathcal{O}^*) & \subseteq & \Gamma(\mathfrak{W},g,\mathfrak{T}) & \subseteq & \Gamma(\mathfrak{W},g) \\ & & \cap| & & \cap| \\ & & \Gamma(\mathfrak{W},g,\mathfrak{M}^*) & \subseteq & \Gamma(\mathfrak{W},g,\mathfrak{M}). \end{array}$$

Here $\Gamma(\mathfrak{W},g)$ is an $\mathfrak{H}(V)$-module. If g and h are meromorphic line bundles on $\mathfrak{W}$, then

$$\left.\begin{array}{l} s \in \Gamma(\mathfrak{W},g,\mathfrak{M}) \\ t \in \Gamma(\mathfrak{W},h,\mathfrak{M}) \end{array}\right\} \quad \text{imply} \quad s\cdot t \in \Gamma(\mathfrak{W},gh,\mathfrak{M})$$

$$\left.\begin{array}{l} s \in \Gamma(\mathfrak{W},g,\mathcal{O}) \\ t \in \Gamma(\mathfrak{W},g,\mathcal{O}) \end{array}\right\} \quad \text{imply} \quad s + t \in \Gamma(\mathfrak{W},g,\mathfrak{M})$$

$$s \in \Gamma(\mathfrak{W},g,\mathfrak{M}^*) \quad \text{implies} \quad \frac{1}{s} \in \Gamma(\mathfrak{W},\tfrac{1}{g},\mathfrak{M}^*)$$

$$s \in \Gamma(\mathfrak{W},g,\mathcal{O}^*) \quad \text{implies} \quad \frac{1}{s} \in \Gamma(\mathfrak{W},\tfrac{1}{g},\mathcal{O}^*).$$

Especially, $\bigoplus_{p=0}^{\infty}\Gamma(\mathfrak{W},g^p)$ carries the structure of a graded ring.

Now, an isomorphism $\lambda\colon \mathfrak{K}(V) \to \Gamma(\mathfrak{W},1,\mathfrak{M})$ shall be defined. For $f \in \mathfrak{K}(V)$, set $\lambda(f)_i = f|V_i$. Obviously $f|V_i = f|V_j$ on V_{ij} if $(i,j) \in I[1]$. Hence, $\lambda(f) \in \Gamma(\mathfrak{W},1,\mathfrak{M})$.

Obviously, λ is an isomorphism with

$$\begin{array}{l} \lambda\colon \mathfrak{H}(V) \to \Gamma(\mathfrak{W},1) \\ \lambda\colon \mathfrak{T}(V) \to \Gamma(\mathfrak{W},1,\mathfrak{T}) \\ \lambda\colon \mathfrak{H}^*(V) \to \Gamma(\mathfrak{W},1,\mathcal{O}^*) \\ \lambda\colon \mathfrak{K}^*(V) \to \Gamma(\mathfrak{W},1,\mathfrak{M}^*). \end{array}$$

Hence, λ can be used to identify these sets and their algebraic structure. Especially, if $s \in \Gamma(\mathfrak{W},g,\mathfrak{M})$ and $t \in \Gamma(\mathfrak{W},g,\mathfrak{M}^*)$ then $s/t \in \Gamma(\mathfrak{W},1,\mathfrak{M}) = \mathfrak{K}(V)$ is a meromorphic function.

Lemma 14.1. Let X be a complex space. Let $\mathfrak{V}$ be a Stein-covering of X. Let $\mathfrak{W}$ be a shrinkage of $\mathfrak{V}$, which is also an open covering of X. Let $f_\lambda \in \mathfrak{K}(X)$ for $\lambda = 1,\ldots,p$ meromorphic functions on X. Then a meromorphic line bundle $g \in Z^1(\mathfrak{W},\mathfrak{M}^*)$ on $\mathfrak{W}$ and holomorphic 0-cochains $s_\lambda \in C^0(\mathfrak{V},\mathfrak{O})$ exist for $\lambda = 0,\ldots,p$ such that $s_\lambda = f_\lambda \cdot s_0$ in $C^0(\mathfrak{V},\mathfrak{O})$ for $\lambda = 1,\ldots,p$ and such that $s_\lambda | \mathfrak{W} \in \Gamma(\mathfrak{V},g)$ are holomorphic sections of g over $\mathfrak{W}$ for $\lambda = 0,1,\ldots,p$ with $s_0 \in \Gamma(\mathfrak{W},g,\mathfrak{T})$.

Proof. By Lemma 3.2, a holomorphic function $t_{\lambda i} \in \mathfrak{H}(V_i)$ exists on V_i for each $i \in I[0,\mathfrak{V}]$ and $\lambda = 1,\ldots,p$ such that $t_{\lambda i} f_\lambda | V_i$ is holomorphic on V_i and such that $t_{\lambda i} | W_i \in \mathfrak{T}(W_i)$ is not a zero divisor in $\mathfrak{H}(W_i)$. Then $s_{0i} = t_{1i},\ldots,t_{pi} \in \mathfrak{H}(V_i)$ and $s_{\lambda i} = s_{0i} f_\lambda | V_i \in \mathfrak{H}(V_i)$ are holomorphic on V_i for $\lambda = 1,\ldots,k$ and each $i \in I[0,\mathfrak{V}]$. Also, $s_{0i} \in \mathfrak{T}(W_i)$ if $i \in I[0,\mathfrak{W}]$. Hence, $s_\lambda = g_\lambda \cdot s_0$ on $\mathfrak{V}$ by definition.

If $(i,j) \in I[0,\mathfrak{W}]$, then $s_{0i} | W_{ij}$ and $s_{0j} | W_{ij}$ are non-zero divisors. Hence $g_{ij} = {}^{s_{0i}|W_{ij}}/{}_{s_{0j}|W_{ij}} \in \mathfrak{K}^*(W_{ij})$ is meromorphic on W_{ij} and not a zero divisor. Obviously $g_{ij} \cdot g_{jk} = g_{ik}$ on W_{ijk} if $(i,j,k) \in I(1,\mathfrak{W})$. Hence, $g = \{g_{ij}\}_{(ij)\in I(2,\mathfrak{W})} \in Z^1(\mathfrak{W},\mathfrak{M}^*)$ is a meromorphic line bundle on $\mathfrak{W}$ with $s_0 | \mathfrak{W} \in \Gamma(\mathfrak{W},g,\mathfrak{T})$. Then $s_\lambda | \mathfrak{W} = f_\lambda \cdot s_0 | \mathfrak{W}$ is a meromorphic section of g over $\mathfrak{W}$; q.e.d.

Let s be a meromorphic section of the meromorphic line bundle g on the open covering $\mathfrak{V}$ of the complex space X. Let N be the set of all $x \in X$ such that an index $i \in I[0]$ exists with $x \in V_i$

and $s_{ix} = 0$ where s_{ix} is the germ of s at x. If $x \in V_j$ then $g_{ijx} \in \mathfrak{M}_x^*$ is not a zero divisor in $\mathfrak{M}_x^*$, hence, $s_{jx} = 0$ if and only if $s_{ix} = g_{ijx}s_{jx} = 0$. Therefore,

$$N = \{x \in X | s_{ix} = 0 \text{ for all } i \in I \text{ with } x \in V_i\}.$$

Obviously, N is open in X and $s|G \equiv 0$ for an open subset $G \neq \emptyset$ of X if and only if $G \subseteq N$. Define $X - N = \text{supp } s$ as the <u>support</u> of s.

<u>Lemma 14.2.</u> Let s be a holomorphic section of the meromorphic line bundle g on the open covering $\mathfrak{V}$ of the complex space X. Let $\mathfrak{B}$ be the set of branches of X. Define $N = X\text{-supp } s$. Then

a) $\mathfrak{B}_0 = \{B \in \mathfrak{B} | B \cap N \neq \emptyset\} = \{B \in \mathfrak{B} | B \subseteq \overline{N}\}$.

b) $\mathfrak{B}_1 = \{B \in \mathfrak{B} | B \subseteq \text{supp } s\} = \mathfrak{B} - \mathfrak{B}_0$.

c) $\overline{N} = \bigcup_{B \in \mathfrak{B}_0} B$ is analytic in X.

d) $\text{supp } s = \bigcup_{B \in \mathfrak{B}_1} B$ is analytic in X.

e) supp S = X if and only if $s \in \Gamma(\mathfrak{V}, g, \mathfrak{T})$.

f) supp S = $\emptyset$ if and only if s = 0.

g) If X is irreducible, then s = 0 if and only if $s_{ix} = 0$ for some $(x,i) \in X \times I$ with $x \in V_i$.

<u>Proof.</u> a) Define $\mathfrak{B}_0 = \{B \in \mathfrak{B} | B \cap N \neq \emptyset\}$. Take $B \in \mathfrak{B}_0$. Let B' be the set of all points of B which are simple points of X. Then B' is open in B and X. Hence, $B' \cap \overline{N}$ is the closure of $B' \cap N$ $= \emptyset$ in B'. Take $x \in B' \cap \overline{N}$. Then $s_{ix} \in \mathcal{O}_x - \mathfrak{T}_x = \{0_x\}$. Hence, $s_{ix} = 0$. Therefore, $x \in N \cap B'$. Therefore $\overline{N} \cap B' = N \cap B' \neq \emptyset$ is

open and closed in the connected manifold B'. Therefore, $N \supseteq B'$, which implies $\overline{N} \supseteq B$.

If $B \in \mathcal{B}$, and if $B \subseteq \overline{N}$, then $B' \subseteq \overline{N}$ and $B' = B' \cap \overline{N} \neq \emptyset$ as the closure of $B' \cap N$. Hence, $B' \cap N \neq \emptyset$, which implies $B \in \mathcal{B}_0$, which proves a). Now, b) is a trivial consequence of a). Now, $\mathcal{B}_0 = \{B \in \mathcal{B} \mid B \cap N \neq \emptyset\}$ implies $N \subseteq \bigcup_{B \in \mathcal{B}_0} B$ because $\mathcal{B}$ is a covering of X. Hence, $\overline{N} \subseteq \bigcup_{B \in \mathcal{B}_0} B$, because any union of branches is analytic. Also, $\mathcal{B}_0 = \{B \in \mathcal{B} \mid B \subseteq \overline{N}\}$ implies $\bigcup_{B \in \mathcal{B}_0} B \subseteq \overline{N}$ which proves $\overline{N} = \bigcup_{B \in \mathcal{B}_0} B$.

Now b) implies $\bigcup_{B \in \mathcal{B}_1} B \subseteq \operatorname{supp} s$. Take $x \in \operatorname{supp} s$. Then $i \in I$ with $x \in V_i$ exists such that $s_{ix} \neq 0$. Hence, $B \in \mathcal{B}$ exists, such that $(s_i|B)_x \neq 0$. Hence, a simple point z of X with $z \in B \cap V_i$ exists such that $s_{iz} \neq 0$. Then an open neighborhood U of z exists such that $s_{iy} \neq 0$ for all $y \in U$ because z is a simple point of X. Hence, $N \subseteq X - U$ and $\overline{N} \subseteq X - U$. Therefore, $z \in B - \overline{N}$. By a) $B \in \mathcal{B} - \mathcal{B}_0 = \mathcal{B}_1$. Hence, $x \in \bigcup_{B \in \mathcal{B}_1} B$. Therefore, $\bigcup_{B \in \mathcal{B}_1} B = \operatorname{supp} s$.

Suppose that supp s = X. Then $s_{ix} \neq 0$ for all $x \in V_i$. Hence, $s \in \mathfrak{T}(V_i)$ if $i \in I[0]$, which implies $s \in \Gamma(\mathfrak{V}, g, \mathfrak{T})$. If $s \in \Gamma(\mathfrak{V}, g, \mathfrak{T})$, then $s_{ix} \neq 0$ for all $x \in V_i$ if $i \in I[0]$. Hence, $N = \emptyset$ by the definition of N. Hence, supp $s = X - N = X$. Therefore, e) is proved.

If $s = 0$, then $N = X$ and supp $s = X - N = \emptyset$. If supp $s = \emptyset$, then $N = X$ and $s = 0$.

g) Let X be irreducible. Assume $x \in V_i$ with $i \in I$ and $s_{ix} = 0$ exist. Then $x \in \ B \in \mathcal{B}_0$. But $\mathcal{B}$ consist only of one element.

Hence, $\mathcal{L}_0 = \mathcal{L}$ and $\mathcal{L}_1 = \emptyset$. Therefore, supp s = $\emptyset$ which implies s = 0; q.e.d.

§15. Preparations

Some lemmata will be proved in this paragraph which shall be needed later.

Lemma 15.1. Let X be a complex space of pure dimension m. Let S_X be the set of non-simple points of X. Let Y be a complex manifold of pure dimension $n \leqq m$. Let $\varphi: X \to Y$ be a holomorphic map. Let N be the set of all $x \in X - S_X$ where the Jacobian of φ has a rank smaller than n. Then $\overline{N}$ is an analytic subset of X.

Proof. If X is normal, then this is true by Remmert [12] Satz 16. Suppose that X is not normal. Let $\rho: \hat{X} \to X$ be the normalization. Define $\psi = \varphi \circ \rho: \hat{X} \to Y$. Let M be the set of all simple points x of $\hat{X}$ such that the Jacobian of ψ at x has a rank smaller than n. Then $\overline{M}$ is analytic. Define $X_0 = X - S_X$ and $\hat{X}_0 = S^{-1}(X_0)$. Then $\rho_0 = \rho: \hat{X}_0 \to X_0$ is biholomorphic and $\hat{X}_0 \cap M = \rho_0^{-1}(N)$. Because ρ_0 is proper, $M' = \rho_0(M)$ is analytic in X with $M' \cap X_0 = N$. Then $\overline{N}$ is the union of all branches B of M' with $B \cap X_0 \neq \emptyset$. Hence, $\overline{N}$ is analytic in X; q.e.d.

Lemma 15.2. Let X be a complex space of pure dimension m. Let S_X be the set of simple points of X. Let $f_1, \ldots, f_k$ be meromorphic functions on X with $k \leqq m$. Let $P = P_{f_1} \cup \ldots \cup P_{f_k}$ be the union of their pole sets. Define $X_0 = X - (S_X \cup P)$ and $\omega = df_1 \wedge \ldots \wedge df_k$ on X_0. Define

$$N = \{x \in X_0 | \omega(x) = 0\}.$$

Then $\overline{N}$ is an analytic subset of X.

Proof. Define $X_1 = X - P$. Define

$$\Gamma_1 = \{(x, f_1(x), \ldots, f_k(x)) \mid x \in X_1\}.$$

The closure $\Gamma = \overline{\Gamma}_1$ of Γ_1 in $X \times \mathbb{P}^k$ is an analytic subset of X. Let $\pi: \Gamma \to X$ and $\varphi: \Gamma \to \mathbb{P}^k$ be the projections. Then $\pi_1 = \pi: \Gamma_1 \to X_1$ is biholomorphic. Let Γ_S be the set of non-simple points of Γ. Define $\Gamma_0 = \Gamma_1 - \Gamma_S$. Then $X_0 = X_1 - X_X$ and $\pi_0 = \pi_1: \Gamma_0 \to X_0$ is biholomorphic. Let M be the set of all points $(x,y) \in \Gamma - \Gamma_S$ such that the rank of φ at (x,y) is smaller than k. By Lemma 15.1, $\overline{M}$ is analytic in Γ. Also $M' = \pi(\overline{M})$ is analytic in X. Because M is analytic in $\Gamma - \Gamma_S$ and because Γ_0 is open in $\Gamma - \Gamma_S$, this implies $\Gamma_0 \cap M = \Gamma_0 \cap \overline{M}$. If $x \in X_0$, then

$$\varphi(\pi_0^{-1}(x)) = (f_1(x), \ldots, f_k(x)).$$

Hence, $N = \pi_0(M \cap \Gamma_0) = M' \cap X_0$. Then $\overline{N}$ is the union of all branches B of M' with $B \cap X_0 \neq \emptyset$. Hence, $\overline{N}$ is analytic in X;

q.e.d.

Lemma 15.3. Let U be a poly-disc with center 0 in $\mathbb{C}^q$. Let $q \geq k$, and $U = Z_1 \times Z_2$ with $Z_1 \subseteq \mathbb{C}^k$ and $Z_2 \subseteq \mathbb{C}^{q-k}$. Let Y be an open, connected subset in $\mathbb{C}^n$ and define $X = U \times Y \subseteq \mathbb{C}^m$ with $m = q + n$. Let $\varphi: X \to Y$ and $F = (f_1, \ldots, f_k): X \to Z_1$ be the projections. Define $S = \{0\} \times Y$ with $0 \in U$. Let $\mathfrak{R}$ be the $\varphi^* \mathfrak{H}(Y)$-module of all holomorphic functuon $f \in \mathfrak{H}(X)$ such that $f_1, \ldots, f_k, f$

are φ-dependent; i.e.,

$$\mathfrak{R} = \mathfrak{G}(X) \cap \mathfrak{S}_\varphi(X, f_1, \ldots, f_k).$$

If h is a non-negative integer, define

$$\mathfrak{R}^h = \{f \in \mathfrak{R} \mid \sigma_f(x) \geqq h \text{ for all } x \in S\}.$$

Then

1. The functions $f_1, \ldots, f_k$ are φ-independent.

2. If f is a holomorphic function on X, then $f \in \mathfrak{R}$ if and only if there exists a series, called a Hartogs series,

$$\text{(15.1)} \qquad \mathfrak{v} = \sum_{\mu_1, \ldots, \mu_k = 0}^{\infty} a_{\mu, \ldots, \mu_k} z_1^{\mu_1} \ldots z_k^{\mu_k}$$

with $a_{\mu_1, \ldots, \mu_k} \in \mathfrak{G}(Y)$ which converges uniformly on every compact subset of $Z_1 \times Y$ such that

$$\text{(15.2)} \qquad f(x) = \sum_{\mu_1, \ldots, \mu_k = 0}^{\infty} a_{\mu_1, \ldots, \mu_k}(\varphi(x)) f_1(x)^{\mu_1} \ldots f_k(x)^{\mu_k}$$

if $x \in X$. This Hartogs series $\mathfrak{v} = \mathfrak{w}(f)$ is uniquely defined.

3. If $f \in \mathfrak{R}$, then $f \in \mathfrak{R}^h$ if and only if $a_{\mu_1, \ldots, \mu_k} \equiv 0$ for all coefficients in $\mathfrak{v} = \mathfrak{w}(f)$ with $\mu_1 + \ldots + \mu_k \leqq h - 1$.

4. The set $\mathfrak{R}^h$ is a $\varphi^* \mathfrak{G}(Y)$-submodule of $\mathfrak{R}$. The $\varphi^* \mathfrak{G}(Y)$-quotient module $\Delta^h = \mathfrak{R}/\mathfrak{R}^{h+1}$ is a free $\varphi^* \mathfrak{G}(Y)$-module of rank $\binom{k+h}{h}$.

Proof. 1. Define $\varphi = (g_1,\ldots,g_n)\colon X \to Y$. Then $df_1 \wedge \cdots \wedge df_k \wedge dg_1 \wedge \cdots \wedge dg_n \not\equiv 0$ on X, because f_μ and g_ν are different projection functions. Hence, $f_1,\ldots,f_k$ are φ-independent by Lemma 2.4.

2. Let $\psi\colon X \to Z_1 \times Y$ be the projection. The algebra $\mathfrak{H}(Z_1 \times Y)$ of all holomorphic functions on $Z_1 \times Y$ is identified with the algebra of all Hartogs series (15.1) which converge uniformly on every compact subset of $Z_1 \times Y$. Then $\psi^*\colon \mathfrak{H}(Z_1 \times Y) \to \mathfrak{H}(X)$ is an injective ring homomorphism such that $\psi^*(\mathfrak{u}) = f$ if is given by (15.1) and f by (15.2). Define $\omega = df_1 \wedge \cdots \wedge df_k \wedge dg_1 \wedge \cdots \wedge dg_n$. The uniform convergence of (15.1) implies

$$df \wedge \omega = \sum_{\mu_1,\ldots,\mu_k=0}^{\infty} d((a_{\mu_1,\ldots,\mu_k})f_1^{\mu_1} \cdots f_k^{\mu_k}) \wedge \omega \equiv 0$$

on X. Hence, $f_1,\ldots,f_k,f$ are φ-dependent. Therefore, $\psi^*(\mathfrak{H}(Z_1 \times Y)) \subseteq \mathfrak{R}$. If $f \in \mathfrak{R}$, then $df \wedge \omega \equiv 0$ on X by Lemma 4.5 which implies $f_{z_\lambda}(z_1,\ldots,z_m) = 0$ for $\lambda = k+1,\ldots,q$. Hence, $\tilde{f} \in \mathfrak{H}_1(Z_1 \times Y)$ exists such that $f(z_1,\ldots,z_m) = \tilde{f}(z_1,\ldots,z_k,z_{q+1},\ldots,z_m)$ or $f = \psi^*(\tilde{f})$. Therefore, $\psi^*\colon \mathfrak{H}(Z_1 \times Y) \to \mathfrak{R}$ is a ring isomorphism. Now 2) is proved with $\mathfrak{M} = (\psi^*)^{-1}$.

3) Take $f \in \mathfrak{R}$. Then

$$f(z_1,\ldots,z_m) = \sum_{\mu_1,\ldots,\mu_k=0}^{\infty} a_{\mu_1,\ldots,\mu_k}(z_{q+1},\ldots,z_u) z_1^{\mu_1} \cdots z_k^{\mu_k}$$

if $(z_1,\ldots,z_m) \in X$. If $f \in \mathfrak{R}^h$, then all partial derivatives of order smaller than h are identically zero on S. Especially

$a_{\mu_1,\dots,\mu_k} \equiv 0$ on Y if $\mu_1 + \dots + \mu_k < h$.

If $x = (0,y)$ with $y \in Y$, and if $a_{\mu_1,\dots,\mu_k} = 0$ for $\mu_1 + \dots + \mu_k < h$, then a poly-disc Y_1 with center zero exists such that

$$f(x+z) = \sum_{\mu_1+\dots+\mu_m=0}^{\infty} b_{\mu_1,\dots,\mu_m} z_1^{\mu_1} \dots z_m^{\mu_m}$$

for $z \in Y$. Here $b_{\mu_1,\dots,\mu_m} = 0$ if $\mu_1 + \dots + \mu_k < h$. Hence, $b_{\mu_1,\dots,\mu_m} = 0$ if $\mu_1 + \dots + \mu_m < h$. Therefore, $\sigma_f(x) \geqq h$. This proves 3.

4. Obviously, $\mathfrak{R}$ and $\mathfrak{R}^h$ are $\varphi^*\mathfrak{G}(Y)$-modules, which can be considered as $\mathfrak{G}(Y)$-module by a change of language (if $g \in \mathfrak{G}(Y)$ and $f \in \mathfrak{R}$ then $g \cdot f = (\varphi^* g) f$). Now,

$$T^h = \{ \sum_{\mu_1+\dots+\mu_k \leqq h} a_{\mu_1,\dots,\mu_k} z_1^{\mu_1} \dots z_k^{\mu_k} \mid a_{\mu,\dots,\mu_k} \in \mathfrak{G}(Y) \}$$

is a free $\mathfrak{G}(Y)$-module of rank $\binom{k+h}{h}$. Define

$$\Sigma^{h+1} = \{ \mathfrak{a} \in \mathfrak{G}(Z_1 \times Y) \mid a_{\mu_1,\dots,\mu_k} \equiv 0 \text{ if } \mu_1 + \dots + \mu_p \leqq h \}.$$

Then Σ^{h+1} is an $\mathfrak{G}(Y)$-module. A surjective $\mathfrak{G}(Y)$-module homomorphism $\tau \colon \mathfrak{G}(Z_1 \times Y) \to T^h$ is defined by

$$\tau\Big(\sum_{\mu_1,\dots,\mu_k=0}^{\infty} a_{\mu_1,\dots,\mu_k} z_1^{\mu_1} \dots z_k^{\mu_k} \Big) = \sum_{\mu_1+\dots+\mu_k \leqq h} a_{\mu_1,\dots,\mu_k} z_1^{\mu_1} \dots z_k^{\mu_k}$$

Then

$$\begin{array}{ccccccccc}
0 & \longrightarrow & \mathfrak{R}^{h+1} & \longrightarrow & \mathfrak{R} & \xrightarrow{\kappa} & \Delta^h & \longrightarrow & 0 \\
 & & \uparrow{\scriptstyle \psi^*} & & \uparrow{\scriptstyle \psi^*} & & & & \\
0 & \longrightarrow & \Sigma^{h+1} & \longrightarrow & \mathfrak{g}(Z_1 \times Y) & \xrightarrow[\tau]{} & T^h & \longrightarrow & 0
\end{array}$$

is defined, where the lines are exact with κ as the residual map and where the columns are isomorphisms. Hence, an $\mathfrak{g}(Y)$-module isomorphism $\psi^*\colon T^h \to \Delta^h$ is defined such that $\psi^* \circ \tau = \kappa \circ \psi^*$. Then Δ^h is a free $\mathfrak{g}(Y)$-module of rank $\binom{k+h}{h}$ which proves 4; q.e.d.

<u>Lemma 15.4.</u> Let $k \leqq q \leqq m$. Let X be a poly-disc with center 0 in $\mathbb{C}^m = \mathbb{C}^k \times \mathbb{C}^{q-k} \times \mathbb{C}^n$ where $n = m - q$. Then $X = Z_1 \times Z_2 \times Y$ where Z_1, Z_2, Y and $U = Z_1 \times Z_2$ are poly-discs with center 0 in $\mathbb{C}^k$, $\mathbb{C}^{q-k}$, $\mathbb{C}^n$ and $\mathbb{C}^q$ respectively. Let $\varphi\colon X \to Y$ and $F = (f_1,\ldots,f_k)\colon X \to Z_1$ be the projections. Define $\mathfrak{R} = \mathfrak{g}(X) \cap \mathfrak{K}_\varphi(X,f_1,\ldots,f_k)$. Take $f \in \mathfrak{K}_\varphi(X,f_1,\ldots,f_k)$. Then $g \in \mathfrak{R}$ and $h \in \mathfrak{R}$ with $h \not\equiv 0$ exist such that $f = {}^g/_h$ on X and such that the germs g_x and h_x of g respectively h are coprime in $\mathcal{O}_x$ for each $x \in X$.

<u>Proof.</u> Let $\psi\colon X \to Z_1 \times Y_1$ be the projection. Then $\varphi = (g_1,\ldots,g_n)\colon X \to Y$ and $\psi = (f_1,\ldots,f_k,g_1,\ldots,g_n)$. Define $\omega = df_1 \wedge \cdots \wedge df_k \wedge dg_1 \wedge \cdots \wedge dg_n$. Let P_f be the pole set of f. Then $df \wedge \omega \equiv 0$ on $X - P_f$ because $f \in \mathfrak{K}_\varphi(X,f_1,\ldots,f_k)$. But $df \wedge \omega \equiv 0$ on $X - P_f$ implies $f \in \mathfrak{K}_\psi(X)$.

If $K \subseteq Z_1 \times Y$ is compact, then $K' = Z_1 \times \{0\} \times Y$ is compact in X. The fiber $\psi^{-1}(z,y) = \{z\} \times Z_2 \times \{y\}$ is irreducible and intersects K' if $(z,y) \in K$. Hence, ψ is quasi-proper with branch number $r_\psi = 1$. Hence, $f \in \psi^* \mathfrak{K}(Z_1 \times Y)$, by Theorem 9.2. Hence, $f' \in \mathfrak{K}(Z_1 \times Y)$ exists such that $\psi^*(f') = f$. Because $Z_1 \times Y$ is a poly-disc, holomorphic functions $g' \in \mathfrak{H}(Z_1 \times Y)$ and $h' \in \mathfrak{H}(Z_1 \times Y)$ exist such that $h' \not\equiv 0$ and $g' = fh'$ and that the germs of g' and h' are coprime at every point of $Z_1 \times Y$. Then $\psi^*(g') = g \in \mathfrak{R}$ and $h = \psi^*(h') \in \mathfrak{R}$ by Lemma 15.3 and $h \cdot f = g$ with $h \not\equiv 0$. The sets $N_g = \{x \in X | g(x) = 0\}$ and $N_h = \{x \in X | h(x) = 0\}$ are analytic with $\psi^{-1}\psi(N_g) = N_g$ and $\psi^{-1}\psi(N_h) = N_h$ where $\psi(N_h)$ and $\psi(N_g)$ do not have a common branch. Hence, N_g and N_h do not have a common branch. Therefore, the germs g_x and h_x of g and h at $x \in X$ are coprime in $\mathcal{O}_x$ if $\mathcal{O}$ is the sheaf of holomorphic functions on X; q.e.d.

<u>Lemma 15.5.</u> Let X and Y be complex spaces of pure dimension m and n respectively with $m - n = q \geqq 0$. Let $\varphi: X \to Y$ be a holomorphic map. Let (α, β, π) be a product representation of φ at $y_0 \in Y$ with $\beta(y_0) = 0$. Suppose that $\beta: U_\beta \to U'_\beta$ is biholomorphic. Define

$$\eta = \alpha: \varphi^{-1}(y_0) \cap U_\alpha \to U''_\alpha \times \{0\} \qquad (0 \in \mathbb{C}^n).$$

Assume that u is a simple point of X and of $\varphi^{-1}(y_0) \cap U_\alpha$ and that φ and η are regular at u. Then α is regular at u.

<u>Proof.</u> A product representation (γ, δ, ψ) of φ at u exists such that $\gamma: U_\gamma \to U'_\gamma = U''_\gamma \times U'_\delta$ and $\delta: U_\delta \to U'_\delta$ are biholomorphic and such that $U_\gamma \subseteq U_\alpha$ and $U_\delta \subseteq U_\beta$ with $\beta | U_\delta = \delta$. Let $\rho: U''_\gamma \times U'_\delta \to U''_\gamma$ be the projection. Then

$$\zeta = \rho\circ\gamma\colon U_\gamma \cap \varphi^{-1}(y_0) \to U''_\gamma$$

is biholomorphic. Then $\zeta(u) = (u_1,\ldots,u_q) \in U''_\gamma$ and $\eta\circ\zeta^{-1}\colon U''_\gamma \to U''_\alpha \times \{0\}$ is regular at $\zeta(u)$. Define $\eta\circ\zeta^{-1} = (f_1,\ldots,f_q,0 \ldots 0)$. Define

$$\frac{\partial(f_1,\ldots,f_q)}{\partial(z_1,\ldots,z_q)} = \Delta(z_1,\ldots,z_q).$$

Then Δ is the Jacobian of $\eta\circ\zeta^{-1}$ and $\Delta(\zeta(u)) \neq 0$. Define

$$\alpha\circ\gamma^{-1} = (h_1,\ldots,h_m).$$

Then

$$(h_{q+1},\ldots,h_m) = \pi\circ\alpha\circ\gamma^{-1} = \beta\circ\varphi\circ\gamma^{-1}$$

$$= \beta\circ\delta^{-1}\circ\psi = \delta\circ\delta^{-1}\circ\psi = \psi.$$

Hence, $h_\mu(z_1,\ldots,z_m) = z_\mu$ for $\mu = q+1,\ldots,m$. Hence, the Jacobian of $\alpha\circ\gamma^{-1}$ is

$$D(z_1,\ldots,z_m) = \frac{\partial(h_1,\ldots,h_m)}{\partial(z_1,\ldots,z_m)} = \frac{\partial(h_1,\ldots,h_q)}{\partial(z_1,\ldots,z_q)}\ .$$

Now,

$$\alpha\circ\gamma^{-1}(z_1,\ldots,z_q,0 \ldots 0) = \alpha\circ\zeta^{-1}(z_1,\ldots,z_q) = \eta\circ\zeta^{-1}(z_1,\ldots,z_q)$$

Hence,

$$h_\mu(z_1,\ldots,z_q,0\ \ldots\ 0) = f_\mu(z_1,\ldots,z_q)$$

for $\mu = 1,\ldots,q$. Therefore, $D(z_1,\ldots,z_q,0\ \ldots\ 0) = \Delta(z_1,\ldots,z_q)$. Now, $\gamma(u) = (\zeta(u),0) = (u_1,\ldots,u_q,0,\ldots,0)$. Hence, $D(\gamma(u)) = \Delta(\zeta(u)) \neq 0$. Therefore, $\alpha\circ\gamma^{-1}$ is regular at $\gamma(u)$. Hence α is regular at u; q.e.d.

§16. Pseudoconcave maps

Let X be a complex space. Let $V \neq \emptyset$ be a subset of the open set U in X. Define the holomorphic convex hull of V in U by

$$\mathfrak{L}(V\|U) = \{x \in U \mid |f(x)| \leq \|f\|_V \text{ for all } f \in \mathfrak{H}(U)\}.$$

Let X and Y be complex spaces of pure dimension m and n respectively with $m - n = q > 0$. Let $\varphi: X \to Y$ be a holomorphic map. Let G be an open subset of Y with $\tilde{G} = \varphi^{-1}(G) \neq \emptyset$. Then φ is said to be pseudoconcave over G if and only if an open subset $\Omega \neq \emptyset$ of $\tilde{G}$ exists such that

1. The restriction $\varphi: \overline{\Omega} \cap \tilde{G} \to G$ is proper.

2. If $y \in G \cap \varphi(X)$ and if B is a branch of $\varphi^{-1}(y)$ then $B \cap \Omega \neq \emptyset$.

3. If $x \in (\overline{\Omega}-\Omega) \cap \tilde{G}$ and if U is an open neighborhood of x, then open neighborhoods A and B with $\overline{A}$ compact exist such that

$$x \in B \subseteq \overline{B} \subset A \subseteq \overline{A} \subset U \cap \tilde{G}$$

$$\varphi(A \cap \Omega) = \varphi(A)$$

$$\mathfrak{L}(A \cap \Omega \cap \varphi^{-1}(y) \| A \cap \varphi^{-1}(y)) \supseteq B \cap \varphi^{-1}(y)$$

with $A \cap \varphi^{-1}(y) \cap \Omega \neq \emptyset$ if $y \in \varphi(A)$.

Obviously, if $G_0 \neq \emptyset$ is an open subset of G then φ is also pseudoconcave over G_0. Condition 1 and 2 imply that the restriction $\varphi_G = \varphi: \tilde{G} \to G$ is quasi-proper.

Now, the following Theorems shall be proven:

Theorem 16.2. Let X and Y be irreducible complex spaces of dimension m and n respectively with $m - n = q > 0$. Let $\varphi: X \to Y$

be a full holomorphic map of rank n which is pseudoconcave over an open subset $G \neq \emptyset$ of Y. Let $f_1,\dots,f_d$ be φ-dependent, meromorphic functions on X. Then $f_1,\dots,f_d$ are algebraically dependent over $\varphi^*\mathfrak{K}(Y)$.

<u>Theorem 16.6.</u> Let X and Y be irreducible complex spaces of dimension m and n respectively with $m - n = q > 0$. Assume that X is normal. Let $\varphi: X \to Y$ be a full holomorphic map of rank n which is pseudoconcave over an open subset $G \neq \emptyset$ of Y. Then the field $\mathfrak{K}(X)$ of meromorphic functions on X is an algebraic function field over the field $\varphi^*\mathfrak{K}(Y)$ of meromorphic functions lifted from Y to X by φ.

These theorems are the third main results of this investigation. Their proofs are complicated and lengthy, but do not require any new concepts or means. They follow the ideas of [2] where Y is a point and Siegel [15] where X is compact and Y is a point. These proofs will be subdivided into a number of steps Σ_ν $(\nu = 1,\dots,27)$. For these steps the following assumptions $A_1,\dots,A_5$ are universally made.

(A_1) An irreducible complex space X of dimension m is given.

(A_2) An irreducible complex space Y of dimension n with $m - n = q > 0$ is given.

(A_3) A holomorphic map $\varphi: X \to Y$ of rank n is given.

(A_4) The map φ is pseudoconcave over an open subset G of Y with $\tilde{G} = \varphi^{-1}(G) \neq \emptyset$.

(A_5) Meromorphic functions $f_1,\dots,f_d$ are given on X such that $f_1,\dots,f_k$ are φ independent. Here $1 \leqq k \leqq d$ and $k \leqq q$.

(A_6) If $M \subseteq Y$, define $\varphi^{-1}(M) = \tilde{M}$. If $\tilde{M} \neq \emptyset$, define $\varphi_M = \varphi: \tilde{M} \to M$ as the restriction.

Observe, that $\varphi_G: \tilde{G} \to G$ is quasi-proper as remarked before. Now, the steps for the proofs of Theorem 16.2 and Theorem 16.6 begin:

(Σ_1) The set $D_0 = \{x \in X \mid \text{rank}_x \varphi \leqq n-1\}$ is analytic and thin in X and $D_0' = \varphi(D_0)$ is almost thin in Y (Proposition 1.24). Because $\varphi_G = \varphi: \tilde{G} \to G$ is quasi-proper, Lemma 1.3 and Proposition 8.4 imply that $D_0'' = D_0' \cap G = \varphi_G(D_0 \cap \tilde{G})$ is analytic in G with $\dim D_0'' \leqq n - 2$. The map $\varphi: X - D_0 \to Y$ is q-fibering. Define $G_0 = G - D_0''$. Then $\tilde{G}_0 = \varphi^{-1}(G_0) \subseteq X - D_0$.

(Σ_2) Let S_Y be the set of non-simple points of Y. Then S_Y and $\tilde{S}_Y = \varphi^{-1}(S_Y)$ are thin analytic subsets of Y and X respectively, because φ has rank n.

(Σ_3) Let S_X be the set of non-simple points of X. Then S_X is a thin analytic subset of X. By Lemma 1.26, the set D_1' of all $y \in \varphi(X)$ such that a branch of $\varphi^{-1}(y)$ is contained in S_X is an almost thin subset of Y. Define $D_1'' = D_1' \cap G_0$.

<u>Lemma:</u> D_1'' is a thin analytic subset of G_0.

<u>Proof.</u> The map $\varphi_{G_0} = \varphi_G: \tilde{G}_0 \to G_0$ is q-fibering, quasi-proper and holomorphic and $S_X \cap \tilde{G}_0$ is the set of non-simple points of $\tilde{G}_0$. Then D_1'' is the set of all $y \in \varphi_{G_0}(S_X \cap \tilde{G}_0)$ such that $S_X \cap \tilde{G}_0$ contains a branch of $\varphi^{-1}(y) = \varphi_{G_0}^{-1}(y)$. By Proposition 8.15, D_1'' is a thin analytic subset of G_0; q.e.d.

(Σ_4) Let $D_2 = P_{f_1} \cup \ldots \cup P_{f_1}$ be the union of the pole sets of $f_1,\ldots,f_d$. Then D_2 is a thin analytic subset of X. By Lemma 1.26, the set D_2' of all $y \in \varphi(X)$ such that a branch of $\varphi^{-1}(y)$ is contained in D_2 is an almost thin subset of Y. As in ($\Sigma 3$), it follows, that $D_2'' = D_2' \cap G_0$ is a thin analytic subset of G_0.

($\Sigma 5$) The map $\varphi_G\colon \tilde{G} \to G$ is quasi-proper. Hence, $\varphi(\tilde{G}) = \varphi_G(\tilde{G})$ is a pure n-dimensional analytic subset of G by Lemma 8.1. Also $D_0'' = G - G_0$ is thin analytic in G and $D_1'' \cup D_2''$ are thin analytic in G_0. Hence, a schlicht chart $\delta\colon U_\delta \to U_\delta'$ of Y exists such that

$$U_\delta \subseteq \varphi(G) \cap (G_0 - (D_1'' \cup D_2'' \cup S_Y)).$$

Then $\tilde{U}_\delta = \varphi^{-1}(U_\delta) \neq \emptyset$ is open in X and $\varphi_{U_\delta}\colon \tilde{U}_\delta \to U_\delta$ is a surjective, quasi-proper, q-fibering holomorphic map. Define $\delta = (g_1,\ldots,g_n)$. Then $dg_1 \wedge \ldots \wedge dg_n \neq 0$ at every point $y \in U_\delta$. On $\tilde{U}_\delta - (S_X \cup D_2)$ define the holomorphic differential form ω by

$$\omega = df_1 \wedge \ldots \wedge df_k \wedge d(g_1 \circ \varphi) \wedge \ldots \wedge d(g_n \circ \varphi).$$

By Lemma 4.7, $\omega \not\equiv 0$ on each branch of $\tilde{U}_\delta - (S_X \cup D_2)$, because $f_1,\ldots,f_k$ are φ-independent. Hence,

$$D_3^0 = \{x \in \tilde{U}_\delta - (S_X \cup D_2) \,|\, \omega(x) = 0\}$$

is a thin analytic subset of $\tilde{U}_\delta - (S_X \cup D_2)$. By Lemma 15.2, the closure $D_3 = \overline{D_3^0} \cap \tilde{U}_\delta$ is an analytic subset of U_δ. Also D_3 is thin because D_3^0 is thin. By Proposition 8.15, the set D_3'' of all $y \in U_\delta$

such that D_3 contains a branch of $\varphi^{-1}(y) = \varphi_{U_\delta}^{-1}(y)$ is a thin analytic subset of U_δ.

(Σ_6) The map φ is pseudoconcave over G. Hence, an open subset $\Omega \neq \emptyset$ of $\tilde{G}$ exists such that $\varphi: \overline{\Omega} \cap \tilde{G} \to G$ is proper, such that for every $y \in G \cap \varphi(X)$ and each branch B of $\varphi^{-1}(y)$ the intersection $B \cap \Omega$ is not empty and such that for every $x \in (\overline{\Omega}-\Omega) \cap \tilde{G}$ and every open neighborhood U of x, open neighborhoods A and B with $\overline{A}$ compact exist such that

$$x \in B \subseteq \overline{B} \subset A \subseteq \overline{A} \subset U \cap \tilde{G}$$
$$\varphi(A \cap \Omega) = \varphi(A)$$
$$\mathfrak{X}(A \cap \Omega \cap \varphi^{-1}(y) \| A \cap \varphi^{-1}(y)) \supseteq B \cap \varphi^{-1}(y)$$

if $y \in \varphi(A)$.

Hence, if f is holomorphic on A and $x \in B$, then $|f(x)| \leqq \|f\|_{A \cap \Omega}$. Also, if $y \in G$ and $A \cap \varphi^{-1}(y) \neq \emptyset$, if f is holomorphic on $A \cap \varphi^{-1}(y)$ and if $x \in B \cap \varphi^{-1}(y)$, then $|f(x)| \leqq \|f\|_{A \cap \Omega \cap \varphi^{-1}(y)}$.

(Σ_7) Take open, locally finite coverings $\mathfrak{W}^\lambda = \{V_i^\lambda\}_{i \in I}$ of X such that $\overline{\mathfrak{W}}^\lambda$ is compact for $\lambda = 0,1,2$, such that $\mathfrak{W}^\lambda$ is a shrinkage of $\mathfrak{W}^{\lambda+1}$ for $\lambda = 0,1$ and such that $\mathfrak{W}^2$ is Stein. Assume $V_i^0 \neq \emptyset$ for each $i \in I$. Then V_i^2 is a Stein space and $\emptyset \neq V_i^0 \subseteq \overline{V}_i^0 \subseteq V_i^1 \subseteq V_i^1 \subseteq V_i^2$ for each $i \in I$. Define $I^\lambda[p] = I[p, \mathfrak{W}^\lambda]$. Then $I = I^\lambda[0]$ for $\lambda = 0,1,2$ and $I^\lambda[p] \subseteq I^{\lambda+1}[p]$ for $\lambda = 0,1$. Define $V_i = V_i^0$ for for $i \in I$, define $\mathfrak{W} = \mathfrak{W}^0$ and $I[p] = I^0[p]$.

By Lemma 14.1, $g \in Z^1(\mathfrak{W}^1, \mathfrak{M}^*)$ and $s_\lambda \in C^0(\mathfrak{W}^2, \mathcal{O})$ exist for $\lambda = 0,1,\ldots,d$, such that $s_\lambda = f_\lambda s_0$ for $\lambda = 1,\ldots,d$ and such that $s_\lambda | \mathfrak{W}^1 \in \Gamma(\mathfrak{W}^1, g)$ for $\lambda = 0,1,\ldots,d$ and such that $s_0 | \mathfrak{W}^1 \in \Gamma(\mathfrak{W}^1, g, \mathfrak{T})$.

Hence, $s_{\lambda i} = f_\lambda s_{0i}$ on V_i^2 and $s_{\lambda i}/s_{0i} = f_\lambda$ on V_i^1 with $s_{0i}|V_i^1 \in \mathfrak{T}(V_i)$ for $i \in I$. Moreover, $s_{\lambda i} = g_{ij}s_{\lambda j}$ on V_{ij}^1 if $(i,j) \in I^1[1]$ for $\lambda = 0,1,\ldots,d$. Especially, ${}^{s_{0i}}/_{s_{0j}} = g_{ij}$ on V_{ij}^1 if $(i,j) \in I^1[1]$. Because $s_{0i}|V_i^1 \in \mathfrak{T}(V_i^1)$, the analytic set

$$E_i = \{x \in V_i^1 | s_{0i}(x) = 0\}$$

is thin in V_i^1 for each $i \in I$. Moreover, g_{ij} is holomorphic on $V_{ij}^1 - E_j$ and $g_{ij}(x) \neq 0$ if $x \in V_{ij}^1 - (E_i \cup E_j)$ when $(i,j) \in I^1[1]$. The set $E_i \cap \overline{V}_i$ is compact and thin in X with $E_i \cap \overline{V}_i \subseteq V_i^1$. Because $\mathfrak{W}^1$ is locally finite

$$D_4 = \bigcup_{i \in I} E_i \cap \overline{V}_i$$

is closed and thin.

For $i \in I$, define $\varphi_i = \varphi\colon V_i^1 \to Y$. By Lemma 1.26, the set E_i' of all $y \in \varphi(V_i^1)$ such that E_i contains a branch of $\varphi_i^{-1}(y) = \varphi^{-1}(y) \cap V_i^1$ is almost thin in Y. Let E_i'' be the set of all $y \in \varphi(V_i^1) \cap G_0$ such that E_i contains a branch B of $\varphi_i^{-1}(y) = \varphi^{-1}(y) \cap V_i^1$ with $B \cap \overline{V}_i \neq \emptyset$. Then $E_i'' \subseteq E_i'$ and E_i'' is almost thin in Y with $E_i'' \subseteq G_0$.

<u>Lemma.</u> E_i'' is closed in G_0.

<u>Proof.</u> Take $a \in \overline{E_i''} \cap G_0 - E_i''$. A sequence of points $\{y_\mu\}_{\mu \in \mathbb{N}}$ converges to a with $y_\mu \in E_i'' \cap G_0$. Then $y_\mu \neq a$ for all $\mu \in \mathbb{N}$. A

branch B_μ of $\varphi_i^{-1}(y_\mu) = \varphi^{-1}(y_\mu) \cap V_i^1$ exists with $B_\mu \subseteq E_i$ and $B_\mu \cap \overline{V}_i \neq \emptyset$. Take $x_\mu \in B_\mu \cap \overline{V}_i$. Because $\overline{V}_i$ is compact, a subsequence $\{x_{\mu_\nu}\}_{\nu\in\mathbb{N}}$ converges to $x \in \varphi^{-1}(a) \cap \overline{V}_i$. The set $C = \bigcup_{\mu=1}^{\infty} B_\mu$ is analytic in $V_i^1 - \varphi^{-1}(a)$ and singular at x. Because $y_\mu \in G_0$, $\varphi^{-1}(y_\mu) \cap V_i^1$ is pure q-dimensional, $\dim B_\mu = q$. Because $a \in G_0$, also $\varphi^{-1}(a) \cap V_i^1$ is pure q-dimensional. By Remmert and Stein [11] a branch B of $\varphi^{-1}(a) \cap V_i^1$ with $x \in B$ exists such that C is singular at every point of B. Hence,

$$B \subseteq \overline{C} \cap V_i^1 \subseteq \overline{E}_i \cap V_i^1 = E_i$$

and $x \in B \cap \overline{V}_i$. Therefore, $a \in E_i''$ which is impossible. Hence, $\overline{E}_i'' \cap G_0 = E_i''$ is closed in G_0; q.e.d.

The union $D_4' = \bigcup_{i\in I} E_i'$ is almost thin in Y. Take $y_0 \in U_\delta - (D_3'' \cup D_4')$. Then $\varphi^{-1}(y_0) \cap \overline{\Omega}$ is compact because $\varphi\colon \overline{\Omega} \cap G \to \tilde{G}$ is proper. Hence, the index set

$$I_* = \{i \in I \mid V_i^2 \cap \varphi^{-1}(y_0) \cap \Omega \neq \emptyset\}$$

if finite. Because $\mathfrak{w}^2$ is a covering, $I_* \neq \emptyset$. Then

$$D_4'' = \bigcup_{i\in I_*} E_i''$$

is closed in G_0. Hence, $G_1 = U_\delta - (D_3'' \cup D_4'')$ is open with

$$y_0 \in (U_\delta - D_3'' \cup D_4') \subseteq U_\delta - (D_3'' \cup \bigcup_{i\in I} E_i'') \subseteq U_\delta - (D_3'' \cup \bigcup_{i\in I_*} E_i'') = G_1.$$

Hence, G_1 is an open neighborhood of y_0 with

$$y_0 \in G_1 \subseteq U_\delta \subseteq (G_0 - D_1'' \cup D_2'' \cup S_Y) \cap \varphi(X).$$

The map $\varphi_{G_1} : \tilde{G}_1 \to G_1$ is surjective, q-fibering, quasi-proper and holomorphic.

Using a parallel translation in C^n, it can be assumed that $\delta(y_0) = 0$.

(Σ_8) Lemma. An open, connected neighborhood G_2 of y_0 exists such that $\overline{G}_2$ is compact, such that $\overline{G}_2 \subset G_1$ and such that an open subset $\hat{G}_2$ of $\tilde{G}_2 = \varphi^{-1}(G_2)$ with $\hat{G}_2 \supseteq \tilde{G}_2 \cap \overline{\Omega}$ exists such that $\hat{G}_2 \cap \overline{V}_i^1 = \emptyset$ if $i \in I - I_*$.

Proof. Because $\mathfrak{V}^2$ is locally finite,

$$V^1 = \bigcup_{i\in I-I_*} \overline{V}_i^1$$

is closed. If $x \in \overline{\Omega} \cap \varphi^{-1}(y_0) \cap V^1$, then $x \in \overline{\Omega} \cap \varphi^{-1}(y_0) \cap \overline{V}_i^1$ for some $i \in I - I_*$. But $x \in \overline{\Omega} \cap \varphi^{-1}(y_0) \cap V_i^2$ implies $i \in I_*$; contradiction. Hence, $\overline{\Omega} \cap \varphi^{-1}(y_0) \subseteq \tilde{G}_1 - V^1$ where $\tilde{G}_1 - V^1$ is open. Now, it is claimed, that an open connected neighborhood G_2 of y_0 with $\overline{G}_2 \subseteq G_1$ and $\overline{G}_2$ compact exists such that $\tilde{G}_2 = \varphi^{-1}(G_2)$ and $\tilde{G}_2 - V^1 \supseteq \tilde{G}_2 \cap \overline{\Omega}$. Otherwise, a sequence $\{z_\mu\}_{\mu\in N}$ of points in G_1 converges to y_0 where

$x_\mu \in \varphi^{-1}(z_\mu) \cap \overline{\Omega} \cap V^1$ exists. Let K be a compact neighborhood of y_0 with $K \subseteq G_1$. A number μ_0 exists such that $z_\mu \in K$ for all $\mu \geqq \mu_0$. Then $\varphi^{-1}(K) \cap \overline{\Omega}$ is compact and $x_\mu \in \varphi^{-1}(K) \cap \overline{\Omega} \cap V^1$ for $\mu \geqq \mu_0$, a subsequence of $\{x_\mu\}_{\mu \in \mathbb{N}}$ converges to a point $x \in \varphi^{-1}(K) \cap \overline{\Omega} \cap V^1$ which implies $x \in \varphi^{-1}(y_0) \cap \overline{\Omega} \cap = \emptyset$. Contradiction! Therefore, an open, connected neighborhood G_2 of y_0 exists such that $\overline{G}_2$ is compact and contained in G_1 and such that $\hat{G}_2 = \tilde{G}_2 - V^1 \supseteq \tilde{G}_2 \cap \overline{\Omega}$. Then $\hat{G}_2$ is an open neighborhood of $G_2 \cap \overline{\Omega}$ with $\hat{G}_2 \subseteq \tilde{G}_2$, such that $\hat{G}_2 \cap \overline{V}_i^1 = \emptyset$ if $i \in I - I_*$; q.e.d.

(Σ_9) <u>Lemma.</u> a) The map $\varphi_{G_2} : \tilde{G}_2 \to G_2$ is surjective, quasi-proper, q-fibering and holomorphic.

b) $\tilde{G}_2 \cap \overline{\Omega} \subseteq \hat{G}_2 \subseteq \tilde{G}_2 \subseteq X - (D_0 \cup \tilde{S}_Y)$

c) If $y \in G_2$, then $\varphi^{-1}(y) \cap (S_X \cup D_2 \cup D_3)$ is a thin analytic subset of $\varphi^{-1}(y)$.

d) If $y \in G_2$, if $i \in I$ and if $\varphi^{-1}(y) \cap V_i \cap \hat{G}_2 \neq \emptyset$, then $i \in I_*$ and $\varphi^{-1}(y) \cap V_i \cap E_i$ is a thin analytic subset of $\varphi^{-1}(y) \cap V_i$.

<u>Proof.</u> a) and b) are true by the construction of G_2 and because $G_2 \subseteq G_1$.

c) Suppose c) is not true. Then a branch B of $\varphi^{-1}(y)$ exists with $B \subseteq S_X \cup D_2 \cup D_3$. Hence, $B \subseteq S_X$ or $B \subseteq D_2$ or $B \subseteq D_3$. If $B \subseteq S_X$ then $y \in D_1' \cap G_2 = D_1' \cap G_0 \cap G_2 = D_1'' \cap G_2 \subseteq D_1'' \cap G_1 = \emptyset$. If $B \in D_2$, then $y \in D_2' \cap G_2 = D_2' \cap G_0 \cap G_2 = D_2'' \cap G_2 \subseteq D_2'' \cap G_1 = \emptyset$. If $B \subseteq D_3$, then $y \in D_3'' \cap G_2 \subseteq D_3'' \cap G_1 = \emptyset$. Hence, a contradiction exists in each case which proves c).

d) If $y \in G_2$, if $i \in I$ and if $\varphi^{-1}(y) \cap V_i \cap \hat{G}_2 \neq \emptyset$, then $\overline{V}_i^1 \cap \hat{G}_2 \neq \emptyset$. Hence, $i \in I_*$ by (Σ_8). Suppose that a branch B of $\varphi^{-1}(y) \cap V_i \cap E_i$ is also a branch of $\varphi^{-1}(y) \cap V_i$. Then a branch B^1 of $\varphi^{-1}(y) \cap V_i^1$ exists such that $B \subseteq B^1$. Then $B \subseteq E_i \cap B^1$ and $E_i \cap B^1$ is not thin but analytic on the irreducible complex space B^1. Hence, $E_i \cap B^1 = B^1$ or $B^1 \subseteq E_i$. Moreover, $B^1 \cap \overline{V}_i \supseteq B \neq \emptyset$. Hence, $y \in E_i''$ with $i \in I_*$ which implies $y \in D_4'' \cap G_2 \subseteq D_4'' \cap G_1 = \emptyset$. This contradiction proves d); q.e.d.

(Σ_{10}). Let ϑ be a metric on X. Let $\vartheta_0 > 0$ be the Lebesgue-number of the covering $\mathfrak{W}$ of the compact set $K_2 = \varphi^{-1}(\overline{G}_2) \cap \overline{\Omega}$. A map $\gamma\colon K_2 \to I$ exists such that for each $x \in K_2$

$$\{z \mid \vartheta(z,x) \leqq \vartheta_0\} \subseteq V_{\gamma(x)}.$$

Take an open connected neighborhood G_3 and G_4 of y_0 with $\overline{G}_4 \subseteq G_3 \subset \overline{G}_3 \subset G_2$. Then $\overline{G}_4$, $\overline{G}_3$, $\varphi^{-1}(\overline{G}_3) \cap \overline{\Omega} = K_3$ and $\varphi^{-1}(\overline{G}_4) \cap \overline{\Omega} = K_4$ are compact with $K_4 \subseteq K_3 \subseteq K_2$ and $K_3 \subseteq \hat{G}_2$.

By (Σ_6), for each $x \in K_3$ open neighborhoods N_x and M_x of x can be taken such that

a) The neighborhood $\overline{N}_x$ is compact with

$$x \in M_x \subseteq \overline{M}_x \subseteq N_x \subseteq \overline{N}_x \subseteq \hat{G}_2$$

$$\operatorname{diam} M_x < \vartheta_0/3, \quad \operatorname{diam} N_x < 2\vartheta_0/3$$

b) If $x \in K_3 - \tilde{G}_3$, then $\varphi(\overline{N}_x) \cap \overline{G}_4 = \emptyset$.

c) If $x \in K_3 \cap \overline{\Omega} \cap \tilde{G}_3$, then $\overline{N}_x \subseteq \Omega \cap \tilde{G}_3$.

d) If $x \in K_3 \cap (\overline{\Omega} - \Omega) \cap \tilde{G}_3$ then

$$\mathcal{L}(N_x \cap \Omega \| N_x) \supseteq M_x$$

$$\mathcal{L}(N_x \cap \varphi^{-1}(y) \cap \Omega \| N_x \cap \varphi^{-1}(y)) \supseteq M_x \cap \varphi^{-1}(y)$$

if $y \in \varphi(N_x) \subseteq G_2$.

(Σ_{11}). <u>Lemma.</u> Points $x_\mu \in \varphi^{-1}(y_0) \cap \overline{\Omega}$ and $u_\mu \in \varphi^{-1}(y_0) \cap \hat{G}_2 - (D_2 \cup D_3 \cup D_4)$ for $\mu = 1,\dots,p$ exist such that

a) A product representation $(\alpha_\mu, \beta, \pi_\mu)$ at u_μ with $\alpha_\mu(u_\mu) = 0$ exists for $\mu = 1,\dots,p$ such that

$$y_0 \in U_\beta \subseteq \overline{U}_\beta \subseteq G_4 \subseteq U_\delta \text{ with } \beta = \delta | U_\beta$$

and such that $\overline{U}_{\alpha_\mu}$ is compact and contained in $\hat{G}_2$. Let κ_μ be the sheet number of α_μ and define

$$\kappa = \sum_{\mu=1}^{p} \kappa_\mu$$

b) For each $\mu = 1,\dots,p$, poly-disc product representations $(P_{\rho\mu}, Q_\rho, \alpha_\mu, \beta)$ exist such that $\alpha_\mu(P_{\rho\mu}) = P'_{\rho\mu} = P''_\mu \times Q'_\rho$ for $\rho = 1,2,3$ such that

$$\overline{Q}_1 \subset Q_2 \subset \overline{Q}_2 \subset Q_3 \subset \overline{Q}_3 \subset U_\beta$$

Observe that Q_ρ is independent of $\mu = 1,\ldots,p$ and that P''_μ is independent of $\rho = 1,2,3$. Then

$$\begin{array}{ccc}
U_{\alpha_\mu} \supseteq \overline{P}_{3\mu} \supset P_{\rho\mu} & \xrightarrow{\alpha_\mu} & P''_\mu \times Q'_\rho = P'_{\rho\mu} \\
\Big\downarrow \varphi & & \Big\downarrow \pi_{\rho\mu} \\
Q_\rho & \xrightarrow[\beta]{} & Q'_\rho
\end{array}$$

is a commutative diagram, where $\pi_{\rho\mu}$ is the projection.

c) For each $\mu = 1,\ldots,p$ and $\rho = 1,2,3$, a poly-disc product representation $(R_{\rho\mu}, Q_\rho, \alpha_\mu, \beta)$ exists which is a retraction of $(P_{\rho\mu}, Q_\rho, \alpha_\mu, \beta)$ with the same condensor θ_μ for $\rho = 1,2,3$ and fixed μ. Define $\theta = \mathrm{Max}(\theta_1,\ldots,\theta_p)$. Let

$$S_{\rho\mu} = \alpha_\mu^{-1}(\{0\} \times Q'_\rho) \qquad 0 \in \mathbb{C}^q$$

be the common center section of $(P_{\rho\mu}, Q_\rho, \alpha_\mu, \beta)$ and $(R_{\rho\mu}, Q_\rho, \alpha_\mu, \beta)$. Also, $\alpha_\mu(R_{\rho\mu}) = R'_{\rho\mu} = R''_\mu \times Q'_\rho$ where the poly-disc R''_μ does not depend on ρ and where $\overline{R}''_\mu \subset P''_\mu$. Then

$$\begin{array}{ccc}
P_{\rho\mu} \supset R_{\rho\mu} & \xrightarrow{\alpha_\mu} & R''_\mu \times Q'_\rho = R'_{\rho\mu} \\
\Big\downarrow \varphi & & \Big\downarrow \pi_{\rho\mu} \\
Q_\rho & \xrightarrow{\beta} & Q'_\rho
\end{array}$$

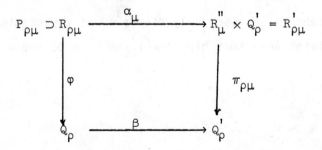

d) The inverse image $\alpha_\mu^{-1}(0)$ consists of κ_μ simple points $u_{\mu 1} = u_\mu, u_{\mu 2}, \ldots, u_{\mu\kappa_\mu}$ of X which are also regular points of α_μ and φ. For each $\mu = 1,\ldots,p$ and $\nu = 1,\ldots,\kappa_\mu$ an open neighborhood $L_{\mu\nu}$ of $u_{\mu\nu}$ in $P_{3\mu}$ and a concentric poly-disc L_μ'' of P_μ'' independent of ν exist such that

$$a_{\mu\nu} = \alpha_\mu | L_{\mu\nu}: L_{\mu\nu} \to L_\mu'' \times Q_3'$$

is biholomorphic such that $f_1,\ldots,f_d$ are holomorphic on $L_{\mu\nu}$. Define

$$f_{\lambda\mu\nu} = f_\lambda \circ \alpha_{\mu\nu}^{-1}: L_\mu'' \times Q_3' \to \mathbb{C}.$$

Then numbers $1 < \lambda_1 < \ldots < \lambda_{q-k} \leqq q$ exists such that

$$\gamma_{\mu\nu}: L_\mu'' \times Q_3' \to L_{\mu\nu}^0$$

is a biholomorphic map onto an open subset $L_{\mu\nu}^0$ of $\mathbb{C}^m$ where

$$\gamma_{\mu\nu}(z) = (f_{1\mu\nu}(z),\ldots,f_{k\mu\nu}(z), z_{\lambda_1},\ldots,z_{\lambda_{k-q}}, z_{q+1},\ldots,z_m)$$

if $z = (z_1,\ldots,z_m) \in L_\mu'' \times Q_3'$. Here $\lambda_\rho = \lambda_\rho(\mu,\nu)$ depends on μ and ν for each $\rho = 1,\ldots,k - q$.

e) For $\mu = 1,\ldots,p$ define $I_\mu = \{i \in I | x_\mu \in \overline{V}_i\}$. Then

$$I_\mu = \{i \in I | \overline{P}_{3\mu} \cap \overline{V}_i \neq \emptyset\}$$

is finite and

$$\overline{P}_{3\mu} \subset M_{x_\mu} \cap \Big(\bigcap_{i \in I_\mu} V_i^1 - \bigcup_{i \in I - I_\mu} \overline{V}_i \Big).$$

f) Define $\partial R_{\rho\mu} = \partial_{\alpha_\mu} R_{\rho\mu}$ for $\mu = 1,\ldots,p$ and $\rho = 1,2,3$. Then $\partial R_{\rho\mu} \cap D_4 = \emptyset$ for $\mu = 1,\ldots,p$ and $\rho = 1,2,3$.

g) For $\mu = 1,\ldots,p$ and $\rho = 1,2,3$

$$\bigcup_{\mu=1}^{p} R_{\rho\mu} \supseteq \overline{\Omega} \cap \varphi^{-1}(Q_\rho)$$

$$\bigcup_{\mu=1}^{p} \overline{R}_{\rho\mu} \supseteq \overline{\Omega} \cap \varphi^{-1}(\overline{Q}_\rho).$$

<u>Proof.</u> Take $x \in \overline{\Omega} \cap \varphi^{-1}(y_0)$. Because $\mathfrak{W}^1$ is locally finite, also $\overline{\mathfrak{W}}$ is locally finite and $I_x = \{i \in I \mid x \in \overline{V}_i\}$ is finite. Hence, $\bigcup_{i \in I - I_x} \overline{V}_i$ is closed and does not contain x. A product representation $(\alpha_x, \beta_x, \pi_x)$ of φ at $x \in \varphi^{-1}(y_0) \cap \Omega \subseteq K_3$ exists such that $\alpha_x(x) = 0$ and such that

$$U_{\alpha_x} \subset \overline{U}_{\alpha_x} \subseteq M_x \cap \Big(\bigcap_{i \in I_x} V_i^1 - \bigcup_{i \in I - I_x} \overline{V}_i^1 \Big) \subseteq \overline{N}_x \subset \hat{G}_2$$

and such that $\overline{U}_{\beta_x} \subset G_4$ with $\beta_x = \delta | U_{\beta_x}$. Then $\overline{U}_{\alpha_x}$ and $\overline{U}_{\beta_x}$ are compact. Moreover,

$$\begin{array}{ccc} U_{\alpha_x} & \xrightarrow{\ \alpha_x\ } & U''_{\alpha_x} \times U'_{\beta_x} = U'_{\alpha_x} \\ \Big\downarrow \varphi & & \Big\downarrow \pi_x \\ U_{\beta_x} & \longrightarrow & U'_{\beta_x} \end{array}$$

commutes where π_x is the projection and where U''_{α_x} is open in $\mathbb{C}^q$.

Define $D_{4x} = U_{\alpha_x} \cap \bigcup_{i \in I_x} E_i$. Because I_x is finite, because E_i is analytic and thin in V_i^1 and because $U_{\alpha_x} \subset V_i^1$ if $i \in I_x$, the set D_{4x} is analytic and thin in U_{α_x}. Because $\alpha_x\colon U_{\alpha_x} \to U'_{\alpha_x}$ is proper, open and holomorphic, $D'_{4x} = \alpha_x(D_{4x})$ is thin and analytic in U'_{α_x}.

Take any $i \in I_x$. Suppose that a branch B of $\varphi^{-1}(y_0) \cap V_i^1$ with $B \subseteq E_i$ exists. Then $y_0 \in E'_i \subseteq D'_4$ which is wrong. Hence, $\varphi^{-1}(y_0) \cap E_i$ is a thin analytic subset of $\varphi^{-1}(y_0) \cap V_i^1$ which implies $\dim \varphi^{-1}(y_0) \cap E_i \leqq q - 1$. Therefore,

$$\dim D_{4x} \cap \varphi^{-1}(y_0) \leqq q - 1.$$

If $w \in D_{4x} \cap \varphi^{-1}(y_0)$, then $\alpha_x(w) = (z,t)$ with $z \in U''_{\alpha_x}$ and $t \in U'_{\beta_x}$. Then $t = \pi_x(z,t) = \pi_x(\alpha_x(w)) = \beta_x(\varphi(w)) = \beta_x(y_0) = 0$. Hence, $\alpha_x(w) = (z,0) \in D'_{4x} \cap (U''_{\alpha_x} \times \{0\})$. If $(z,0) \in D'_{4x} \cap (U''_{\alpha_x} \times \{0\})$. Then $w \in D_{4x}$ exists such that $\alpha_x(w) = (z,0)$. Then $\beta_x(\varphi(w)) = \pi_x(\alpha_x(w)) = \pi_x(z,0) = 0 = \beta_x(y_0)$. Because β_x is biholomorphic $\varphi(w) = y_0$ and $w \in D_{4x} \cap \varphi^{-1}(y_0)$. Therefore,

$$D'_{4x} \cap (U''_{\alpha_x} \times \{0\}) = \alpha_x(D_{4x} \cap \varphi^{-1}(y_0)).$$

Because $\alpha_x\colon \varphi^{-1}(y_0) \cap U_{\alpha_x} \to U''_{\alpha_x} \times \{0\}$ is proper, surjective, open, light and holomorphic, and because $D_{4x} \cap \varphi^{-1}(y_0)$ is thin in

$\varphi^{-1}(y_0) \cap U_{\alpha_x}$, also $D'_{4x} \cap (U''_{\alpha_x} \times \{0\})$ is thin in $U''_{\alpha_x} \times \{0\}$. Hence, $\dim D'_{4x} \cap (U''_{\alpha_x} \times \{0\}) < q$.

By a linear transformation on $\mathbb{C}^q$ only, the map α_x can be chosen such that

$$\{(\underbrace{0_1 \ldots 0}_{(q-1)})\} \times \mathbb{C} \times \{(\underbrace{0 \ldots 0}_{n-q})\} \cap D'_{4x}$$

contains $0 \in \mathbb{C}^m$ as an isolated point.

A poly-disc Q'_x with center 0 in $\mathbb{C}^n$ and with $\overline{Q}'_x \subset U'_{\beta_x}$ and a poly-disc

$$P''_x = \prod_{\nu=1}^{q} \{z_\nu \mid |z_\nu| < r_\nu(x)\} \subseteq \overline{P}''_x \subset U''_{\alpha_x}$$

exists such that

$$\emptyset = D'_{4x} \cap \left(\prod_{\nu=1}^{q-1} \{z_\nu \mid |z_\nu| \leqq r_\nu(x)\} \times \{z_q \mid |z_q| = r_q(x)\} \times \overline{Q}'_x\right).$$

A number $\varepsilon(x)$ exists such that

$$0 < \varepsilon(x) < \tfrac{1}{4}\mathrm{Min}\{r_1(x),\ldots,r_q(x)\}$$

$$\prod_{\nu=1}^{q} \{z_\nu \mid |z_\nu| \leqq r_\nu(x) + 2\varepsilon(x)\} \subseteq U''_{\alpha_x}$$

and such that for each $(a_1,\ldots,a_q) \in \mathbb{C}^q$ with $|a_\nu| \leqq \varepsilon(x)$ for $\nu = 1,\ldots,q$ and for each t_q with $0 < r_q(x) - \varepsilon(x) \leqq t_q \leqq r_q(x) + \varepsilon(x)$

$$\emptyset = D'_{4x} \cap \prod_{\nu=1}^{q-1} \{z_\nu \mid |z_\nu - z_\nu| \leqq r_\nu(x)\} \times \{z_q \mid |z_q - a_q| = t_q\} \times \overline{Q}^1_x .$$

Define

$$A'_x = \prod_{\nu=1}^{q} \{z_\nu \mid |z_\nu| < \varepsilon(x)\} \subseteq \overline{A}^1_x \subseteq U''_{\alpha_x} .$$

Then $\overline{A}^1_x \times \overline{Q}^1_x \subseteq U''_{\alpha_x} \times U'_{\beta_x}$. Define

$$x \in A_x = \alpha_x^{-1}(A_x \times Q'_x) \subseteq \overline{A}_x \subset U_{\alpha_x} .$$

Then A_x is open and defined for every point x of the compact set $\varphi^{-1}(y_0) \cap \overline{\Omega}$.

Finitely many points $x_1, \ldots, x_p$ exist in $\overline{\Omega} \cap \varphi^{-1}(y_0)$ such that

$$\overline{\Omega} \cap \varphi^{-1}(y_0) \subseteq A_{x_1} \cup \ldots \cup A_{x_p} .$$

Define $A_\mu = A_{x_\mu}$ and $A'_\mu = A'_{x_\mu}$. Define $\varepsilon_\mu = \varepsilon(x_\mu)$ and $r_{\nu\mu} = r_\nu(x_\mu)$ for $\nu = 1, \ldots, q$ and $\mu = 1, \ldots, p$. Then

$$U'_\beta = Q'_{x_1} \cap \ldots \cap Q'_{x_p}$$

is a poly-disc in $\mathbb{C}^n$ with center 0 and $\overline{U}'_\beta \subset U'_\delta$. Define $U_\beta = \delta^{-1}(U'_\beta) = \beta_{x_\mu}^{-1}(U'_\beta)$. Then U_β is an open connected neighborhood of y_0 with $\overline{U}_\beta$ compact. Moreover, $\overline{U}_\beta \subseteq G_4 \subset U_\delta$. Define $\beta = \delta | U_\beta = \beta_{x_\mu} | U_\beta$. Then $\beta\colon U_\beta \to U'_\beta$ is a schlicht chart at y_0 with $\beta(y_0) = 0$.

Define

$$I_\mu = I_{x_\mu} = \{i \in I \mid x_\mu \in \overline{V}_i\}.$$

Define $D_{4x_\mu} = D_{4\mu}$ and $D'_{4x_\mu} = D'_{4\mu}$ for $\mu = 1,\dots,p$.

Now, it is claimed that an open neighborhood Q of y_0 exists such that $\overline{Q} \subseteq U_\beta$, such that $\beta(Q) = Q'$ is a poly-disc in $\mathbb{C}^n$ with center 0, and such that

$$\varphi^{-1}(\overline{Q}) \cap \overline{\Omega} \subseteq A_1 \cup \dots \cup A_p.$$

Suppose the claim is wrong. A sequence $\{z_\lambda\}_{\lambda \in \mathbb{N}}$ of points $z_\lambda \in U_\beta$ converges to y_0 such that $\xi_\lambda \in \varphi^{-1}(z_\lambda) \cap \overline{\Omega}$ exists with $\xi_\lambda \notin A_\mu$ for $\mu = 1,\dots,p$ and all $\lambda \in \mathbb{N}$. Because $\varphi^{-1}(\overline{U}_\beta) \cap \overline{\Omega}$ is compact, it can be assumed that $\{\xi_\lambda\}_{\lambda \in \mathbb{N}}$ converges to $\xi \in \varphi^{-1}(\overline{U}_\beta) \cap \overline{\Omega}$. Then $\varphi(\xi) = y_0$. Hence, $\xi \in \varphi^{-1}(y_0) \cap \overline{\Omega}$. An index μ with $1 \leqq \mu \leqq p$ exists such that $\xi \in A_\mu$. Hence, $\lambda \in \mathbb{N}$ exists such that $\xi_\lambda \in A_\mu$, which is wrong. Hence, the claim is correct, and Q, Q' can be taken as indicated before.

The set

$$\Phi_\mu = \varphi^{-1}(y_0) \cap (S_x \cup D_2 \cup D_3 \cup D_{4\mu})$$

is analytic in $Y_{\alpha_{x_\mu}}$. By (Σ_9), $\varphi^{-1}(y_0) \cap (S_x \cup D_2 \cup D_3)$ is a thin analytic subset of $\varphi^{-1}(y_0)$. Because

$$\dim(D_{4\mu} \cap \varphi^{-1}(y_0)) \leqq q - 1$$

also $D_{4\mu} \cap \varphi^{-1}(y_0)$ is a thin analytic subset of $\varphi^{-1}(y_0) \cap U_{\alpha_{x_\mu}}$. Hence, Φ_μ is a thin analytic subset of $\varphi^{-1}(y_0) \cap U_{\alpha_{x_\mu}}$. Because

$$\alpha_{\mu *} = \alpha_{x_\mu} : \varphi^{-1}(y_0) \cap U_{\alpha_{x_\mu}} \to U''_{\alpha_{x_\mu}} \times \{0\}$$

is a light, proper, surjective, open, holomorphic map, $\Phi'_\mu = \alpha_{x_\mu}(\Phi_\mu)$ and $\hat{\Phi}_\mu = \alpha_{\mu *}^{-1}(\Phi'_\mu)$ are thin analytic subsets of $U''_{\alpha_{x_\mu}} \times \{0\}$ respectively $\varphi^{-1}(y_0) \cap U_{\alpha_{x_\mu}}$. Moreover, a thin analytic subset ψ'_μ of $U''_{\alpha_\mu} \times \{0\}$ exists such that $\psi_\mu = \alpha_{\mu x}^{-1}(\psi'_\mu)$ is thin and analytic in $\varphi^{-1}(y_0) \cap U_{\alpha_{x_\mu}}$ and such that $\alpha_{\mu *}$ is locally biholomorphic on $\varphi^{-1}(y_0) \cap U_{\alpha_{x_\mu}} - \psi_\mu$.

Because A_μ is an open neighborhood of x_μ with $A_\mu \subseteq U_{\alpha_{x_\mu}}$, a point

$$u_\mu \in \varphi^{-1}(y_0) \cap A_\mu - (\hat{\Phi}_\mu \cup \psi_\mu) \subseteq U_{\alpha_{x_\mu}}$$

can be taken for each $\mu = 1,\ldots,p$. Let $u_{\mu 1} = u_\mu, u_{\mu 2}, \ldots, u_{\mu \kappa_\mu}$ be the different points of $\alpha_{\mu *}^{-1}(\alpha_{x_\mu}(u_\mu))$. Obviously, $\kappa_\mu < \infty$ and

$$u_{\mu\nu} \in \varphi^{-1}(y_0) \cap U_{\alpha_{x_\mu}} - (\hat{\Phi}_\mu \cup \psi_\mu).$$

Now, it is claimed that the following 6 statements are true for $\nu = 1,\ldots,\kappa_\mu$ and $\mu = 1,\ldots,p$.

1. Statement: The functions $f_1,\ldots,f_d$ are holomorphic at $u_{\mu\nu}$.

2. Statement: $u_{\mu\nu}$ is a simple point of X.

3. Statement: $\omega(u_{\mu\nu}) \neq 0$.

4. Statement: $u_{\mu\nu}$ is a regular point of φ.

5. Statement: $u_{\mu\nu}$ is a regular point of α_{x_μ}.

6. Statement: κ_μ is the sheet number of α_{x_μ}.

Proof of the Statements. 1. Because $u_\mu \notin \hat{\Phi}_\mu$, also $\alpha_{x_\mu}(u_\mu) \notin \Phi'_\mu$ and $u_{\mu\nu} \notin \Phi_\mu$. Hence, $u_{\mu\nu} \in X - D_2$. Therefore, $f_1,\ldots,f_d$ are holomorphic at $u_{\mu\nu}$.

2. Because $u_{\mu\nu} \notin \Phi_\mu$, also $u_{\mu\nu} \notin S_X$. Hence, $u_{\mu\nu}$ is a simple point of X.

3. Because $u_{\mu\nu} \notin \Phi$, also $u_{\mu\nu} \in \tilde{U}_\delta - (S_X \cup D_2 \cup D_3)$. By (Σ_5) this implies $\omega(u_{\mu\nu}) \neq 0$.

4. Because $u_{\mu\nu} \in \tilde{U}_\delta - (S_X \cup D_2)$ and $\omega(u_{\mu\nu}) \neq 0$, also $d(g_1 \circ \varphi) \wedge \ldots \wedge d(g_n \circ \varphi) \neq 0$ at $u_{\mu\nu}$. Hence, $u_{\mu\nu}$ is a regular point of $\delta \circ \varphi$. Because $\delta\colon U_\delta \to U'_\delta$ is biholomorphic, $u_{\mu\nu}$ is a regular point of φ.

5. Because $u_\mu \notin \psi_\mu$, also $\alpha_{x_\mu}(u_\mu) \notin \psi_\mu'$. Hence, $u_{\mu\nu} \notin \psi_\mu$. Hence, $\alpha_{\mu x}$ is regular at $u_{\mu\nu}$ which is a simple point of X and $\varphi^{-1}(y_0) \frown U_{\alpha_{x_\mu}}$. By Lemma 15.5, α_{x_μ} is regular at $u_{\mu\nu}$.

6. Because α_{x_μ} is regular at each point of $\alpha_{x_\mu}^{-1}(\alpha_{x_\mu}(u_\mu)) = \alpha_{\mu x}^{-1}(\alpha_{x_\mu}(u_\mu))$, the number κ_μ of points in this set is the sheet number of α_{x_μ}; q.e.d.

Define $\kappa = \kappa_1 + \dots + \kappa_p$. Define

$$\alpha_{x_\mu}(u_{\mu\nu}) = \alpha_{x_\mu}(u_\mu) = (a_{1\mu},\dots,a_{q\mu},0 \dots 0).$$

Then $(a_{1\mu},\dots,a_{q\mu}) \in A_\mu'$ especially $|a_{\lambda\mu}| < \varepsilon_\mu$ for $\lambda = 1,\dots,q$. Therefore,

$$U_\mu'' = \prod_{\lambda=1}^{q} \{z_\lambda \mid |z_\lambda - a_{\lambda\mu}| < r_{\lambda\mu} + \varepsilon_\mu\} \subseteq \overline{U}_\mu'' \subseteq U_{\alpha_{x_\mu}}'' .$$

Define

$$U_{\alpha_\mu}'' = \prod_{\lambda=1}^{q} \{z_\lambda \mid |z_\lambda| < r_{\lambda\mu} + \varepsilon_\mu\}$$

$$U_{\alpha_\mu} = \alpha_{x_\mu}^{-1}(U_\mu'' \times U_\beta').$$

A biholomorphic map

$$\ell_\mu : U''_\mu \times U'_\beta \to U''_{\alpha_\mu} \times U'_\beta = U'_{\alpha_\mu}$$

is defined by

$$\ell_\mu(z_1,\ldots,z_m) = (z_1-a_{1\mu},\ldots,z_q-a_{q\mu},z_{q+1},\ldots,z_m)$$

such that $\ell_\mu(\alpha_{x_\mu}(u_{\mu\nu})) = 0 \in \mathbb{C}^m$. Then

$$\alpha_\mu = \ell_\mu \circ \alpha_{x_\mu} : U_{\alpha_\mu} \longrightarrow U''_{\alpha_\mu} \times U'_\beta$$

is a chart of X with sheet number κ_μ such that

$$\alpha_\mu^{-1}(0) = \{u_{\mu 1},\ldots,u_{\mu,\kappa_\mu}\}$$

and such that α_μ is regular at every point $u_{\mu\nu}$. Let $\pi_\mu : U''_{\alpha_\mu} \times U'_\beta \to U'_\beta$ be the projection. Then $\pi_\mu \circ \ell_\mu = \pi_{x_\mu}$ and

$$\pi_\mu \circ \alpha_\mu = \pi_\mu \circ \ell_\mu \circ \alpha_{x_\mu} = \pi_{x_\mu} \circ \alpha_{x_\mu} = \beta \circ \varphi$$

on U_{α_μ}. Hence, $(\alpha_\mu,\beta,\pi_\mu)$ is a product representation at u_μ over y_0. Here $\overline{U}_{\alpha_\mu}$ is compact and contained in $\overline{U}_{\alpha_{x_\mu}} \subseteq \hat{G}_2$. Hence, a) is proved.

Open subsets T'_1 of $\mathbb{C}^q$, T'_2 of $\mathbb{C}^n$ and $T_{\mu\nu}$ of U_{α_μ} exist such that $0 \in T'_1 \subseteq U''_{\alpha_\mu}$ and $0 \in T'_2 \subseteq U'_\beta$ and $u_{\mu\nu} \in T_{\mu\nu}$ for $\nu = 1,\ldots,\kappa_\mu$ and $\mu = 1,\ldots,p$, such that

$$\alpha_\mu | T_{\mu\nu}: T_{\mu\nu} \to T_1' \times T_2'$$

is biholomorphic. Define $\eta_{\mu\nu} = (\alpha_\mu | T_{\mu\nu})^{-1}$. Moreover, the neighborhood $T_{\mu\nu}$ can be taken such that $f_1,\ldots,f_d$ are holomorphic on $T_{\mu\nu}$ and such that $\omega(x) \neq 0$ for all $x \in T_{\mu\nu}$. Define $f_{\lambda\mu\nu} = f_\lambda \circ \eta_{\mu\nu}$ for $\lambda = 1,\ldots,d$, for $\nu = 1,\ldots,\kappa_\mu$ and $\mu = 1,\ldots,p$. Then

$$0 \neq \eta_{\mu\nu}^*(\omega) = \sum_{1 \leqq \rho_1 < \ldots < \rho_k \leqq q} \frac{\partial(f_{1\mu\nu},\ldots,f_{k\mu\nu})}{\partial(z_{\rho_1},\ldots,z_{\rho_k})} dz_{\rho_1} \wedge \cdots \wedge dz_{\rho_1} \wedge$$

$$dz_{q+1} \wedge \cdots \wedge dz_m$$

on $T_1' \times T_2'$. Hence, $(\lambda_1,\ldots,\lambda_{q-k})$ with $1 \leqq \lambda_1 < \ldots < \lambda_{q-k} \leqq q$ exists such that

$$\frac{\partial(f_{1\mu\nu},\ldots,f_{k\mu\nu})}{\partial(z_{\rho_1},\ldots,z\rho_k)}(0) \neq 0$$

where $1 \leqq \rho_1 < \ldots < \rho_k \leqq q$ and $(\rho_1,\ldots,\rho_q,\lambda_1,\ldots,\lambda_{q-k})$ is a permutation of $(1,\ldots,q)$. Of course, $\lambda_1,\ldots,\lambda_{q-k}$ depend on μ,ν. Then

$$\frac{\partial(f_{1\mu\nu},\ldots,f_{k\mu\nu},z_{\lambda_1},\ldots,z_{\lambda_{q-k}},z_{q+1},\ldots,z_m)}{\partial(z_1,\ldots\ldots\ldots\ldots\ldots\ldots\ldots\ldots,z_m)}(0) \neq 0.$$

A poly-disc L_3' with center $0 \in \mathbb{C}^q$ and $\overline{L}_3' \subseteq T_1'$ and a poly-disc Q_3' with center $0 \in \mathbb{C}^n$ and $\overline{Q}_3' \subseteq Q' \cap T_2' \subset U_\beta'$ exist such that the map

$$\gamma_{\mu\nu}: L_3' \times Q_3' \to \hat{L}_{\mu\nu} \subset \mathbb{C}^m$$

is biholomorphic where

$$\gamma_{\mu\nu}(z) = (f_{1\mu\nu}(z), \dots, f_{k\mu\nu}(z), z_{\lambda_1}, \dots, z_{\lambda_{q-k}}, z_{q+1}, \dots, z_m)$$

for $(z_1, \dots, z_m) \in L_3' \times Q_3'$ and where $\hat{L}_{\mu\nu} = \gamma_{\mu\nu}(L_3' \times Q_3')$ is open in $\mathbb{C}^m$.

Take poly-discs Q_1' and Q_2' with center $0 \in \mathbb{C}^n$ and

$$\overline{Q}_1' \subset Q_2' \subset \overline{Q}_2' \subset Q_3' \subset \overline{Q}_3' \subset Q' \cap T_2' \subset U_\beta'.$$

Define $Q_\rho = \beta^{-1}(Q_\rho')$. Then

$$\overline{Q}_1 \subset Q_2 \subset Q_2 \subset Q_3 \subset \overline{Q}_3 \subset Q \subset \overline{Q} \subset U_\beta.$$

Define

$$P_\mu'' = P_{\alpha_{x_\mu}}'' = \prod_{\lambda=1}^{q} \{z_\lambda \mid |z_\lambda| < r_{\lambda_\mu}\} \subseteq \overline{P}_\mu'' \subset U_{\alpha_\mu}''$$

$$P_{\rho\mu} = \alpha_\mu^{-1}(P_\mu'' \times Q_\rho') = \alpha_{x_\mu}^{-1}(\ell_\mu^{-1}(P_\mu'' \times Q_\rho').$$

Then $(P_{\rho\mu}, \alpha_\mu)$ is a poly-disc-chart. Define

$$\emptyset \neq R''_\mu = \prod_{\lambda=1}^{q} \{z_\lambda \mid |z_\lambda| < r_{\lambda\mu} - \tfrac{1}{2}\varepsilon_\mu\} \subset \overline{R}''_\mu \subset P''_\mu$$

$$R_{\rho\mu} = \alpha_\mu^{-1}(R''_\mu \times Q'_\rho).$$

Then $(R_{\rho\mu}, \alpha_\mu)$ is a poly-disc chart. Define $P'_{\rho\mu} = P''_\mu \times Q'_\rho$ and $R'_{\rho\mu} = R''_\mu \times Q'_\rho$. Let $\pi_{\rho\mu}\colon P''_\mu \times Q'_\rho \to Q'_\rho$ be the projection. Then $(P_{\rho\mu}, Q_\rho, \alpha_\mu, \beta)$ and $(R_{\rho\mu}, Q_\rho, \alpha_\mu, \beta)$ are poly-disc product representations and $(R_{\rho\mu}, Q_\rho, \alpha_\mu, \beta)$ is a retraction of $(P_{\rho\mu}, Q_\rho, \alpha_\mu, \beta)$ with the common center section

$$S_{\rho\mu} = \alpha_\mu^{-1}(\{0\} \times Q'_\rho).$$

The condensor $\theta_{\mu\rho}$ is

$$\theta_{\mu\rho} = \left(\mathrm{Max}\left(\frac{r_{1\mu} - \frac{1}{2}\varepsilon_\mu}{r_{1\mu}}, \ldots, \frac{r_{q\mu} - \frac{1}{2}\varepsilon_\mu}{r_{q\mu}}\right)\right)^{\frac{1}{\kappa_\mu}} = \theta_\mu.$$

Hence, $\theta_{\mu\rho} = \theta_\mu$ independent of $\rho = 1,2,3$. Define

$$\theta = \mathrm{Max}(\theta_1, \ldots, \theta_p).$$

Then $0 < \theta < 1$. Now, b) and c) are proved.

e) If $1 \leqq \mu \leqq p$, then

$$\overline{P}_{3\mu} \subset U_{\alpha_\mu} \subset U_{\alpha_{x_\mu}} \subset M_{x_\mu} \cap \left(\bigcap_{i \in I_\mu} V_i^1 - \bigcup_{i \in I - I_\mu} \overline{V}_i\right) \subseteq N_{x_\mu}.$$

If $i \in I$ and $\overline{P}_{3\mu} \cap \overline{V}_i \neq \emptyset$, then $i \in I_\mu$. If $i \in I_\mu$, then $x_\mu \in \overline{V}_i$.

Now, $x_\mu \in P_{3\mu}$ is claimed.

$$\ell_\mu(\alpha_{x_\mu}(x_\mu)) = \ell_\mu(0) = (-a_{1\mu},\ldots,-a_{q\mu},0 \ldots 0)$$

with $|a_{\lambda\mu}| < \varepsilon_\mu < r_{\lambda\mu}$ for $\lambda = 1,\ldots,q$. Hence,

$$x_\mu \in \alpha_{x_\mu}^{-1}\ell_\mu^{-1}(P_\mu'' \times Q_3') = P_{3\mu}.$$

Hence, $x_\mu \in \overline{P}_{3\mu} \cap \overline{V}_i$. Therefore, $\overline{P}_{3\mu} \cap V_i \neq \emptyset$. Hence,

$$I_\mu = \{i \in I | \overline{P}_{3\mu} \cap \overline{V}_i \neq \emptyset\}.$$

This proves e).

f) Define $\partial R_{\rho\mu} = \partial_{\alpha_\mu} R_{\rho\mu}$. If $x \in \partial R_{\rho\mu} \cap D_4$, then $x \in \overline{P}_{3\mu} \cap E_i \cap \overline{V}_i$ for some $i \in I$. Hence, $i \in I_\mu$. Now, $\alpha_\mu(x) = z = (z_1,\ldots,z_m)$ with $|z_\lambda| = r_{\lambda\mu} - \frac{1}{2}\varepsilon_\mu$ for $\lambda = 1,\ldots,q$ and $z' = (z_{q+1},\ldots,z_m) \in \overline{Q}_\rho'$. Then

$$\ell_\mu^{-1}(z) = (z_1+a_{1\mu},\ldots,z_q+a_{q\mu},z_{q+1},\ldots,z_m) = (x_1,\ldots,x_m) = x'$$

where $x' = \alpha_{x_\mu}(x)$ and $|x - a_{\lambda\mu}| = r_{\lambda\mu} - \frac{1}{2}\varepsilon_\mu$ for $\lambda = 1,\ldots,q$. Because $i \in I_\mu$,

$$x \in U_{\alpha_{x_\mu}} \cap \bigcup_{j \in I_\mu} E_j = D_{4\mu}$$

$x' \in \alpha_{x_\mu}(D_{4\mu}) = D'_{4\mu}$. Define $t_q = r_{q\mu} - \frac{1}{2}\varepsilon_\mu$. Then

$$x' \in D'_{4\mu} \cap \prod_{\lambda=1}^{q-1} \{z_\lambda \mid |z_\lambda - a_{\lambda\mu}| \leqq r_{\lambda\mu}\} \times \{z_q \mid \quad |z_q - a_q| = t_q\} \times \overline{Q}'_{x_\mu} = \emptyset$$

because $r_{q\mu} - \varepsilon_\mu < t_q < r_{q\mu} + \varepsilon_\mu$. Contradiction! Hence, $\partial R_{\rho\mu} \cap D_4 = \emptyset$, which proves f.

g) Take $z = (z_1, \ldots, z_m) \in A'_\mu \times Q$. Then $|z_\lambda| < \varepsilon_\mu$ for $\lambda = 1, \ldots, q$. Hence,

$$|z_\lambda - a_{\lambda\mu}| \leqq |z_\lambda| + |a_{\lambda\mu}| < 2\varepsilon_\mu < 3\varepsilon_\mu - \frac{1}{2}\varepsilon_\mu < r_{\lambda_\mu} - \frac{\varepsilon_\mu}{2}$$

for $\mu = 1, \ldots, q$. Hence,

$$\ell_\mu(z) = (z_1 - a_{1\mu}, \ldots, z_q - a_{q\mu}, z_{q+1}, \ldots, z_m) \in R''_\mu \times Q'_\rho$$

which implies $z \in \ell_\mu^{-1}(R''_\mu \times Q_\rho)$. Hence, $A'_\mu \times Q'_\rho \subseteq \ell_\mu^{-1}(R''_\mu \times Q'_\rho)$. Therefore,

$$\alpha_{x_\mu}^{-1}(A'_\mu \times Q'_\rho) \subseteq \alpha_{x_\mu}^{-1} \ell_\mu^{-1}(R''_\mu \times Q'_\rho) = R_{\rho\mu}.$$

Take $x \in \overline{\Omega} \cap \varphi^{-1}(Q_\rho)$. Then $x \in \overline{\Omega} \cap \varphi^{-1}(\overline{Q})$. Hence, $x \in A_\mu$ for some index μ. Hence, $\alpha_{x_\mu}(x) = z \in (A'_\mu \times Q'_{x_\mu})$ with $\pi_{x_\mu}(\alpha_{x_\mu}(x)) \in Q'_\rho$. Hence, $\alpha_{x_\mu}(x) = z \in A'_\mu \times Q'_\rho$. Therefore, $x \in \alpha_{x_\mu}^{-1}(A'_\mu \times Q'_{\rho\mu}) \subseteq R_{\rho\mu}$. Therefore,

$$\overline{\Omega} \cap \varphi^{-1}(Q_\rho) \subseteq \bigcup_{\mu=1}^{p} R_{\rho\mu}.$$

Replacing Q_ρ' by $\overline{Q}_\rho'$, it follows as above that

$$\alpha_{x_\mu}^{-1}(A_\mu \times \overline{Q}_\rho') = \alpha_{x_\mu}^{-1}(\ell_\mu^{-1}(R_\mu'' \times \overline{Q}_\rho') \subseteq \overline{R}_{\rho\mu}.$$

Take $x \in \overline{\Omega} \cap \varphi^{-1}(\overline{Q}_\rho) \subseteq \overline{\Omega} \cap \varphi^{-1}(\overline{Q})$. Hence, $x \in A_\mu$ for some index μ. Therefore, $\alpha_{x_\mu}(x) \in A_\mu' \times Q_{x_\mu}'$ and $\pi_{x_\mu}(\alpha_{x_\mu}(x)) = \beta(\varphi(x)) \in \beta(\overline{Q}_\rho) = \overline{Q}_\rho'$. Hence, $\alpha_{x_\mu}(x) \in A_\mu' \times \overline{Q}_\rho'$.and $x \in \alpha_{x\mu}^{-1}(A' \times \overline{Q}_{\rho\mu}') \subseteq \overline{R}_{\rho\mu}$. Therefore,

$$\overline{\Omega} \cap \varphi^{-1}(\overline{Q}_\rho) \subseteq \bigcup_{\mu=1}^{p} \overline{R}_{\rho\mu}$$

which proves g).

d) The intersection $L_\mu'' = L_3' \cap P_\mu''$ is a poly-disc with center 0 in $\mathbb{C}^q$. Then $\eta_{\mu\nu}(L_\mu'' \times Q_3') = L_{\mu\nu}$ is an open subset of $P_{3\mu} \cap T_{\mu\nu}$ and $f_1,\ldots,f_d$ are holomorphic on $L_{\mu\nu}$. Moreover, $u_{\mu\nu} = \eta_{\mu\nu}(0) \in L_{\mu\nu}$. Then $\alpha_{\mu\nu} = \alpha_\mu | L_{\mu\nu} = \eta_{\mu\nu}^{-1} | L_{\mu\nu}: L_{\mu\nu} \to L_\mu'' \times Q_3'$ is biholomorphic. The image $\gamma_{\mu\nu}(L_\mu'' \times Q_3') = L_{\mu\nu}^0$ is open in $\mathbb{C}^m$ and $\gamma_{\mu\nu}: L_\mu'' \times Q_3' \to L_{\mu\nu}^0$ is biholomorphic, which proves d); q.e.d.

(Σ_{12}) Define $J = \{\mu \in \mathbb{N} | 1 \leqq \mu \leqq p\}$ and $\tau: J \to I$ by $\tau(\mu) = \gamma(x_\mu)$.

<u>Lemma.</u> a) If $\mu \in J$, then $\overline{P}_{\rho\mu} \subseteq M_{x_\mu} \subseteq N_{x_\mu} \subset \overline{N}_{x_\mu} \subset V_{\tau(\mu)}$ for $\rho = 1,2,3$.

b) Define

$$P_\rho = \{P_{\rho\mu}\}_{\mu\in J} \qquad \overline{P}_\rho = \{\overline{P}_{\rho\mu}\}_{\mu\in J}$$

$$R_\rho = \{R_{\rho\mu}\}_{\mu\in J} \qquad \overline{R}_\rho = \{\overline{R}_{\rho\mu}\}_{\mu\in J}$$

$$\partial R_\rho = \{\partial R_{\rho\mu}\}_{\mu\in J}.$$

The $\overline{P}_\rho$, $\overline{R}_\rho$, ∂R_ρ are finite compact families, with $\overline{R}_\rho$ contained in $\overline{P}_\rho$ and ∂R_ρ contained in $\overline{R}_\rho$. Also $(\overline{P}_\rho,\tau)$, $(\overline{R}_\rho,\tau)$ and $(\partial R_\rho,\tau)$ are refinements of $\mathfrak{W}$.

<u>Proof.</u> a) Take $\mu \in J$. Define $i = \tau(\mu) = \gamma(x_\mu)$. By construction

$$\overline{P}_{\rho\mu} \subseteq M_{x_\mu} \subseteq N_{x_\mu} \subseteq \overline{N}_{x_\mu} \qquad \text{and} \qquad x_\mu \in M_{x_\mu}$$

$$\{z \in X|\ \ (z,x_\mu) \leqq \vartheta_0\} \subset V_{\gamma(x\ )} = V_i.$$

Also, diam $N_{x_\mu} \leqq \frac{2}{3}\vartheta_0$. Hence, if $z \in \overline{N}_{x_\mu}$ then $\vartheta(z,x_\mu) \leqq \frac{2}{3}\vartheta_0 < \vartheta_0$. Therefore, $\overline{N}_{x_\mu} \subset V_i$, which proves a). b) is an immediate consequence; q.e.d.

(Σ_{13}) <u>Lemma.</u> Take $\mu \in J$ and $y \in Q_\rho$. Define $i = \tau(\mu)$. Suppose that f is holomorphic on $\varphi^{-1}(y) \cap V_i$. Suppose that $x \in \overline{P}_{\rho\mu} \cap \varphi^{-1}(y)$ is given. Then $x' \in \overline{\Omega} \cap \varphi^{-1}(y) \cap V_i$ exists such that $|f(x)| \leqq |f(x')|$.

<u>Proof.</u> If $x \in \overline{\Omega}$, take $x = x'$. If $x_\mu \in \Omega$, then $M_{x_\mu} \subset \Omega$ Hence, $x \in \overline{P}_{\rho\mu} \subseteq M_{x_\mu} \subseteq \Omega$ and $x = x'$. If $x_\mu \in \overline{\Omega} - \Omega$, then

$x_\mu \in K_3 \cap (\overline{\Omega}-\Omega) \cap \tilde{G}_3$ and

$$x \in \overline{P}_{\rho\mu} \subseteq M_{x_\mu} \cap \varphi^{-1}(y) \subseteq \mathcal{L}(N_{x_\mu} \cap \Omega \cap \varphi^{-1}(y) \| N_{x_\mu} \cap \varphi^{-1}(y))$$

because $y = \varphi(x) \in \varphi(P_{\rho\mu}) \subseteq \varphi(N_{x_\mu})$. Here $\overline{N}_{x_\mu} \subseteq V_1$ by (Σ_{12}). Define $A = N_{x_\mu} \cap \Omega \cap \varphi^{-1}(y)$. Then $|f(x)| \leqq \|f\|_A$. Now, $C = \overline{N}_{x_\mu} \cap \overline{\Omega} \cap \varphi^{-1}(y)$ is compact with $A \subseteq C \subseteq V_1 \cap \varphi^{-1}(y)$. Hence, $x' \in C \subseteq \Omega \cap V_1 \cap \varphi^{-1}(y)$ exists such that $\|f\|_A \leqq |f(x')|$. Hence, $|f(x)| \leqq |f(x')|$; q.e.d.

(Σ_{14}) <u>Lemma.</u> If $y \in \overline{Q}_\rho$, each branch of $\varphi^{-1}(y)$ intersects $\bigcup_{\mu=1}^{p} \overline{R}_{\rho\mu}$. If $y \in Q_\rho$, each branch of $\varphi^{-1}(y)$ intersects $\bigcup_{\mu=1}^{p} R_{\rho\mu}$.

<u>Proof.</u> Leb B be a branch of $\varphi^{-1}(y)$. Then $B \cap \Omega \neq \emptyset$ hence,

$$\emptyset \neq B \cap \overline{\Omega} \subseteq \varphi^{-1}(\overline{Q}_\rho) \cap \overline{\Omega} \subseteq \bigcup_{\mu=1}^{p} \overline{R}_{\rho_\mu} \text{ if } y \in \overline{Q}_\rho$$

$$\emptyset \neq B \cap \Omega \subseteq \varphi^{-1}(Q\ ) \cap \overline{\Omega} \subseteq \bigcup_{\mu=1}^{p} R_{\rho\mu} \text{ if } y \in Q_\rho; \quad \text{q.e.d.}$$

(Σ_{15}) <u>Lemma.</u> Each branch of $\tilde{Q}_\rho = \varphi^{-1}(Q_\rho)$ intersects $\bigcup_{\mu=1}^{p} R_{\rho\mu}$. Moreover, $\varphi_{Q_\rho} = \varphi\colon \tilde{Q}_\rho \to Q_\rho$ is quasi-proper.

<u>Proof.</u> The map $\varphi_G = \varphi\colon \tilde{G} \to G$ is quasi-proper and Q_ρ is open in G. Hence, $\varphi_{Q_\rho} = \varphi_G\colon \tilde{Q}_\rho \to Q_\rho$ is quasi-proper. Let B be a

branch of $\tilde{Q}_\rho$. Take a simple point x of $\tilde{Q}_\rho$ and define $y = \varphi(x)$. Then $y \in Q_\rho$. Let B_0 be a branch of $\varphi^{-1}(y)$ with $x \in B_0$. Because x is a simple point of $\tilde{Q}_\rho$ and $x \in B$, this implies $B_0 \subseteq B$. By (Σ_{14})

$$\emptyset = B_0 \cap \bigcup_{\mu=1}^{p} R_{\rho\mu} \subseteq B \cap \bigcup_{\mu=1}^{p} R_{\rho\mu}.$$

(Σ_{16}) <u>Lemma.</u> If $\mu \in J$, if $j = \tau(\mu)$ and if $i \in I_\mu$, then $\partial R_{1\mu} \subseteq V_i^1 \cap V_j \subseteq V_{ij}^1$. Moreover, g_{ij} is holomorphic on $\partial R_{1\mu}$ and

$$\Gamma = \sup\{\|g_{i\tau(\mu)}\|_{\partial R_{1\mu}} \mid i \in I_\mu \text{ and } \mu \in J\}$$

is finite with $1 \leqq \Gamma < \infty$.

<u>Proof.</u> If $i \in I_\mu$, then $\partial R_{1\mu} \subseteq \overline{P}_{3\mu} \subseteq V_i^1$ by $(\Sigma_{11}e)$. Moreover, $\partial R_{1\mu} \subseteq \overline{P}_{3\mu} \subseteq V_{\tau(\mu)} = V_j$ by (Σ_{12}). Hence $\partial R_{1\mu} \subseteq V_i^1 \cap V_j \subseteq V_{ij}^1$ and g_{ij} is meromorphic on V_{ij}^1 and holomorphic on $V_{ij}^1 - E_j$. Now, $\partial R_{1\mu} \cap E_j = \partial R_{1\mu} \cap V_j \cap E_j \subseteq \partial R_{1\mu} \cap D_4 = \emptyset$. Hence, g_{ij} is holomorphic on the compact set $\partial R_{1\mu}$. Hence, $\|g_{i\tau(\mu)}\|_{\partial R_{1\mu}} < \infty$ for $\mu \in J$. Therefore, $\Gamma < \infty$. Take any $\mu \in J$. Define $j = \tau(\mu)$. Then $x_\mu \in M_{x_\mu} \subseteq V_{\tau(\mu)} = V_j$. Hence, $j \in I_\mu$. Hence,

$$\Gamma \geqq \|g_{jj}\|_{\partial R_{1\mu}} = 1; \quad \text{q.e.d.}$$

(Σ_{17}). Define $H = Q_2$ and $\tilde{H} = \varphi^{-1}(H)$. Define $\mathfrak{W}^\lambda = \mathfrak{V}^\lambda \cap \tilde{H} = \{V_i^\lambda \cap \tilde{H}\}_{i \in I}$ and $W_i^\lambda = V_i^\lambda \cap \tilde{H}$ for $\lambda = 0,1,2$. Define $\mathfrak{W} = \mathfrak{W}^0$ and $W_i = W_i^0 = V_i \cap \tilde{H}$. Define $I[p,\lambda] = I[p,\mathfrak{W}^\lambda]$. For $\lambda = 0,1$, write

$$\Gamma(\mathfrak{w}^\lambda, g^t | \mathfrak{w}^\lambda) = \Gamma(\mathfrak{w}^\lambda, g^t) \quad \text{if} \quad t \in \mathbb{N}.$$

Take $s \in C^0(\mathfrak{w}^\lambda, \mathfrak{O})$ where $\lambda \in \{0,1,2\}$. Then $s = \{s_i\}_{i\in I[0,\lambda]}$ where s_i is holomorphic on W_i^λ. If $\mu \in J$. Then

$$\overline{R}_{1\mu} \subseteq \overline{P}_{1\mu} \subset V_{\tau(\mu)} \cap \tilde{Q}_2 = V_{\tau(\mu)} \cap \tilde{H} = W^0_{\tau(\mu)} = W_{\tau(\mu)}.$$

Hence, $s_{\tau(\mu)}$ is holomorphic on the open neighborhood $W^0_{\tau(\mu)}$ of the compact sets $\overline{R}_{1\mu}$ and $\overline{P}_{1\mu}$. Hence,

$$\|s\| = \sup\{\|s_{\tau(\mu)}\|_{\overline{P}_{1\mu}} | \mu \in J\} < \infty$$

$$\langle s\rangle = \sup\{\|s_{\tau(\mu)}\|_{\overline{R}_{1\mu}} | \mu \in J\} < \infty$$

are defined, finite and are semi-norms on $C^0(\mathfrak{w}^\lambda, \mathfrak{O})$. Obviously

$$\langle s\rangle \leqq \|s\|.$$

(Σ_{18}). <u>Lemma</u>. $\langle\ \rangle$ and $\|\ \|$ are norms on $\Gamma(\mathfrak{w}^\lambda, g^t)$ if $\lambda = 0,1$ and if $t \in \mathbb{N}$.

<u>Proof.</u> Take $s \in \Gamma(\mathfrak{w}^\lambda, g^t)$ with $\langle s\rangle = 0$. Then $s_{\tau(\mu)} \equiv 0$ on $R_{1\mu}$. Define $N = \tilde{H} - \operatorname{supp} s$. Then $R_{1\mu} \subseteq N$ for $\mu = 1,\ldots,p$. Let $\mathfrak{L}$ be the set of branches of $\tilde{H}$. Define

$$\mathfrak{L}_0 = \{B \in \mathfrak{L} \,|\, B \cap N \neq \emptyset\}$$

and $\mathcal{B}_1 = \mathcal{B} - \mathcal{B}_0$. Then $\operatorname{supp} S = \bigcup_{B \in \mathcal{B}_1} B$. Each branch of $R_{2\mu}$ intersects $R_{1\mu}$ because $\alpha_\mu : R_{2\mu} \to R'_{2\mu}$ is proper, light, surjective and open, and because $R_{1\mu} = \alpha_\mu^{-1}(R'_{1\mu})$. Hence, $s_{\tau(\mu)} \equiv 0$ on $R_{2\mu}$. Therefore, $R_{2\mu} \subseteq N$. Each branch $B \in \mathcal{B}$ intersects some $R_{2\mu}$, i.e., $B \cap R_{2\mu} \neq \emptyset$ for some μ by (Σ_{15}). Hence, $B \cap N \neq \emptyset$ for each $B \in \mathcal{B}$. Therefore, $\mathcal{B}_0 = \mathcal{B}$ and $\mathcal{B}_1 = \emptyset$. Hence, $\operatorname{supp} s = \emptyset$. By Lemma 14.2 $s = 0$. If $s \in \Gamma(\mathfrak{M}^\lambda, g)$ with $\|s\| = 0$, then $\langle s\rangle \leqq \|s\|$ implies $\langle s\rangle = 0$; q.e.d.

(Σ_{19}) <u>Lemma.</u> Take $t \in \mathbb{N}$. Take $s \in \Gamma(\mathfrak{M}^1, g^t)$. Then

$$\|s\| \leqq \Gamma^t \langle s\rangle.$$

<u>Proof.</u> Take $\nu \in J$. Define $i = \tau(\nu)$. Take $x \in \overline{P}_{1\nu}$. Define $y = \varphi(x) \in H$. Then $\overline{P}_{1\nu} \subseteq W_i$ and

$$x \in \varphi^{-1}(y) \cap W_i = \varphi^{-1}(y) \cap \tilde{H} \cap V_i = \varphi^{-1}(y) \cap V_i$$

and s_i is holomorphic on $\varphi^{-1}(y) \cap V_j$ with $x \in \overline{P}_{1\nu} \cap \varphi^{-1}(y)$. By (Σ_{13}), $x' \in \overline{\Omega} \cap \varphi^{-1}(y) \cap V_i$ exists such that $|s_i(x)| \leqq |s_i(x')|$. Hence, $x' \in \overline{\Omega} \cap \varphi^{-1}(Q_1)$. By $(\Sigma_{11}g)$, $x' \in R_{1\mu}$ for some index $\mu \in J$. Define $j = \tau(\mu)$. Then $\overline{R}_{1\mu} \subseteq V_j \cap \tilde{H} = W_j$. Also, $x' \in \overline{P}_{3\mu} \cap V_i$. Hence, $i \in I_\mu$ and $\overline{P}_{3\mu} \subseteq V_i^1$. Therefore, $\overline{R}_{1\mu} \subset V_i^1 \cap \tilde{H} = W_i^1$. Hence, s_i is holomorphic on an open neighborhood of $\overline{R}_{1\mu}$. By Lemma 13.4, $x'' \in \partial R_{1\mu}$ exists such that $|s_i(x')| \leqq |s_i(s'')|$. Now $\overline{R}_{1\mu} \subseteq W_i^1 \cap W_j^1 = W_{ij}^1$. Hence $s_i = g_{ij}^t s_j$. Now, $\partial R_{1\mu} \subseteq V_i \cap V_j$ and $i \in I_\mu$.

By (Σ_{16}) g_{ij} is holomorphic on $\partial R_{1\mu}$ and $|g_{ij}(x'')| \leqq \Gamma$. Hence,

$$|s_i(x)| \leqq |s_i(x')| \leqq |s_i(x'')| = |g_{ij}(x'')|^t |s_j(x'')| \leqq \Gamma^t |s_j(x'')|$$

$$\leqq \Gamma^t \langle s \rangle$$

for each $x \in \overline{P}_{1\nu}$. Hence, $\|s_{\tau(\nu)}\|_{\overline{P}_{1\nu}} \leqq \Gamma^t \langle s \rangle$ for each $\nu \in J$.

Therefore, $\|s\| \leqq \Gamma\langle s \rangle$; q.e.d.

Observe that Γ is independent of the choice of $t \in \mathbb{N}$ and of the choice of $s \in \Gamma(\mathfrak{W}^1, g^t)$.

(Σ_{20}). Let $0 \leqq h \in \mathbb{Z}$ be integers. Let $s \in C^0(\mathfrak{W}^\lambda, \mathcal{O})$ with $\lambda \in \{0,1,2\}$. Suppose that $h \leqq \sigma_{s_j}(x)$ for all $x \in S_{1\mu}$ with $j = \tau(\mu)$ and for all $\mu \in J$. Then

$$\langle s \rangle \leqq \theta^h \|s\|.$$

Recall that $S_{1\mu}$ is the common central section of $P_{1\mu}$ and $R_{1\mu}$ defined in $(\Sigma_{11}c)$ and that θ was defined in $(\Sigma_{11}c)$ and does not depend on the choices of h and $s \in C^0(\mathfrak{W}^\lambda, \mathcal{O})$, and that $0 < \theta < 1$.

<u>Proof.</u> Take $\mu \in J$. Define $j = \tau(\mu)$. Then

$$\overline{R}_{1\mu} \subseteq \overline{P}_{1\mu} \subseteq V_j \cap \tilde{H} = W_j \subseteq W_j^\lambda.$$

Hence, s_j is holomorphic in an open neighborhood of $\overline{P}_{1\mu}$, with

$\sigma_{s_j}(x) \geqq h$ for all $x \in S_{1\mu}$. By Proposition 13.6

$$\|s_j\|_{\overline{R}_{1\mu}} \leqq \theta_\mu^h \|s_j\|_{\overline{P}_{1\mu}} \leqq \theta^h \|s\|$$

for all $\mu \in J$. Hence, $\langle s \rangle \leqq \theta^h \|s\|$; q.e.d.

(Σ_{21}) Lemma. Take $t \in \mathbb{N}$ and $h \in \mathbb{N}$ such that

$$h > t \frac{\log \Gamma}{\log {}^1/_\theta} .$$

Suppose that $s \in \Gamma(\mathfrak{W}^1, g^t)$ and that $\sigma_{s_j}(x) \geqq h$ for all $x \in S_{1\mu}$ with $j = \tau(\mu)$ and all $\mu \in J$. Then $s \equiv 0$.

Proof. Obviously

$$h \log \frac{1}{\theta} > t \log \Gamma$$

$$\left(\frac{1}{\theta}\right)^h > \Gamma^t$$

$$1 > \Gamma^t \theta^h > 0.$$

Suppose that $\langle s \rangle > 0$. Then (Σ_{14}), (Σ_{19}) and (Σ_{20}) imply

$$\langle s \rangle \leqq \theta^h \|s\| \leqq \Gamma^t \theta^h \langle s \rangle < \langle s \rangle.$$

Contradiction! Hence, $\langle s \rangle = 0$. By (Σ_{18}), $s = 0$; q.e.d.

(Σ_{22}) Now, the maps $\gamma_{\mu\nu}$ shall be modified. Take $\mu \in J$ and

$\nu \in \mathbb{N}$ with $1 \leqq \nu \leqq \kappa_\mu$. An injective holomorphic map

$$\varphi_{\mu\nu}: Q_3 \to \tilde{Q}_3$$

is defined by

$$\varphi_{\mu\nu}(y) = \alpha_{\mu\nu}^{-1}(0,\beta(y)).$$

Define $S_{\rho\mu\nu} = \alpha_{\mu\nu}^{-1}(\{0\} \times Q'_\rho)$. Then $S_{\rho\mu} = \bigcup_{\nu=1}^{\kappa_\mu} S_{\rho\mu\nu}$ and $S_{\rho\mu\nu} = \varphi_{\mu\nu}(Q_\rho)$. Here $S_{\rho\mu\nu}$ is a smooth n-dimensional complex submanifold of $L_{\mu\nu}$ and $\varphi_{\mu\nu}: Q_\rho \to S_{\rho\mu\nu}$ is biholomorphic for $\rho = 1,2,3$. If $y \in Q_3$, define $x = \varphi_\mu\ (y)$. Then $\alpha_\mu(x) = \alpha_{\mu\nu}(x) = (0,\beta(y)$ and

$$\beta(\varphi(x)) = \pi_\mu(\alpha_\mu(x)) = \pi_\mu(0,\beta(y)) = \beta(y).$$

Because β is injective, $y = \varphi(x) = \varphi(\varphi_{\mu\nu}(y))$. Hence,

$$\varphi \circ \varphi_{\mu\nu} = \mathrm{Id}: Q_3 \to Q_3$$

is the identity.

The functions $f_1,\ldots,f_d$ are holomorphic on $L_{\mu\nu}$. Hence, $g_{\lambda\mu\nu} = f_\lambda \circ \varphi_{\mu\nu}$ is holomorphic on Q_3 and $g'_{\lambda\mu\nu} = f_\lambda \circ \varphi_{\mu\nu} \circ \beta^{-1}$ is holomorphic on Q'_3. Take $x \in S_{3\mu\nu}$. Then $\alpha_\mu(x) = \alpha_{\mu\nu}(x) = (0,z)$ with $z \in Q'_3$. Then $\beta(\varphi(x)) = \pi_\mu(\alpha_\mu(x)) = x \in Q'_3$. Hence,

$$x = \alpha_{\mu\nu}^{-1}(0,\beta(\varphi(x)) = \varphi_{\mu\nu}(\varphi(x)).$$

Therefore,

$$\mathfrak{g}_{\lambda\mu\nu}(\varphi(x)) = f_\lambda(\varphi_{\mu\nu}(\varphi(x)) = f_\lambda(x),$$

if $x \in S_{\mathfrak{z}\mu\nu}$. The function

$$F_{\lambda\mu\nu} = f_\lambda - \mathfrak{g}_{\lambda\mu\nu}\circ\varphi$$

is meromorphic on $\tilde{Q}_{\mathfrak{z}}$ and holomorphic on $L_{\mu\nu}$ with

$$F_{\lambda\mu\nu}(x) = 0 \qquad \text{if } x \in S_{\mathfrak{z}\mu\nu}.$$

The function $F'_{\lambda\mu\nu} = F_{\lambda\mu\nu}\circ\alpha^{-1}_{\mu\nu}$ is holomorphic on $L''_\mu \times Q'_{\mathfrak{z}}$ with

$$F'_{\lambda\mu\nu} = f_\lambda\circ\alpha^{-1}_{\mu\nu} - \mathfrak{g}_{\lambda\mu\nu}\circ\varphi\circ\alpha^{-1}_{\mu\nu} = f_{\lambda\mu\nu} - \mathfrak{g}_{\lambda\mu\nu}\circ\beta^{-1}\circ\pi_\mu$$

$$= f_{\lambda\mu\nu} - \mathfrak{g}'_{\lambda\mu\nu}\circ\pi_\mu.$$

Define

$$\gamma'_{\mu\nu}\colon L''_\mu \times Q'_{\mathfrak{z}} \to L^*_{\mu\nu} = \gamma'_{\mu\nu}(L''_\mu \times Q'_{\mathfrak{z}})$$

by

$$\gamma'_{\mu\nu}(z) = (F'_{1\mu\nu}(z),\ldots,F'_{k\mu\nu}(z),z_{\lambda_1},\ldots,z_{\lambda_{q-k}},z_{q+1},\ldots,z_m)$$

for $z = (z_1,\ldots,z_m) \in L'' \times Q'_{\mathfrak{z}}$. Because $\gamma_{\mu\nu}$ is biholomorphic and because $\mathcal{g}'_{\lambda\mu\nu}\circ\pi_\mu$ depends only on the variables $z_{q+1},\ldots,z_m$, the definition of $\gamma_{\mu\nu}$ shows that $\gamma'_{\mu\nu}$ is injective. Hence, $\gamma'_{\mu\nu}$ is an injective holomorphic map of an open subset of $\mathbb{C}^m$ into $\mathbb{C}^m$. Hence, the image set $L^*_{\mu\nu}$ of $\gamma'_{\mu\nu}$ is open and $\gamma'_{\mu\nu}\colon L''_\mu \times Q'_3 \to L^*_{\mu\nu}$ is biholomorphic. Observe that $\gamma'_{\mu\nu}(0,z) = (0,z)$ if $z \in Q'_3$ and $0 \in L''_\mu$.

$$\gamma'_{\mu\nu}\colon \{0\} \times Q'_3 \to \{0\} \times Q'_3$$

is the identity map and

$$\{0\} \times Q'_3 \subseteq L^*_{\mu\nu}.$$

Hence, $\{0\} \times \overline{Q}'_2$ is a compact subset of $L^*_{\mu\nu}$. A positive number r exists such that

$$\prod_{\nu=1}^{q} \{z_\nu \mid |z_\nu| \leqq r\} \times \overline{Q}'_2 \subseteq L^*_{\mu\nu}.$$

Define

$$Z^*_1 = \prod_{\nu=1}^{k} \{z_\nu \mid |z_\nu| < r\}$$

$$Z^*_2 = \prod_{\nu=k+1}^{q} \{z_\nu \mid |z_\nu| < r\}.$$

Then $\overline{Z}^*_1 \times \overline{Z}^*_2 \times \overline{Q}'_2 \subseteq L^*_{\mu\nu}$ and

$$Z'_{\mu\nu} = (\gamma'_{\mu\nu})^{-1}(Z^*_1 \times Z^*_2 \times Q'_2)$$

is an open connected neighborhood of $0 \in \mathbb{C}^m$ with $\overline{Z}'_{\mu\nu} \subseteq L'' \times Q'_3$. Then $Z_{\mu\nu} = \alpha^{-1}_{\mu\nu}(Z'_{\mu\nu})$ is an open connected neighborhood of $u_{\mu\nu}$ with $\overline{Z}_{\mu\nu}$ compact and $\overline{Z}_{\mu\nu} \subset L_{\mu\nu}$. Then

$$\hat{\alpha}_{\mu\nu} = \gamma'_{\mu\nu} \circ \alpha_{\mu\nu} \colon Z_{\mu\nu} \to Z_1^* \times Z_2^* \times Q'_2$$

is biholomorphic with $S_{2\mu\nu} = \hat{\alpha}^{-1}_{\mu\nu}(\{0\} \times Q_2)$ where $0 \in \mathbb{C}^q$. Hence, $Z_{\mu\nu}$ is an open neighborhood of $S_{2\mu\nu}$. Let $\pi\colon \mathbb{C}^m \to \mathbb{C}^n$ be the projection defined by $\pi(z_1,\ldots,z_m) = (z_{q+1},\ldots,z_m)$. Then

$$\pi \circ \hat{\alpha}_{\mu\nu} = \pi \circ \gamma'_{\mu\nu} \circ \alpha_{\mu\nu} = \pi_\mu \circ \alpha_{\mu\nu} = \beta \circ \varphi .$$

Hence,

$$\begin{array}{ccc} \tilde{H} \supseteq Z_{\mu\nu} & \xrightarrow{\ \hat{\alpha}_{\mu\nu}\ } & Z_1^* \times Z_2^* \times Q'_2 \\ \downarrow \varphi & & \downarrow \pi \\ H = Q_2 & \xrightarrow{\ \beta\ } & Q'_2 \end{array}$$

is a smooth product representation of φ at $u_{\mu\nu}$ and over y_0 where

$$\hat{\alpha}_{\mu\nu}(x) = (F_{1\mu\nu}(x),\ldots,F_{k\mu\nu}(x), z_{\lambda_1},\ldots,z_{\lambda_{q-k}}, z_{q+1},\ldots,z_m)$$

if $\alpha_\mu(x) = (z_1,\ldots,z_m)$ where $g_\lambda(\varphi(x)) = z_{q+\lambda}$.

(Σ_{23}) Take $\mu \in J$ and $\nu \in \mathbb{N}$ with $1 \leqq \nu \leqq \kappa_\mu$. Define $\psi_{\mu\nu} = \varphi\colon Z_{\mu\nu} \to H$. Let $\mathfrak{R}_{\mu\nu}$ be the $\psi^*_{\mu\nu}\mathfrak{H}(H)$-module of all holomorphic functions f on $Z_{\mu\nu}$ such that $f_1|Z_{\mu\nu},\ldots,f_k|Z_{\mu\nu},f$ are $\psi_{\mu\nu}$-dependent on $Z_{\mu\nu}$. Hence,

$$\mathfrak{R}_{\mu\nu} = \mathfrak{H}(Z_{\mu\nu}) \cap \mathfrak{K}_{\psi_{\mu\nu}}(Z_{\mu\nu},f_1|Z_{\mu} ,\ldots,f_k|Z_{\mu\nu}).$$

If $0 \leqq h \in \mathbb{Z}$, define

$$\mathfrak{R}^h_{\mu\nu} = \{f \in \mathfrak{R}_{\mu\nu} | \sigma_f(x) \geqq h \text{ for all } x \in S_{2\mu\nu}\}$$

<u>Lemma.</u> a) $\mathfrak{R}_{\mu\nu}$ is the $\psi^*_{\mu\nu}\mathfrak{H}(H)$-module of all holomorphic functions f on $Z_{\mu\nu}$ such that $F_{1\mu\nu},\ldots,F_{k\mu\nu},f$ are $\psi_{\mu\nu}$-dependent on $Z_{\mu\nu}$, i.e.,

$$\mathfrak{R}_{\mu\nu} = \mathfrak{H}(Z_{\mu\nu}) \cap \mathfrak{K}_{\psi_{\mu\nu}}(Z_{\mu\nu},F_{1\mu\nu},\ldots,F_{k\mu\nu}).$$

Moreover,

$$\mathfrak{K}_{\psi_{\mu\nu}}(Z_{\mu\nu},f_1|Z_{\mu\nu},\ldots,f_k|Z_{\mu\nu}) = \mathfrak{K}_{\psi_{\mu\nu}}(Z_{\mu\nu},F_{1\mu\nu},\ldots,F_{k\mu\nu}).$$

b) If $0 \leqq h \in \mathbb{Z}$, then $\mathfrak{R}^h_{\mu\nu}$ is a $\psi^*_{\mu\nu}\mathfrak{H}(H)$-submodule of $\mathfrak{R}_{\mu\nu}$ and

$$\Delta^h_{\mu\nu} = \mathfrak{R}_{\mu\nu}/\mathfrak{R}^{h+1}_{\mu\nu}$$

is a free $\psi^*_{\mu\nu}\mathfrak{H}(H)$-module of rank $\binom{k+h}{h}$.

c) If f is meromorphic on $\tilde{H}$ and if $f_1|\tilde{H},\ldots,f_k|\tilde{H},f$ are φ_H-dependent, then $f|Z_{\mu\nu} \in \mathfrak{K}_{\psi_{\mu\nu}}(Z_{\mu\nu},F_{1\mu\nu},\ldots,F_{k\mu\nu})$. If f is also holomorphic on $Z_{\mu\nu}$ then $f|Z_{\mu\nu} \in \mathfrak{R}_{\mu\nu}$.

d) If $f \in \mathfrak{K}_{\psi_{\mu\nu}}(Z_{\mu\nu},F_{1\mu\nu},\ldots,F_{k\mu\nu})$, then $f^0 \in \mathfrak{R}_{\mu\nu}$ and and $f^1 \in \mathfrak{R}_{\mu\nu}$ exist such that $f^0 \not\equiv 0$ and $f = {f^1}/{f^0}$ and such that the germs f^0_x and f^1_x of f^0 and f^1 respectively are coprime at every $x \in Z_{\mu\nu}$.

<u>Proof.</u> Observe that

$$\omega = df_1 \wedge \cdots \wedge df_k \wedge dg_1 \circ \varphi \wedge \cdots \wedge dg_n \circ \varphi$$

is holomorphic on $Z_{\mu\nu}$ and that $\omega \not\equiv 0$ on $Z_{\mu\nu}$. Define

$$\omega_{\mu\nu} = dF_{1\mu\nu} \wedge \cdots \wedge dF_{k\mu\nu} \wedge dg_1 \circ \varphi \wedge \cdots \wedge dg_n \circ \varphi$$

on $Z_{\mu\nu}$. By Lemma 15.3.1, $F_{1,\mu\nu},\ldots,F_{k,\mu\nu}$ are $\psi_{\mu\nu}$-independent. Hence, $\omega_{\mu\nu} \not\equiv 0$ on $Z_{\mu\nu}$. (Obviously, $\omega_{\mu\nu}(x) \neq 0$ for $x \in Z_{\mu\nu}$.) By Lemma 4.5 and Lemma 4.6

$$\mathfrak{K}_{\psi_{\mu\nu}}(Z_{\mu\nu},f_1|Z_{\mu\nu},\ldots,f_1|Z_{\mu\nu}) = \{f \in \mathfrak{K}(Z_{\mu\nu})\,|\,df \wedge \omega = 0 \text{ on } Z_{\mu\nu} - P_f\}$$

$$\mathfrak{K}_{\psi_{\mu\nu}}(Z_{\mu\nu},F_{1\mu\nu},\ldots,F_{k\mu\nu}) = \{f \in \mathfrak{K}(Z_{\mu\nu})\,|\,df \wedge \omega_{\mu\nu} = 0 \text{ on } Z_{\mu\nu} - P_f\}$$

where P_f is the pole set of f. Now,

$$dF_{\lambda\mu\nu} \wedge dg_1 \circ \varphi \wedge \cdots \wedge dg_n \circ \varphi$$

$$= (df_\lambda - d(\mathfrak{g}_{\lambda\mu\nu} \circ \varphi)) \wedge dg_1 \circ \varphi \wedge \cdots \wedge dg_n \circ \varphi$$

$$= df_\lambda \wedge dg_1 \circ \varphi \wedge \cdots \wedge dg_n \circ \varphi - \varphi^*(d\mathfrak{g}_{\lambda\mu\nu} \wedge dg_1 \wedge \cdots \wedge dg_n)$$

$$= df_\lambda \wedge dg_1 \circ \varphi \wedge \cdots \wedge dg_n \circ \varphi,$$

because H has dimension n. Hence, $\omega | Z_{\mu\nu} = \omega_{\mu\nu}$. Therefore,

$$\mathfrak{S}_{\psi_{\mu\nu}}(Z_{\mu\nu}, f_1 | Z_{\mu\nu}, \ldots, f_k | Z_{\mu\nu}) = \psi_{\mu\nu}(Z_{\mu\nu}, F_{1\mu\nu}, \ldots, F_{k\mu\nu})$$

which proves a).

Now, a) and Lemma 15.3 imply b) and d). If f is meromorphic on $\tilde{H}$ and if $f_1 | \tilde{H}, \ldots, f_k | \tilde{H}, f$ are φ_H-dependent, then $\omega_{\mu\nu} \wedge df = \omega \wedge df = 0$ on $Z_{\mu\nu} - P_f$ by Lemma 4.5, which implies c); q.e.d.

The map $\psi^*_{\mu\nu}$: $\mathfrak{H}(H) \to \psi^*_{\mu\nu}\mathfrak{H}(H)$ is an isomorphism. If this is regarded as an identification isomorphism, then $\mathfrak{R}_{\mu\nu}, \mathfrak{R}^h_{\mu\nu}, \Delta^h_{\mu\nu}$ are $\mathfrak{H}(H)$-modules and their direct sum can be taken:

$$\mathfrak{R} = \bigoplus_{\mu=1}^{\mu} \bigoplus_{\nu=1}^{\kappa_\mu} \mathfrak{R}_{\mu\nu}$$

$$\mathfrak{R}^h = \bigoplus_{\mu=1}^{p} \bigoplus_{\nu=1}^{\kappa_\mu} {}^{h}_{\mu\nu}$$

$$\Delta^h = \bigoplus_{\mu=1}^{p} \bigoplus_{\nu=1}^{\kappa_\mu} \Delta^h_{\mu\nu} .$$

Let $\mathfrak{r}^h_{\mu\nu}: \mathfrak{R}_{\mu\nu} \to \Delta^h_{\mu\nu}$ be the residual map. Then

$$\mathfrak{r}^h = \bigoplus_{\mu=1}^{p} \bigoplus_{\nu=1}^{\kappa_\mu} \mathfrak{r}^h_{\mu\nu}: \mathfrak{R} \to \Delta^h$$

is a surjective, $\mathfrak{H}(H)$-module homomorphism with the kernel $\mathfrak{R}^{h+1}$. Here Δ^h is a free $\mathfrak{H}(H)$-module of rank $\kappa\binom{k+h}{h}$ where

$$\kappa = \kappa_1 + \ldots + \kappa_p.$$

(Σ_{24}) Let $\mathfrak{m}$ be a non-negative integer. Let $\xi_0,\ldots,\xi_k$ be indeterminants over $\mathfrak{H}(H)$. Let $\mathcal{H}(\mathfrak{m})$ be the $\mathfrak{H}(H)$-module of all homogeneous polynomials of degree $\mathfrak{m}$ over $\mathfrak{H}(H)$ in the variables $\xi_0,\ldots,\xi_k$, i.e.,

$$A(\xi_0,\ldots,\xi_k) = \sum_{\mu_0+\ldots+\mu_k=\mathfrak{m}} a_{\mu_0 \ldots \mu_k} \xi_0^{\mu_0} \ldots \xi_k^{\mu_k}$$

where $a_{\mu_0 \ldots \mu_k} \in \mathfrak{H}(H)$. Then $\mathcal{H}(\mathfrak{m})$ is a free $\mathfrak{H}(H)$-module of rank $\binom{k+\mathfrak{m}}{k}$. Let $\mathfrak{n}$ be a non-negative integer. Let η_0 and η_1 be additional indeterminants over $\mathfrak{H}(H)$. Then they are also indeterminants over $\mathcal{H}(\mathfrak{m})$. Let $\mathcal{H}(\mathfrak{m},\mathfrak{n})$ be the $\mathfrak{H}(H)$-module of all homogeneous polynomials of degree $\mathfrak{n}$ over $\mathcal{H}(\mathfrak{m})$ in the variables η_0,η_1, i.e., of

$$A(\xi_0,\ldots,\xi_k,\eta_0,\eta_1) = \sum_{\lambda=0}^{\mathfrak{n}} A_\lambda(\xi_0,\ldots,\xi_k)\eta_0^{\mathfrak{n}-\lambda}\eta_1^{\lambda}$$

where $A_\lambda = A_\lambda(\xi_0,\ldots,\xi_k) \in \mathcal{H}(\mathfrak{m})$. Then $\mathcal{H}(\mathfrak{m},\mathfrak{n})$ is a free $\mathfrak{H}(H)$-module of rank $(\mathfrak{n}+1)\binom{k+\mathfrak{m}}{k}$ with $\mathcal{H}(\mathfrak{m},0) = \mathcal{H}(\mathfrak{m})$.

Take $t \in \mathbb{N}$ and $h \in \mathbb{Z}$ with $h \geqq 0$. Suppose that sections

$$\upsilon_1 \in \Gamma(\mathfrak{W}^1, g^t) \qquad \upsilon_0 \in \Gamma(\mathfrak{W}^1, g^t, \mathfrak{r})$$

are given. Then $f = {}^{\upsilon_1}/_{\upsilon_0} \in \mathfrak{K}(\tilde{H})$ is a meromorphic function on $\tilde{H}$. Suppose that $f_1|\tilde{H}, \ldots, f_k|\tilde{H}, f$ are φ_H-dependent. Take $\mu \in J$ and $\nu \in \mathbb{N}$ with $1 \leqq \nu \leqq \kappa_\mu$. Lemma (Σ_{23}) c) and d) $f^0_{\mu\nu} \in \mathfrak{O}_{\mu\nu}$ and $f^1_{\mu\nu} \in \mathfrak{O}_{\mu\nu}$ exist such that $f^0_{\mu\nu} \not\equiv 0$ on $Z_{\mu\nu}$, such that the germs of $f^0_{\mu\nu}$ and $f^1_{\mu\nu}$ are coprime at every point of $Z_{\mu\nu}$ and such that

$$f|Z_{\mu\nu} = {}^{f^1_{\mu\nu}}/_{f^0_{\mu\nu}} .$$

Take

$$A = A(\xi_0, \ldots, \xi_k, \eta_0, \eta_k) = \sum_{\lambda=0}^{n} A_\lambda(\xi_0, \ldots, \xi_k)\eta_0^{-\lambda}\eta_1^\lambda$$

in $\mathcal{H}(m, n)$ with

$$A_\lambda(\xi_0, \ldots, \xi_k) = \sum_{\mu_0+\ldots+\mu_k=m} a_{\mu_0 \ldots \mu_k\lambda}\xi_0^{\mu_0} \ldots \xi_k^{\mu_k}$$

in $\mathcal{L}(m)$. Define

$$A_\lambda(1, f_1, \ldots, f_k) = \sum_{\mu_0+\ldots+\mu_k=m} (a_{\mu_0 \ldots \mu_k\lambda} \circ \varphi) f_1^{\mu_1} \ldots f_k^{\mu_k}$$

on $\tilde{H}$. Then $A_\lambda(1,f_1,\ldots,f_k)$ is meromorphic on $\tilde{H}$ and holomorphic on $Z_{\mu\nu}$. Moreover, $f_1|\tilde{H},\ldots,f_k|\tilde{H}$, $A_\lambda(1,f_1,\ldots,f_k)$ are φ_H-dependent. Especially,

$$A_\lambda(1,f_1,\ldots,f_k)|Z_{\mu\nu} \in \mathfrak{R}_{\mu\nu}.$$

Hence,

$$\mathfrak{y}_{\mu\nu}(A) = \sum_{\lambda=0}^{\mathfrak{w}} A_\lambda(1,f_1,\ldots,f_k)\cdot(f^0_{\mu\nu})^{\mathfrak{w}-\lambda}(f^1_{\mu\nu})^\lambda \in \mathfrak{R}_{\mu\nu}.$$

Obviously

$$\mathfrak{y}_{\mu\nu}\colon \mathcal{H}(\mathfrak{m},\mathfrak{w}) \to \mathfrak{R}_{\mu\nu}$$

is an $\mathfrak{h}(H)$-module homomorphism. Hence

$$\chi^h_{\mu\nu} = \mathfrak{r}^h_{\mu\nu}\circ\mathfrak{y}_{\mu\nu}\colon \mathcal{H}(\mathfrak{m},\mathfrak{w}) \to \Delta^h_{\mu\nu}$$

and

$$\chi^h = \bigoplus_{\mu=1}^{p}\ \bigoplus_{\nu=1}^{\kappa_\mu} \chi^h_{\mu\nu}\colon \mathcal{H}(\mathfrak{m},\mathfrak{w}) \to \Delta^h$$

and $\mathfrak{h}(H)$-modules. Observe that $\chi^h_{\mu\nu}$ and χ^h depend on $\mathfrak{m}$, $\mathfrak{w}$, t, h, $\mathfrak{v}_0$ and $\mathfrak{v}_1$.

<u>Lemma:</u> Define $\Gamma_0 \in \mathbb{N}$ by

$$\Gamma_0 - 1 < \mathrm{Max}\left(k,\frac{\log\Gamma}{\log\frac{1}{\theta}}\right) \leq \Gamma_0.$$

Suppose that $h \in \mathbb{N}$ with $h \geqq (\mathfrak{w} + \mathfrak{w}t)\Gamma_0$. Suppose that $A \in \mathfrak{X}(\mathfrak{w}, \mathfrak{w})$. Define

$$A_\lambda(s_0,\ldots,s_k) = \sum_{\mu_1+\ldots+\mu_k=\mathfrak{w}} (a_{\mu_0\ldots\mu_k} \circ \varphi) s_0^{\mu_0} \ldots s_k^{\mu_k} \in \Gamma(\mathfrak{M}^1, g^{\mathfrak{w}})$$

and

$$\mathfrak{r} = A(s_0,\ldots,s_k,\upsilon_0,\upsilon_1) = \sum_{\lambda=0}^{\mathfrak{w}} A_\lambda(s_0,\ldots,s_k)\upsilon_0^{\mathfrak{w}-\lambda}\upsilon_1^\lambda \in \Gamma(\mathfrak{M}^1, g^{\mathfrak{w}+t\mathfrak{w}}).$$

Then $A_\lambda(s_0,\ldots,s_k)$ are holomorphic sections of $g^{\mathfrak{w}}|\mathfrak{M}^1$ and $\mathfrak{r}$ is a holomorphic section of $g^{\mathfrak{w}+t\mathfrak{w}}|\mathfrak{M}^1$. If $\chi^h(A) = 0$, then $\mathfrak{r} = 0$.

<u>Proof.</u> Because $\chi^h(A) = 0$, also $\chi^h_{\mu\nu}(A) = 0$ for $\nu = 1,\ldots,\kappa_\mu$ and $\mu = 1,\ldots,p$. Hence $\mathfrak{y}_{\mu\nu}(A) \in \mathfrak{R}^{h+1}_{\mu\nu}$ which implies

$$\sigma_{\mathfrak{y}_{\mu\nu}(A)}(x) \geqq h + 1 > h \text{ for all } x \in S_{2\mu\nu}.$$

If $W_i^1 \neq \emptyset$, then

$$A_i = A(s_{0i},\ldots,s_{ki},\upsilon_{0i},\upsilon_{1i})$$

is a holomorphic function on W_i^1.

Take $\mu \in J$ and $\nu \in \mathbb{N}$ with $1 \leqq \nu \leqq \kappa_\mu$. Then $Z_{\mu\nu} \subseteq P_{2\mu} \subseteq V_{\tau(\mu)} \cap \tilde{H} = W_i \subseteq W_i^1$ if $i = \tau(\mu)$. On $Z_{\mu\nu}$

$$f^1_{\mu\nu}/f^0_{\mu\nu} = f = \upsilon_{1i}/\upsilon_{0i}.$$

Hence

$$\upsilon_{0i} f^1_{\mu\nu} = \upsilon_{1i} f^0_{\mu\nu} \qquad \text{on } Z_{\mu\nu}.$$

Because $f^0_{\mu\nu}$ and $f^1_{\mu\nu}$ are coprime at every point of the manifold $Z_{\mu\nu}$, the function

$$\Phi_{\mu\nu} = {}^{\upsilon_{0i}}/_{f^0_{\mu\nu}} \not\equiv 0$$

holomorphic on $Z_{\mu\nu}$ and $\upsilon_{1i} = \Phi_{\mu\nu} f^1_{\mu\nu}$ on $Z_{\mu\nu}$. Also $s_{\lambda i} = f_\lambda s_{0i}$ on $Z_{\mu\nu}$ for $\lambda = 1,\ldots,k$. Hence

$$A_i = s_{0i}^{\mathfrak{w}} \cdot \Phi_{\mu\nu}^{\mathfrak{v}} \, \mathfrak{y}_{\mu\nu}(A)$$

by the definition of $\mathfrak{y}_{\mu\nu}(\mathfrak{a})$. If $x \in S_{2\mu\nu}$, then

$$\sigma_{A_i}(x) \geqq \sigma_{\mathfrak{y}_{\mu\nu}}(A)(x) \geqq h + 1.$$

Hence $\sigma_{A_i}(x) \geqq h + 1$ for all $x \in S_{2\mu}$ if $i = \tau(\mu)$ for $\mu = 1,\ldots,p$.

Now, $\mathfrak{a} \in \Gamma(\mathfrak{M}^1, g^{\mathfrak{w}+t})$ and

$$(h+1) \geq (\mathfrak{w} + t\,\mathfrak{v})\Gamma_0 \geq (\mathfrak{w}+t\,\mathfrak{v}) \, \frac{\log \Gamma}{\log \frac{1}{\theta}} .$$

Hence $\mathfrak{a} \equiv 0$ on $\tilde{H}$ by (Σ_{21}); q.e.d.

(Σ_{25}) Theorem. Define

$$\mathscr{w} = \kappa 2^k (\Gamma_0+1)^k \leqq N.$$

Take $t \in \mathbb{N}$ and define

$$\mathscr{W} = \mathscr{w} t \Gamma_0 \in \mathbb{N}.$$

Suppose that holomorphic sections of $g^t | \mathfrak{M}^1$ are given.

$$\upsilon_1 \in \Gamma(\mathfrak{M}^1, g^t) \text{ and } \upsilon_0 \in \Gamma(\mathfrak{M}^1, g^t, \mathfrak{r})$$

such that $f = \upsilon_1/\upsilon_0 \in \mathfrak{K}(\tilde{H})$ and such that $f_1|\tilde{H},\ldots,f_k|\tilde{H},f$ are φ_H-dependent. Let $\xi_0,\ldots,\xi_k$ be indeterminants over $\mathfrak{H}(H)$.

Then there exist homogeneous polynomials of degree

$$A_\lambda = A_\lambda(\xi_0,\ldots,\xi_k) = \sum_{\mu_0,\ldots,\mu_k=\mathscr{w}} a_{\mu_0+\ldots+\mu_k,\lambda}\,\xi_0^{\mu_0}\cdots\xi_k^{\mu_k}$$

over $\mathfrak{H}(H)$, i.e., $a_{\mu_0\ldots\mu_k,\lambda} \in \mathfrak{H}(H)$, for $\lambda = 0,\ldots,\mathscr{W}$, such that the following conditions are satisfied.

1. At least one of the polynomials $A_0,\ldots,A_{\mathscr{W}}$ is not the zero polynomial.

2. Define $A_\lambda(s_0,\ldots,s_k) \in \Gamma(\mathfrak{M}^1, g^{w})$ by

$$A_\lambda(s_0,\ldots,s_k) = \sum_{\mu_0+\ldots+\mu_k=w} (a_{\mu_0\ldots\mu_k,\lambda}\circ\varphi) s_0^{\mu_0}\ldots s_k^{\mu_k}$$

then at least one of these sections is not the zero section, but

$$\sum_{\lambda=0}^{w} A_\lambda(s_0,\ldots,s_k)\upsilon_0^{w-\lambda}\upsilon_1^{\lambda} \equiv 0 \text{ on } \tilde{H}$$

is the zero section in $\Gamma(\mathfrak{M}^1, g^{w+tw})$.

3. Define a meromorphic function $A_\lambda(1\ f_1\ \ldots\ f_k)$ on $\tilde{H}$ by

$$A_\lambda(1,f_1,\ldots,f_k) = \sum_{\mu_0+\ldots+\mu_k=w} (a_{\mu_0,\ldots,\mu_k,\lambda}\circ\varphi) f_1^{\mu_1}\ldots f_k^{\mu_k}.$$

Then at least one of these functions is not a zero divisor in $\mathfrak{K}(\tilde{H})$, but

$$\sum_{\lambda=0}^{w} A_\lambda(1,f_1,\ldots,f_k) f^\lambda \equiv 0$$

on $\tilde{H}$.

4. The functions $f_1|\tilde{H},\ldots,f_k|\tilde{H},f$ are algebraically dependent $\varphi_H^*\mathfrak{K}(H)$ with

$$[f\colon \varphi_H^*\mathfrak{K}(H)(f_1|\tilde{H},\ldots,f_k|\tilde{H})] \leq w$$

where $\varphi_H^*\mathfrak{K}(H)(f_1|\tilde{H},\ldots,f_k|\tilde{H}))$ is a field and a pure transcendental extension of $\varphi_H^*\mathfrak{K}(H)$ of transcendency degree k.

Remark: Observe, that w does not depend on t, υ_0, υ_1. However, $\mathfrak{w}$ depends on $f_1, f_2, \ldots, f_d$ and on the choices of P_ρ, R_ρ, α_μ, β, $s_\lambda, \mathfrak{v}^\lambda, \mathfrak{m}^\lambda$, H. The number $\mathfrak{w}$ depends in addition on t, but not on the individual choice of $\mathfrak{v}_0$ and $\mathfrak{v}_1$.

Proof. For $h \in \mathbb{N}$ is

$$\frac{(h+1)^k}{k!} \leqq \frac{(h+k)\ldots(h+1)}{k!} = \binom{k+h}{k} \leqq \frac{(k+h)^k}{k!}$$

Take $h = (\mathfrak{w}+t\mathfrak{w})\Gamma_0$. Then

$$h = \mathfrak{w}\Gamma_0 + t\mathfrak{w}\Gamma_0 = \mathfrak{w}(\Gamma_0+1) > \Gamma_0 > k.$$

Then

$$\begin{aligned} k!\ \text{rank}\ \Delta^h &= \kappa k!\binom{k+h}{h} \leqq \kappa(h+k)^k < \kappa 2^k h^k \\ &= \kappa 2^k(\Gamma_0+1)^k \mathfrak{w}^k = w\ \mathfrak{w}^k \\ &< (w+1)(\mathfrak{w}+1)^k \leqq (w+1)\binom{\mathfrak{w}+k}{k}\cdot k! = \\ &= k!\ \text{rank}\ \mathfrak{K}(\mathfrak{w}, w). \end{aligned}$$

Hence rank $\Delta^h <$ rank $\mathfrak{K}(\mathfrak{w}, w)$. Suppose that

$$\chi^h\colon \mathfrak{K}(\mathfrak{w}, w) \to \Delta^h$$

is injective. Then rank $\mathfrak{K}(\mathfrak{w}, w) <$ rank Δ^h which is wrong. Hence $A \in \mathfrak{K}(\mathfrak{w}, w)$ exists such that $A \neq 0$ and $\chi^h(A) = 0$. Define

$$A = \sum_{\lambda=0}^{w} A_\lambda(\xi_0, \ldots, \xi_k)\eta_0^{w-\lambda}\eta_1^\lambda$$

with

$$A_\lambda(\xi_0,\dots,\xi_k) = \sum_{\mu_0+\dots+\mu_k=\mathfrak{m}} a_{\mu_0\dots\mu_k,\lambda}\,\xi_0^{\mu_0}\cdots\xi_k^{\mu_k}$$

with $a_{\mu_0,\dots,\mu_k,\lambda} \in \mathfrak{H}(H)$. Because H is a connected complex manifold, $\mathfrak{H}(H)$ is an integral domain and $\mathfrak{K}(H)$ is a field. Because $A \neq 0$, then $a_{\mu_0,\dots,\mu_k,\lambda} \not\equiv 0$ for some $\mu_0\dots\mu_k,\lambda$. Hence $A_\lambda(\xi_\lambda,\dots,\xi_k)$ is not the zero polynomial. Since $\chi^h(A) = 0$, Lemma (Σ_{24}) implies that

$$\sum_{\lambda=0}^{\mathfrak{w}} A_\lambda(s_0,\dots,s_k)\upsilon_0^{\mathfrak{w}-\lambda}\upsilon_1^{\lambda} \equiv 0$$

is the zero section of $\Gamma(\mathfrak{M}^1, g^{\mathfrak{m}+t\mathfrak{w}})$. Now,

$$s_0^{\mathfrak{m}} A_\lambda(1,f_1,\dots,f_k) = A_\lambda(s_0,\dots,s_k) \qquad \text{on } \mathfrak{M}^1$$

with $s_0|\mathfrak{M}^1 \in \Gamma(\mathfrak{M}^1,g,\mathfrak{T})$. Take the index λ such that $A_\lambda(\xi_0,\dots,\xi_k) \not\equiv 0$. If $A_\lambda(1,f_1,\dots,f_k) \equiv 0$ on any open subset U of $\tilde{H}$, define $\varphi_0 = \varphi\colon U \to H$. Then $f_1|U_1,\dots,f_k|U$ are algebraically dependent over $\varphi_0^*\,\mathfrak{K}(H)$. By Corollary 4.24, $f_1,\dots,f_k$ are φ-dependent on X, which is wrong. Hence $A_\lambda(1,f_1,\dots,f_k)$ is not a zero divisor in $\mathfrak{K}(\tilde{H})$ and $A_\lambda(s_0,\dots,s_k)$ is not the zero section (even $A_\lambda(s_0,\dots,s_k) \in \Gamma(\mathfrak{M}^1,g^{\mathfrak{m}},\mathfrak{T})$ is true). Now

$$s_0^{\mathfrak{m}}\upsilon_0^{\mathfrak{w}}\sum_{\lambda=0}^{\mathfrak{w}} A_\lambda(k,f_1,\dots,f_k)f^{\lambda} = \sum_{\lambda=0}^{\mathfrak{w}} A_\lambda(1_0,\dots,1_k)\upsilon_0^{\mathfrak{w}-\lambda}\upsilon_1^{\lambda} \equiv 0$$

is the zero section. Because $s_0^{\mathfrak{m}}\upsilon_0^{\mathfrak{w}} \in \Gamma(\mathfrak{M}^1,g^{\mathfrak{m}+t\mathfrak{w}},\mathfrak{T})$

$$\sum_{\lambda=0}^{w} A_\lambda(1,f_1,\ldots,f_k)f^\lambda \equiv 0$$

on $\tilde{H}$. Hence the functions $f_1|\tilde{H},\ldots,f_k|\tilde{H},f$ are algebraically dependent over $\varphi_H^* \mathfrak{H}(H)$ and hence over $\varphi_H^* \mathfrak{K}(H)$ with

$$[f\colon \varphi_H^* \mathfrak{K}(H)(f_1|H,\ldots,f_k|H)] \leqq w .$$

The map $\varphi_H^*\colon \mathfrak{K}(H) \to \varphi_H^* \mathfrak{K}(H)$ is an isomorphism. Hence $\varphi_H^* \mathfrak{K}(H)$ is a subfield of $\mathfrak{K}(\tilde{H})$. By Lemma 4.26, $\varphi_H^* \mathfrak{K}(H)(f_1|\tilde{H},\ldots,f_k|\tilde{H})$ is a field and a pure transcendental extension of $\varphi_H^* \mathfrak{K}(H)$ of transcendency degree k; q.e.d.

(Σ_{26}) Corollary. If $f_\lambda \in \mathfrak{E}_\varphi(X,f_1,\ldots,f_k)$ and if $1 \leqq \lambda \leqq d$, then $f_1|\tilde{H},\ldots,f_k|\tilde{H},f_\lambda|\tilde{H}$ are algebraically dependent over $\varphi_H^* \mathfrak{K}(H)$ with

$$[f_\lambda|\tilde{H}\colon \varphi^* \mathfrak{K}(H)(f_1|\tilde{H},\ldots,f_k|\tilde{H})] \leqq w.$$

Remark: Observe that w depends on f_λ by the construction of the coverings and sections $s_0,\ldots,s_d$. Therefore, the estimate of the degree is of almost no value.

Proof. Take $\mathfrak{v}_1 = s_\lambda$ and $\mathfrak{v}_0 \in s_0$. Then the assumptions of (Σ_{25}) are satisfied with $t = 1$, q.e.d.

Now, the construction series (Σ_n) will be interrupted to prove the following theorems:

Theorem 16.1. Let X and Y be irreducible complex spaces of dimension m and n respectively with $m - n = q > 0$. Let $\varphi: X \to Y$ be a holomorphic map of rank n, which is pseudoconcave over an open subset $G \neq \emptyset$ of Y with $\tilde{G} = \varphi^{-1}(G) \neq \emptyset$. Let $f_1, \ldots, f_d$ be φ-dependent meromorphic functions on X. Let U be any open, irreducible subset of G. Let $\tilde{U} = \varphi^{-1}(U) \neq \phi$. Then $f_1|\tilde{U}, \ldots, f_d|\tilde{U}$ are algebraically dependent over $\varphi_U^* \mathfrak{K}(U)$ where $\varphi_U = \varphi: \tilde{U} \to U$.

Proof. The set $D_0 = \{x \in X | \text{rank}_x \varphi \leqq n-1\}$ is thin and analytic in X. Hence

$$D_0 \cap \tilde{U} = \{x \in \tilde{U} | \text{rank}_x \varphi_U \leqq n-1\}$$

is thin and analytic in $\tilde{U}$. The restriction $\varphi_U = \varphi_G: \tilde{U} \to U$ is quasi-proper and has strict rank n. By Proposition 8.6 $D_0^* = \varphi(D_0 \cap \tilde{U})$ is an analytic subset of U with $\dim D_0^* \leqq n - 2$.

Hence $V^* = U - D_0^*$ is an open, irreducible subset of U and Y. The map $\varphi_V = \varphi_U: \tilde{V} \to V$ is q-fibering and quasi-proper. By Lemma 8.20, the set $\mathfrak{L}$ of branches of $\tilde{V}$ is finite. If $B \in \mathfrak{L}$, then $\psi_B = \varphi_V|B: B \to V$ is q-fibering and quasi-proper by Proposition 8.10'.

Define $F = \{f_1, \ldots, f_d\}$ and $k = \varphi - \dim F$. A free subset F' of F exists such that $\mathfrak{K}_\varphi(X,F) = \mathfrak{K}_\varphi(X,F')$ and such that $k = \#F'$. The notation can be taken such that $F' = \{f_1, \ldots, f_k\}$ with $k = 0$ meaning that $F' = \phi$. Then $f_1, \ldots, f_k$ are φ-independent and $f_1, \ldots f_1, \ldots, f_k, f_\lambda$ are φ-dependent for $\lambda = 1, \ldots, d$. By Lemma 4.14, $f_1|\tilde{V}, \ldots, f_k|\tilde{V}$ are φ_V-independent and $f_1|\tilde{V}, \ldots, f_k|\tilde{V}, f_\lambda|\tilde{V}$ are φ_V-

dependent. By Lemma 4.4, $f_1|B,\ldots,f_k|B$ are ψ_B-independent and $f_1|B,\ldots,f_k|B,f_\lambda|B$ are ψ_B-dependent.

Now, it is claimed, that $f_1|B,\ldots,f_k|B,f_\lambda|B$ are algebraically dependent over $\psi_B^*\mathfrak{K}(V)$ for $\lambda = k+1,\ldots,d$. If $k = 0$, this is true by Theorem 9.2. Assume $k > 0$. Then conditions (A_1) - (A_6) hold with V instead of G. By (Σ_{20}) an open irreducible subset H of V exists such that $f_1|\tilde{H},\ldots,f_k|\tilde{H},f_\lambda|\tilde{H}$ are algebraically dependent over $\varphi_H^*\mathfrak{K}(H)$ for $\lambda = k+1,\ldots,d$. Now, $B \cap \tilde{H}$ is a union of branches of $\tilde{H}$. Hence, the restrictions $f_1|\tilde{H}\cap B,\ldots,f_k|\tilde{H}\cap B$, $f_\lambda|\tilde{H}\cap B$ exist and are algebraically dependent over

$$(\varphi|\tilde{H}\cap B)^*\mathfrak{K}(H) = (\psi_B|B\cap\tilde{H})^*\mathfrak{K}(H).$$

Because $\psi_B\colon B \to V$ is quasi-proper, this map ψ_B is full. By Theorem 12.7, $f_1|B,\ldots,f_k|B,f_\lambda|B$ are algebraically dependent over $\psi_B^*\mathfrak{K}(V)$. Hence the claim is proved for $k = 0$ and $k > 0$.

Let $\xi_1,\ldots,\xi_d$ be indeterminants over $\mathfrak{K}(V)$. Then

$$A_{\lambda B} = \sum_{\nu=0}^{r(\lambda,B)} A_{\lambda B\nu}(\xi_1,\ldots,\xi_k)\xi_\lambda^\nu \in \mathfrak{K}(V)[\xi_1,\ldots,\xi_k,\xi_\lambda]$$

with

$$A_{\lambda B\nu} = \sum_{\mu_1+\ldots+\mu_k \leqq m(\lambda,B)} a_{\mu_1\ldots\mu_k\lambda B\nu}\,\xi_1^{\mu_1}\cdots\xi_k^{\mu_k}$$

and with $a_{\mu,\ldots,\mu_k\lambda B\nu} \in \mathfrak{K}(V)$ exist such that $A_{\lambda B,r(\lambda,B)} \not\equiv 0$ in $\mathfrak{K}(V)[\xi_1,\ldots,\xi_k]$ and such that

$$A_{\lambda B}(f_1,\ldots,f_k f_\lambda) = \sum_{\nu=0}^{r(\lambda,B)} A_{\lambda B\nu}(f_1,\ldots,f_k) f_\lambda^\nu$$

if meromorphic on $\tilde{V}$ with $A_{\lambda B}(f_1,\ldots,f_k,f_\lambda)|B \equiv 0$ on B. Here

$$A_{\lambda B\nu}(f_1,\ldots,f_k) = \sum_{\mu_1+\ldots+\mu_k \leqq m(\lambda,B)} \varphi_V^*(a_{\mu_1\ldots\mu_p\lambda B\nu}) f_1^{\mu_1}\ldots f_k^{\mu_k}.$$

Since $\mathfrak{B}$ is finite, $r_\lambda = \sum_{B\in\mathfrak{B}} r(\lambda,B) < \infty$ and

$$A_\lambda = \prod_{B\in\mathfrak{B}} A_{\lambda B} = \sum_{\nu=0}^{r_\lambda} C_{\lambda\nu}(\xi_1,\ldots,\xi_k)\xi_\lambda^\nu$$

with

$$C_{\lambda\nu}(\xi_1,\ldots,\xi_k) = \sum_{\mu_1+\ldots+\mu_k \leqq m_\lambda} C_{\mu_1\ldots\mu_k\lambda\nu}\xi_1^{\mu_1}\ldots\xi_k^{\mu_k}$$

and with $C_{\mu_1\ldots\mu_k\lambda,\nu} \in \mathfrak{K}(V)$ exist. Here

$$C_{\lambda r_\lambda} = \prod_{B\in\mathfrak{B}} A_{\lambda B,r(\lambda,B)} \not\equiv 0 \text{ in } \mathfrak{K}(V)[\xi_1,\ldots,\xi_k].$$

Hence $A_\lambda \not\equiv 0$ in $\mathfrak{K}(V)[\xi_1,\ldots,\xi_k,\xi_\lambda]$ and

$$A_\lambda(f_1,\ldots,f_k,f_\lambda) = \prod_{B\in\mathfrak{B}} A_{\lambda B}(f_1,\ldots,f_k,f_\lambda) \equiv 0 \text{ on } \tilde{V}$$

with

$$A_\lambda(f_1,\ldots,f_k f_\lambda) = \sum_{\lambda=0}^{r_\lambda} C_{\lambda\nu}(f_1,\ldots,f_k) f_\lambda^\nu$$

and

$$C_{\lambda\nu}(f_1,\ldots,f_k) = \sum_{\mu_1+\ldots+\mu_k \leqq m_\lambda} \varphi_V^*(C_{\mu_1\ldots\mu_k\lambda\nu}) f_1^{\mu_1}\ldots f_k^{\mu_k}.$$

Observe that $V = U - D_0^*$ where D_0^* is an analytic subset of U with $\dim D_0^* \leqq n - 2$. By Lemma 3.11, meromorphic functions $e_{\mu_1\ldots\mu_k\lambda\nu}$ exist on U such that $e_{\mu_1\ldots\mu_k\lambda\nu}|V = C_{\mu_1\ldots\mu_k\lambda\nu}$.

Then

$$E_{\nu\lambda}(\xi_1,\ldots,\xi_k) = \sum_{\mu_1+\ldots+\mu_k \leqq m_\lambda} e_{\mu_1\ldots\mu_k,\lambda,\nu}\, \xi_1^{\mu_1}\ldots\xi_k^{\mu_k} \in \mathfrak{F}(U)[\xi_1\ldots\xi_k].$$

and

$$0 \neq E_\lambda = \sum_{\nu=0}^{r_\lambda} E_{\lambda\nu}(\xi_1,\ldots,\xi_k)\xi_\lambda^\nu \in \mathfrak{F}(V)[\xi_1,\ldots,\xi_k,\xi_\lambda].$$

By Lemma 1.25, $\varphi_U^{-1}(D_0^*)$ is a thin analytic subset of U. Hence

$$\sum_{\nu=0}^{r_\lambda} E_{\lambda\nu}(f_1,\ldots,f_k)f_\lambda^\nu = E\,(f_1,\ldots,f_k,f_\lambda) \equiv 0 \text{ on } \tilde{U}$$

where

$$E_{\lambda\nu}(f_1,\ldots,f_k) = \sum_{\mu_1+\ldots+\mu_k \leqq m_\lambda} \varphi_U^*(e_{\mu_1\ldots\mu_k,\lambda,\nu}) f_1^{\mu_1}\ldots f_k^{\mu_k}$$

on $\tilde{U}$. Therefore $f_1|\tilde{U},\ldots,f_k|\tilde{U},f_\lambda|\tilde{U}$ are algebraically dependent over $\varphi_U^* \mathfrak{K}(U)$; hence $f_1|\tilde{U},\ldots,f_d|\tilde{U}$ are algebraically dependent over $\varphi_U^* \mathfrak{K}(U)$; q.e.d.

Theorem 16.2. Let X and Y be irreducible complex spaces of dimension m and n respectively with $m - n = q > 0$. Let $\varphi: X \to Y$ be a full holomorphic map of rank n, which is pseudoconcave over an open subset $G \neq \phi$ of Y. Let $f_1,\ldots,f_d$ be φ-dependent meromorphic functions on X. Then $f_1,\ldots,f_d$ are algebraically dependent over $\varphi^* \mathfrak{K}(Y)$.

Remark: The local irreducibility of X required in [2] Theorem 4 is not needed here. Actually, Theorem 4 holds without this requirement by virtue of Theorem 16.2 for $n = 0$.

Proof. By Corollary 11.11, φ is surjective. Hence $\varphi^{-1}(G) = \tilde{G} \neq \emptyset$. Let U be any open, non-empty irreducible subset of G. By Theorem 17.1, $f_1|\tilde{U},\ldots,f_d|\tilde{U}$ are algebraically dependent over $\varphi_U^* \mathfrak{K}(U)$. By Theorem 12.7, $f_1,\ldots,f_d$ are algebraically dependent over $\varphi^* \mathfrak{K}(Y)$; q.e.d.

Now, the assumptions (A_1) - (A_6) are made again and the construction (Σ_1) - (Σ_{26}) is resumed.

(Σ_{27}) Theorem. Let $\tilde{H}$ be normal. Take $f \in \mathfrak{K}_\varphi(X,f_1,\ldots,f_k)$. Then holomorphic sections $\upsilon_0 \in \Gamma(\mathfrak{W}^1,g^t,\mathfrak{T})$ and $\upsilon_1 \in \Gamma(\mathfrak{W}^1,g^t)$ for some $t \in \mathbb{N}$ exist such that $f|\tilde{H} = \upsilon_1/\upsilon_0$.

Proof. By Theorem 16.1, $f_1|\tilde{H},\ldots,f_k|\tilde{H},f|\tilde{H}$ are algebraically dependent over $\varphi_H^* \mathfrak{K}(H)$ because $f_1,\ldots,f_k,f$ are φ-dependent and because H is open and irreducible in G. Let $\xi_0,\ldots,\xi_k,\xi_{k+1}$ indeterminants over $\mathfrak{K}(H)$. Because $\mathfrak{K}(H)$ is the field of quotients of $\mathfrak{H}(H)$, holomorphic functions $a_{\mu_0\ldots\mu_k,\rho} \in \mathfrak{H}(H)$ exist such that

$$A_\rho(\xi_0,\ldots,\xi_k) = \sum_{\mu_0+\ldots+\mu_k=t} a_{\mu_0\ldots\mu_k\rho}\xi_0^{\mu_0}\ldots\xi_k^{\mu_k}$$

with $t > 0$ and

$$A(\xi_0,\ldots,\xi_{k+1}) = \sum_{\rho=0}^{r} A_\rho(\xi_0,\ldots,\xi_k)\xi_{p+1}^{\rho}$$

with $A_r(\xi_0,\ldots,\xi_k) \not\equiv 0$ in $\mathfrak{H}(H)[\xi_0,\ldots,\xi_k]$ and such that

$$\sum_{\rho=0}^{r} A_\rho(1,f_1,\ldots,f_k)\cdot f^\rho \equiv 0$$

on $\tilde{H}$ with

$$A_\rho(1,f_1,\ldots,f_k) = \sum_{\mu_0+\ldots+\mu_k=t}(a_{\mu_0\ldots\mu_k\rho}\circ\varphi)f_1^{\mu_1}\ldots f_k^{\mu_k}$$

on $\tilde{H}$. Define

$$A_{\rho i} = A_\rho(s_{0i},\ldots,s_{ki}) = \sum_{\mu_0+\ldots+\mu_k=t}(a_{\mu_1\ldots\mu_k}\circ\varphi)s_{0i}^{\mu_0}s_{1i}^{\mu_1}\ldots s_{ki}^{\mu_k}$$

$$= s_{0i}^{t}A_\rho(1,f_1,\ldots,f_k)$$

on W_i^1. Suppose that $A_{ri} \equiv 0$ on an open subset U of $\tilde{H}$. Then

$A_r(1,f_1,\ldots,f_k) \equiv 0$ on an open subset U of $\tilde{H}$. Define $\psi = \varphi\colon U \to H$. Then $f_1|U,\ldots,f_k|U$ are algebraically dependent over $\psi^*\mathfrak{K}(H)$. By Corollary 4.24, $f_1,\ldots,f_k$ are φ-dependent, which is wrong. Hence $A_{ri} \in \mathfrak{O}(W_i^1)$ is not a zero divisor of $\mathfrak{F}(W_i^1)$ whenever $W_i^1 \neq \emptyset$. If so, then

$$\sum_{\rho=0}^{r} A_{\rho i} f^\rho = s_{0i}^t \sum_{\rho=0}^{r} A_\rho(1,f_1,\ldots,f_k) f^\rho \equiv 0$$

on W_i^1. Hence

$$\sum_{\rho=0}^{r} A_{\rho i}(A_{ri})^{r-\rho-1}(A_{ri}f)^\rho + (A_{ri}f)^r \equiv 0$$

on W_i^1. Therefore, the germ of $\upsilon_{1i} = A_{ti}f$ at every $x \in W_i^1$ is integral over $\mathfrak{O}_x$. Because $\tilde{H}$ is normal, also W_i^1 is normal and the meromorphic function υ_{1i} is holomorphic on W_i^1. Define $\upsilon_{0i} = A_{ri} \in \mathfrak{O}(W_i^1)$. If $W_i^1 \cap W_j^1 \neq \emptyset$ then

$$\upsilon_{0i} = A_{ri} = s_{0i}^t A_r(1,f_1,\ldots,f_k) = g_{ij}^t s_{0j}^t A_r(1,f_1,\ldots,f_k)$$

$$= g_{ij}^t A_{rj} = g_{ij}^t \upsilon_{0j}$$

$$\upsilon_{1i} = \upsilon_{0i} f = g_{ij}^t \upsilon_{0j} f = g_{ij}^t \upsilon_{1j}.$$

Hence $\upsilon_0 = \{\upsilon_{0i}\}_{i\in I} \in \Gamma(\mathfrak{M}^1, g^t, \mathfrak{O})$ and $\upsilon_1 = \{\upsilon_{1i}\}_{i\in I} \in \Gamma(\mathfrak{M}^1, g^t)$ are holomorphic sections of $g^t|\mathfrak{M}^1$ with $f = {}^{\upsilon_1}/_{\upsilon_0}$; q.e.d.

<u>Theorem 16.3.</u> Let X and Y be irreducible complex spaces of dimension m and n respectively with $m - n = q > 0$. Let $\varphi: X \to Y$ be a holomorphic map of rank n which is pseudoconcave over the open subset $G \neq \emptyset$ of Y. Suppose that $\tilde{G} = \varphi^{-1}(G)$ is not empty and normal. Let $f_1, \ldots, f_k$ φ-independent meromorphic functions on X. Then a positive integer $e \in \mathbb{N}$ and an open subset H of G with $\tilde{H} \in \varphi^{-1}(H) \neq \emptyset$ exist such that H is biholomorphically equivalent to a poly-disc in $\mathbb{C}^n$ and such that the following property holds.

(P) Let $f \in \mathfrak{K}_\varphi(X, f_1, \ldots, f_k)$ be a meromorphic function on X such that $f_1, \ldots, f_k, f$ are φ-dependent. Then holomorphic functions $a_{\mu_0 \ldots \mu_k, \rho} \in \mathfrak{H}(H)$ exist such that

$$A_\rho = \sum_{\mu_0 + \ldots + \mu_k = s} (a_{\mu_0 \ldots \mu_k, \rho} \circ \varphi) f_1^{\mu_1} \ldots f_k^{\mu_k} \in \mathfrak{K}(\tilde{H})$$

$$\sum_{\rho=0}^{r} A_\rho f^\rho \equiv 0 \text{ on } \tilde{H}$$

where $A_r \in \mathfrak{K}^*(\tilde{H})$ is not a zero divisor in $\mathfrak{K}(\tilde{H})$ with $1 \leqq r \leqq e$.

Especially, $f_1|\tilde{H}, \ldots, f_k|\tilde{H}, f|\tilde{H}$ are algebraically dependent over $\varphi_H^* \mathfrak{K}(H)$ where $\varphi_H = \varphi: \tilde{H} \to H$ and $f|\tilde{H}$ is integral over $\varphi_H^* \mathfrak{K}(H)(f_1|\tilde{H}, \ldots, f_k|\tilde{H})$ with

$$[f|\tilde{H}: \varphi_H^* \mathfrak{K}(H)(f_1|\tilde{H}, \ldots, f_k|\tilde{H})] \leqq e.$$

Here $\varphi_H^* \mathfrak{K}(H)(f_1|\tilde{H}, \ldots, f_k|\tilde{H})$ is a field in $\mathfrak{K}(\tilde{H})$ and is a pure transcendental extension of $\varphi_H^* \mathfrak{K}(H)$ of transcendency degree k.

Proof. Determine H and $\tilde{H}$ as an (Σ_{17}). Determine $e = \mathfrak{w}$ as in (Σ_{25}). Take $f \in \mathfrak{K}_\varphi(X, f_1, \ldots, f_k)$. By Theorem (Σ_{27}) $t \in \mathbb{N}$ and $\upsilon_0 \in \Gamma(\mathfrak{W}^1, g^t, \mathfrak{T})$ and $\upsilon_1 \in \Gamma(\mathfrak{W}^1, g^t)$ with $f|\tilde{H} = \upsilon_1/\upsilon_0$ exist. By Lemma 4.14, $f_1|\tilde{H}, \ldots, f_k|\tilde{H}, f|\tilde{H}$ are φ_H-dependent. By Theorem (Σ_{25}) $a_{\mu_0 \ldots \mu_k, \rho} \in \mathfrak{H}(H)$ exist such that

$$A_\rho = \sum_{\mu_0 + \ldots + \mu_k = s} (a_{\mu_0 \ldots \mu_k, \rho} \circ \varphi) f_1^{\mu_1} \ldots f_k^{\mu_k}$$

$$\sum_{\rho=0}^{e} A_\rho f^\rho \equiv 0 \text{ on } \tilde{H}.$$

where at least one $A_\rho \in \mathfrak{K}^*(H)$ is not a zero divisor of $\mathfrak{K}(H)$. Hence a largest index r exists such that $A_r \not\equiv 0$ on $\tilde{H}$. Then $A_{r+1} \equiv \ldots \equiv A_e \equiv 0$ and $r \leqq e$. Obviously $r = 0$, implies $A_0 \equiv 0$ which is wrong. Hence $1 \leqq r \leqq e$. By Lemma 12.9,

$$K = \varphi_H^* \mathfrak{K}(H)(f_1|\tilde{H}, \ldots, f_k|\tilde{H})$$

is a field and the field of quotients of $\varphi_H^* \mathfrak{K}(H)[f_1|\tilde{H}, \ldots, f_k|\tilde{H}]$. Moreover, K is a pure transcendental extension of $\varphi_H^* \mathfrak{K}(H)$ of transcendence degree k. Obviously

$$0 \not\equiv A_r \in \varphi_H^* \mathfrak{K}(H)[f_1|\tilde{H}, \ldots, f_k|\tilde{H}] \subseteq K \subseteq \mathfrak{K}(\tilde{H}).$$

Because K is a field and $A_r \not\equiv 0$ in K, the function A_r is not a zero divisor of $\mathfrak{K}(\tilde{H})$, i.e., $A_r \in \mathfrak{K}^*(\tilde{H})$. Especially, $f_1|\tilde{H}, \ldots, f_k|\tilde{H}, f|\tilde{H}$ are algebraically dependent over $\varphi_H^* \mathfrak{K}(H)$ and $f|\tilde{H}$

is integral over K with $[f|\tilde{H}: K] = r \leqq e$; q.e.d.

Theorem 16.4. Let X and Y be irreducible complex spaces of dimension m and n respectively with $m - n = q > 0$. Let $\varphi: X \to Y$ be a holomorphic map of rank n which is pseudoconcave over the open subset $G \neq \emptyset$ of Y. Suppose that $\tilde{G} = \varphi^{-1}(G)$ is not empty and normal. Let H be any open, irreducible subset of G with $\varphi^{-1}(H) = \tilde{H} \neq \emptyset$. Define $\varphi_H = \varphi: \tilde{H} \to H$. Then an integer $e \in \mathbb{N}$ exists such that the following property holds:

(P_0) If $f \in \mathfrak{K}_\varphi(X)$ is a φ-dependent meromorphic function on X, then $f|\tilde{H}$ is algebraic over the subfield $\varphi_H^* \mathfrak{K}(H)$ of $\mathfrak{K}(\tilde{H})$ with degree

$$[f|\tilde{H}: \varphi_H^* \mathfrak{K}(H)] \leqq e.$$

Proof. The set $D = \{x \in X | \operatorname{rank}_x \varphi \leqq n-1\}$ is thin and analytic in X. Hence

$$D \cap \tilde{H} = \{x \in \tilde{H} | \operatorname{rank}_x \varphi_H \leqq n-1\}$$

is thin and analytic in $\tilde{H}$. The restriction $\varphi_H: \tilde{H} \to H$ is quasi-proper and has strict rank n. Hence φ_H is surjective. By Proposition 8.4, $D^* = \varphi(D \cap \tilde{H})$ is an analytic subset of H with $\dim D^* \leqq n-2$. Hence $V = H - D^*$ is an open, irreducible subset of H and Y. Because φ_H has strict rank n, the open subset $\tilde{V} = \varphi^{-1}(V)$ is dense in $\tilde{H}$. The restriction $\varphi_V = \varphi_H: \tilde{V} \to V$ is quasi-proper, surjective and q-fibering. By Lemma 8.20, the set $\mathfrak{B}$ of branches of $\tilde{V}$ is finite. If $B \in \mathfrak{B}$, then $\psi_B = \varphi_V|B: B \to V$ is quasi-proper and q-fibering by Proposition 8.10' and surjective by Lemma 8.1, since V is irreducible and n-dimensional. By Theorem 9.2, the branch number e(B) of ψ_B is finite and $\mathfrak{K}_{\psi_B}(B)$ is finite algebraic over $\psi_B^* \mathfrak{K}(V)$ with

$$[\mathfrak{K}_{\psi_B}(B) : \psi_B^* \mathfrak{K}(V)] \leqq e(B).$$

Because $\mathfrak{B}$ is finite,

$$e = \sum_{B \in \mathfrak{B}} e(B) < \infty$$

is a positive integer.

Take $f \in \mathfrak{K}_\varphi(X)$. By Lemma 4.14, $f|\tilde{V} \in \mathfrak{K}_{\varphi_V}(\tilde{V})$. By Lemma 4.4, $f|B \in \mathfrak{K}_{\psi_B}(B)$ because $\psi_B = \varphi_V|B$ for each $B \in \mathfrak{B}$. Meromorphic functions $a_{B\mu} \in \mathfrak{K}(V)$ for $\mu = 0,\ldots,r(B)$ exist such that $1 \leqq r(B) \leqq e(B)$, such that $a_{Br(B)} \not\equiv 0$ on V and such that

$$\sum_{\mu=0}^{r(B)} \varphi_V^*(a_{B\mu}) \cdot f^\mu$$

is meromorphic on $\tilde{V}$ and identically zero on B. Let ξ be an indeterminant over $\mathfrak{K}(V)$ and define

$$A_B(\xi) = \sum_{\mu=0}^{r(B)} a_{B\mu} \cdot \xi^\mu .$$

Then $A_B(\xi) \not\equiv 0$ in the integral domain $\mathfrak{K}(V)[\xi]$. Hence

$$0 \not\equiv A(\xi) = \prod_{B \in \mathfrak{B}} A_B(\xi) = \sum_{\mu=0}^{r} q_\mu \xi^\mu \in \mathfrak{K}(V)[\xi]$$

with

$$1 \leqq r = \sum_{B \in \mathcal{B}} r(B) \leqq \sum_{B \in \mathcal{B}} e(B) = e < \infty$$

and with

$$0 \not\equiv a_r = \prod_{B \in \mathcal{B}} a_{B\mu} \in \mathfrak{K}^*(V).$$

Moreover,

$$A(f) = \sum_{\mu=0}^{r} \varphi_V^*(a_\mu) f^\mu \equiv 0$$

on $\tilde{V}$.

Observe that $V = H - D^*$ and that D^* is analytic in H with $\dim D^* \leqq n-2$. By Lemma 2.7, meromorphic functions $b_\mu \in \mathfrak{K}(H)$ exist such that $b_\mu | V = a_\mu$. Because $\tilde{V}$ is dense in $\tilde{H}$, this implies

$$\sum_{\mu=0}^{r} \varphi_{\tilde{H}}^*(b_\mu) \cdot f^\mu \equiv 0$$

on $\tilde{H}$ with $b_r \not\equiv 0$ on H and $r \leqq e$. Hence $f|\tilde{H}$ is algebraic over $\varphi_H^* \mathfrak{K}(H)$ with $[f|\tilde{H}: \varphi_H^* \mathfrak{K}(H)] \leqq e$. Since

$$\varphi_H^*: \mathfrak{K}(H) \to \varphi_H^* \mathfrak{K}(H)$$

is an isomorphism, $\varphi_H^* \mathfrak{K}(H)$ is a subfield of $\mathfrak{K}(\tilde{H})$; q.e.d.

Remark: The pseudoconcavity in Theorem 16.4 was not really needed. It is enough to require that $\varphi_G = \varphi: \tilde{G} \to G$ is quasi-proper.

<u>Theorem 16.5.</u> Let X and Y be irreducible complex spaces of dimension m and n respectively with $m - n = q > 0$. Let $\varphi: X \to Y$ be a full holomorphic map of rank n which is pseudoconcave over an open subset $G \neq \emptyset$ of Y. Suppose that $\tilde{G} = \varphi^{-1}(G)$ is a normal complex space. Then the field $\mathfrak{K}(X)$ of meromorphic functions on X is an algebraic function field over the field $\varphi^*\mathfrak{K}(Y)$ of meromorphic functions lifted from Y ot X by φ.

<u>Proof.</u> By Corollary 11.11, φ is surjective. Hence $\tilde{G} \neq \emptyset$. A free finite subset F of $\mathfrak{K}(X)$ exists such that

$$\mathfrak{K}(X) = \mathfrak{K}_\varphi(X;F).$$

If $F = \emptyset$, take any open, irredicuble subset $H \neq \emptyset$ of G. Then $\tilde{H} = \varphi^{-1}(H) \neq \emptyset$. Determine $e \in \mathbb{N}$ by Theorem 16.4. If $f \in \mathfrak{K}(X) = \mathfrak{K}_\varphi(X)$, then $f|\tilde{H}$ is algebraic over $\varphi_H^*\mathfrak{K}(H)$ where $\varphi_H = \varphi: \tilde{H} \to H$ and where $[f|\tilde{H}: \varphi_H^*\mathfrak{K}(H)] \leqq e$. By Theorem 12.6, f is algebraic over $\varphi^*\mathfrak{K}(Y)$ with $[f: \varphi^*\mathfrak{K}(Y)] \leqq e$. Hence $\mathfrak{K}(X)$ is a finite algebraic extension of $\varphi^*\mathfrak{K}(Y)$ with

$$[\mathfrak{K}(X): \varphi^*\mathfrak{K}(Y)] \leqq e.$$

Now, consider the case $F \neq \emptyset$. Then $F = \{f_1,\ldots,f_k\}$ where $f_1,\ldots,f_k$ are φ-independent. Determine $e \in \mathbb{N}$ and H as in Theorem 16.3. If $f \in \mathfrak{K}(X) = \mathfrak{K}_\varphi(X,F)$, then $f_1|\tilde{H},\ldots,f_k|\tilde{H},f|\tilde{H}$ are algebraically dependent over $\varphi_H^*\mathfrak{K}(H)$ and $f_1|\tilde{H},\ldots,f_k|\tilde{H}$ are algebraically independent over $\varphi_H^*\mathfrak{K}(H)$ with

$$[f|\tilde{H}: \varphi_H^*\mathfrak{K}(H)(f_1|\tilde{H},\ldots,f_k|\tilde{H})] \leqq e.$$

By Theorem 12.7, $f_1,\dots,f_k,f$ algebraically dependent over $\varphi^*\mathfrak{K}(Y)$ and $f_1,\dots,f_k$ are algebraically independent over $\varphi^*\mathfrak{K}(Y)$ with

$$[f\colon \varphi^*\mathfrak{K}(Y)(f_1,\dots,f_k)] \leqq e.$$

Hence $\varphi^*\mathfrak{K}(Y)(f_1,\dots,f_k)$ is a pure transcendental extension of $\varphi^*\mathfrak{K}(Y)$ with transcendence degree k and $\mathfrak{K}(X)$ is a finite algebraic extension of the field $\varphi^*\mathfrak{K}(Y)(f_1,\dots,f_k)$ with

$$[\mathfrak{K}(X)\colon \varphi^*\mathfrak{K}(Y)(f_1,\dots,f_k)] \leqq e.$$

Therefore, $\mathfrak{K}(X)$ is an algebraic function field over $\varphi^*\mathfrak{K}(Y)$;

q.e.d.

This implies immediately Theorem 16.6 as formulated on page 299 . Theorems 6.1 to 6.6 represent the third main result of these investigations.

§17. A counter example by Kas

Let D be the unit disc. Kas pointed out to the authors, a construction by Kodaira which gives an example of a connected complex manifold X of dimension 2 and a surjective, regular, full, 1-fibering, holomorphic map $\varphi: X \to D$ with $\varphi^*(\mathfrak{K}(D)) = \mathfrak{K}(X)$ such that there are two φ-dependent meromorphic functions ξ and η which are algebraically independent over $\varphi^*(\mathfrak{H}(D)) = \mathfrak{H}(X)$. Especially, $\mathfrak{K}(X)$ is not the field of quotients of $\mathfrak{H}(X)$.

For the construction of X, the following theorem is needed.

<u>Theorem 17.1.</u> Let $\{M_\lambda\}_{\lambda\in\Lambda}$ be a family of pure n-dimensional complex manifolds[11)]. Suppose that for each pair $(\lambda,\rho) \in \Lambda \times \Lambda$ an open subset $A_{\lambda\rho}$ of M_ρ and, if $A_{\lambda\rho} \neq \emptyset$, a biholomorphic map $\gamma_{\lambda\rho}: A_{\lambda\rho} \to A_{\rho\lambda}$ are given such that

1) $\gamma_{\rho\sigma}^{-1}(A_{\lambda\rho} \cap A_{\sigma\rho}) \subseteq A_{\lambda\sigma}$

2) $\gamma_{\lambda\rho}\circ\gamma_{\rho\sigma} = \gamma_{\lambda\sigma}$ on $\gamma_{\rho\sigma}^{-1}(A_{\lambda\rho} \cap A_{\sigma\rho})$

3) $A_{\rho\rho} = M_\rho$ and $\gamma_{\rho\rho}: M_\rho \to M_\rho$ is the identity.

4) If $x \in M_\rho$ and $y \in M_\sigma$ and if in the case $x \in A_{\sigma\rho}$ also $y \neq \gamma_{\sigma\rho}(x)$, then open neighborhoods U of x in M_ρ and V of y in M_σ exist such that

$$V \cap \gamma_{\rho\sigma}(U \cap A_{\sigma\rho}) = \emptyset.$$

Then a complex manifold X of pure dimension n and biholomorphic maps $\alpha_\lambda: M_\lambda \to X_\lambda$ of M_λ onto an open subset of X_λ of X exist such that

$$X = \bigcup_{\lambda \in \Lambda} X_\lambda$$

$$\alpha_\rho(A_{\lambda\rho}) = \alpha_\lambda(A_{\rho\lambda}) = X_\lambda \cap X_\rho \qquad \text{if } A_{\lambda\rho} \neq \emptyset$$

$$\alpha_\lambda^{-1} \circ \alpha_\rho = \gamma_{\lambda\rho}: A_{\lambda\rho} \to A_{\rho\lambda} \qquad \text{if } A_{\lambda\rho} \neq \emptyset.$$

Moreover, if each M_λ has a countable base of open sets and if Λ is at most countable, then X has a countable base of open sets. If each M_λ is connected, and if $A_{\lambda\rho} \neq \emptyset$ for each $(\lambda,\rho) \in \Lambda \times \Lambda$, then X is connected.

This theorem was stated without proof in [19] Satz 10-11. Therefore, the proof shall be given here.

Proof. Let M be the disjoint union of $\{M_\lambda\}_{\lambda\in\Lambda}$

$$M = \dot{\bigcup_{\lambda\in\Lambda}} M_\lambda .$$

Then M_λ is an open subset of the pure n-dimensional complex M with $M_\lambda \cap M_\rho = \emptyset$ if $\lambda \neq \rho$.

If $x \in M$ and $y \in M$, say $x \sim y$ if and only if $x \in A_{\lambda\rho}$ and $y \in A_{\rho\lambda}$ and $\gamma_{\lambda\rho}(x) = y$ for some $(\lambda,\rho) \in \Lambda \times \Lambda$. It is claimed that ~ defines an equivalence relation on M.

If $x \in M$, then $x \in M_\rho = A_{\rho\rho}$ and $x = \gamma_{\rho\rho}(x)$. Hence $x \sim x$. If $x \in M$ and $y \in M$ and $x \sim y$, then $x \in A_{\lambda\rho}$ and $y \in A_{\lambda\rho}$ with $y = \gamma_{\lambda\rho}(x)$

By 2) and 3) $\gamma_{\lambda\rho}^{-1} = \gamma_{\rho\lambda}$. Hence $x = \gamma_{\rho\lambda}(y)$ with $y \in A_{\rho\lambda}$ and $x \in A_{\rho\lambda}$. Therefore $y \sim x$. Assume $(x,y,z) \in M \times M \times M$ and $x \sim y$ and $y \sim z$. Then $x \in A_{\rho\sigma}$ and $y = \gamma_{\rho\sigma}(x) \in A_{\sigma\rho}$. Also $y \in A_{\lambda\tau}$ and $z \in A_{\tau\lambda}$ with $z = \gamma_{\sigma\lambda}(y)$. Because $y \in A_{\sigma\tau} \cap A_{\lambda\rho} \subseteq M_\tau \cap M_\rho$ and because the union was disjoint, $\tau = \rho$. Hence

$$x \in \gamma_{\rho\sigma}^{-1}(A_{\sigma\rho} \cap A_{\lambda\rho}) \subseteq A_{\lambda\sigma}$$

$$\gamma_{\lambda\sigma}(x) = \gamma_{\lambda\rho}(\gamma_{\rho\sigma}(x)) = \gamma_{\lambda\rho}(y) = z.$$

Hence $x \sim z$. An equivalence relation $\mathcal{R}$ on M is defined. Let $X = M/\mathcal{R}$ be the quotient space and let $\alpha: M \to X$ be the residual map. Then X is topological space and α is continuous and surjective. Define $X_\lambda = \alpha(M_\lambda)$ and

$$\alpha_\lambda = \alpha: M_\lambda \to X_\lambda.$$

Obviously, α_λ is surjective and continuous. If $x \in M_\lambda$ and $y \in M_\lambda$ with $\alpha_\lambda(x) = \alpha_\lambda(y)$, then $y = \gamma_{\lambda\rho}(x)$ where $x \in A_{\lambda\rho}$ and $y \in A_{\rho\lambda}$. Hence $x \in M_\lambda \cap M_\rho$ which implies $\lambda = \rho$. Therefore, $y = \gamma_{\lambda\lambda}(x) = x$. The map $\alpha_\lambda: M_\lambda \to X_\lambda$ is bijective and continuous. Obviously, $X = \bigcup_{\lambda\in\Lambda} X_\lambda$.

If $U \subseteq M_\lambda$, then

$$\alpha^{-1}(\alpha(U)) = \bigcup_{\rho\in\Lambda} \gamma_{\rho\lambda}(U \cap A_{\rho\lambda})$$

is claimed. If $y = \gamma_{\rho\lambda}(x)$ with $x \in U \cap A_{\rho\lambda}$, then $y \sim x$ and $\alpha(y) = \alpha(x) \in \alpha(U)$. Hence $y \in \alpha^{-1}(\alpha(U))$. If $y \in \alpha^{-1}(\alpha(U))$. Then

$\alpha(y) = \alpha(x)$ with $x \in U$. Hence $x \in A_{\rho\lambda}$ and $y \in A_{\lambda\rho}$ with $\gamma_{\rho\lambda}(x) = y$ since $U \subseteq M_\lambda$. Hence the claim is proved.

If $U \subseteq M_\lambda$ is open, then U is open in M. Also $U \cap A_{\rho\lambda}$ is open in M_λ and $\gamma_{\rho\lambda}(U \cap A_{\rho\lambda})$ is open in M. Hence $\alpha^{-1}(\alpha(U))$ is open in M. Therefore, $\alpha(U)$ is open. Because $\{M_\lambda\}_{\lambda\in\Lambda}$ is an open covering of M, the map α is open. Hence $X_\lambda = \alpha(M_\lambda)$ is open. Also $\alpha_\lambda \doteq \alpha : M_\lambda \to X_\lambda$ is open. Therefore $\alpha_\lambda : M_\lambda \to X_\lambda$ is a topological map.

If $x \in X$ and $y \in X$ with $x \neq y$, then $x' \in M_\lambda$ and $y' \in M_\rho$ exist such that $\alpha(x') = x$ and $\alpha(y') = y$. Because $x \neq y$, also $x' \nsim y'$ and $x' \in M_\lambda - A_{\rho\lambda}$ or $x' \in A_{\rho\lambda}$ with $y' \neq \gamma_{\rho\lambda}(x')$. In either case, open neighborhoods U of x in M_λ and V of y in M_ρ exist such that $\gamma_{\rho\lambda}(U \cap A_{\rho\lambda}) \cap V = \emptyset$. Then $\alpha(U)$ and $\alpha(V)$ are open in X with $x \in \alpha(U)$ and $y \in \alpha(V)$. If $z \in \alpha(U) \cap \alpha(V)$ then $z = \alpha(z_\rho) = \alpha(z_\lambda)$ with $z_\lambda \in U$ and $z_\rho \in V$. Because $z_\lambda \sim z_\rho$, this implies $z_\rho = \gamma_{\rho\lambda}(z_\lambda) \in V \cap \gamma_{\rho\lambda}(U \cap A_{\rho\lambda}) = \emptyset$. This contradiction shows, that $\alpha(U) \cap \alpha(V) = \emptyset$. The space X is a Hausdorff space.

If $x \in A_{\rho\lambda}$, then $\gamma_{\rho\lambda}(x) \sim x$. Hence $\alpha(\gamma_{\rho\lambda}(x)) = \alpha(x)$. Moreover,

$$\alpha_\lambda(A_{\rho\lambda}) = \alpha(\gamma_{\rho\lambda}(A_{\rho\lambda}) = \alpha(A_{\lambda\rho}) = \alpha\ (A_{\lambda\rho}) \subseteq X_\lambda \cap X_\rho.$$

If $x \in X_\lambda \cap X_\rho$, then $x' \in A_{\lambda\rho}$ and $x'' \in A_{\rho\lambda}$ with $\alpha(x') = x = \alpha(x'')$ exist. Hence $x'' = \gamma_{\lambda\rho}(x')$ and

$$\alpha_\lambda(A_{\rho\lambda}) = \alpha_\rho(A_{\lambda\rho}) = X_\lambda \cap X_\rho.$$

If $x \in A_{\lambda\rho}$, then $\gamma_{\lambda\rho}(x) = y \in A_{\rho\lambda}$ and $x{\sim}y$. Hence $\alpha_\rho(x) = \alpha(x) = \alpha(y)$

$= \alpha_\lambda(y)$ or $\gamma_{\lambda\rho}(x) = y = \alpha_\lambda^{-1}(\alpha_\rho(x))$. Therefore

$$\alpha_\lambda^{-1} \circ \alpha_\rho = \gamma_{\lambda\rho}\colon A_{\lambda\rho} \to A_{\rho\lambda}.$$

Let $\mathfrak{A}_\lambda$ be the set of all schicht charts of the complex manifold M_λ. Then $\mathfrak{A}_\lambda$ is the complex structure of M_λ. If $\beta \in \mathfrak{A}_\lambda$, define $U_\beta^\lambda = \alpha_\lambda(U_\beta)$. Then U_β^λ is open on X and $\alpha_\lambda\colon U_\beta \to U_\beta^\lambda$ is a topological map. Define

$$\tilde{\mathfrak{A}}_\lambda = \{\beta \circ \alpha_\lambda^{-1}\colon U_\beta^\lambda \to U_\beta' \mid \beta \in \mathfrak{A}_\lambda\}$$

$$\mathfrak{A} = \bigcup_{\lambda \in \Lambda} \tilde{\mathfrak{A}}_\lambda ,$$

Then it is claimed that $\mathfrak{A}$ is an atlas of X. If $\beta \circ \alpha_\lambda^{-1} \in \tilde{\mathfrak{A}}_\lambda$, then $\beta \cdot \alpha_\lambda^{-1}\colon U_\beta^\lambda \to U_\beta'$ is topological. Because $\{U_\beta\}_{\beta \in \mathfrak{A}_\lambda}$ is an open covering of M_λ,

$$\bigcup_{\lambda \in \Lambda} \bigcup_{\beta \in \mathfrak{A}_\lambda} U_\beta^\lambda = \bigcup_{\lambda \in \Lambda} X_\lambda = X.$$

If $\beta \in \mathfrak{A}_\lambda$ and $\delta \in \mathfrak{A}_\rho$ with $U_\beta^\lambda \cap U_\delta^\rho \neq \emptyset$, then $U_\beta^\lambda \cap U_\delta^\rho \subseteq X_\lambda \cap X_\rho$ and

$$\begin{aligned}(\beta \circ \alpha_\lambda^{-1}) \circ (\delta \circ \alpha_\rho^{-1})^{-1} &= \beta \circ \alpha_\lambda^{-1} \circ \alpha_\rho \circ \delta^{-1} \\ &= \beta \circ \gamma_{\lambda\rho} \circ \delta^{-1}\end{aligned}$$

on $\delta(\alpha_\rho^{-1}(U_\beta^\lambda \cap U_\delta^\rho)$ is holomorphic. Hence $\mathfrak{A}$ is a complex atlas on X which defines uniquely a complex structure.

If $\beta \in \mathfrak{A}_\lambda$, then $\beta \circ \alpha_\lambda^{-1} \in \mathfrak{A}$ and

$$(\beta\circ\alpha_\lambda^{-1})\circ\alpha_\lambda\circ\beta^{-1} = \mathrm{Id}: U'_\beta \to U'_\beta.$$

Hence $\alpha_\lambda: M_\lambda \to X_\lambda$ is a bijective holomorphic map. Hence α_λ is biholomorphic.

Clearly, if M_λ has a countable base of open sets, then X_λ has a countable base of open sets. If Λ is at most countable, then $X = \bigcup_{\lambda\in\Lambda} X_\lambda$ has a countable base of open sets.

If each M_λ is connected, then each X_λ is connected. If $A_{\lambda\rho} \neq \emptyset$ for each pair $(\lambda,\rho) \in \Lambda \times \Lambda$, then $X_\lambda \cap X_\rho \neq \emptyset$ for each pair $(\lambda,\rho) \in \Lambda \times \Lambda$. Hence $X = \bigcup_{\lambda\in\Lambda} X_\lambda$ is connected; q.e.d.

A special case of Theorem 17.1 shall be constructed now[12]: Define

$$D = \{t \in \mathbb{C} \mid |t| < 1\} \qquad \mathbb{C}^* = \mathbb{C} - \{0\}$$

$$D^* = D - \{0\} = D \cap \mathbb{C}^*$$

$$\Lambda = \mathbb{Z} \qquad M_\lambda = D \times \mathbb{C}^* \qquad \text{for each } \lambda \in \mathbb{Z}$$

$$A_{\rho\lambda} = \begin{cases} D \times \mathbb{C}^* & \text{if } \lambda = \rho \\ D^* \times \mathbb{C}^* & \text{if } \lambda \neq \rho \end{cases}$$

$$\gamma_{\rho\rho} = \mathrm{Id}: A_{\rho\rho} \to A_{\rho\rho} \qquad \text{if } \rho \in \mathbb{Z}$$

$$\gamma_{\lambda\rho}(t,w) = (t, t^{\rho-\lambda}w) \qquad \text{if } (t,w) \in D^* \times \mathbb{C}^*$$

and if $\lambda \neq \rho$ with $\lambda \in \mathbb{Z}$ and $\rho \in \mathbb{Z}$. This definition applies also for $\lambda = \rho$ on $D^* \times \mathbb{C}^*$ and even on $D \times \mathbb{C}^*$ if $0^0 = 1$. Obviously $\gamma_{\rho\lambda}: A_{\rho\lambda} \to A_{\lambda\rho}$ is a biholomorphic map. Moreover

$$\gamma_{\rho\sigma}^{-1}(A_{\lambda\rho} \cap A_{\sigma\rho}) \subseteq A_{\lambda\sigma} .$$

If $(t,w) \in D^* \times \mathbb{C}^*$, then

$$\gamma_{\lambda\rho}(\gamma_{\rho\sigma}(t,w)) = \gamma_{\lambda\rho}(t,t^{\sigma-\rho}w) = (t,t^{\rho-\lambda}t^{\sigma-\rho}w)$$

$$= (t,t^{\sigma-\lambda}w) = \gamma_{\lambda\sigma}(t,w)$$

which is also correct if $t = 0$ and $\lambda = \rho = \sigma$. Therefore assumptions 1) - 3) of Theorem 17.1 are satisfied.

4) Take $(t,w) \in M_\rho$ and $(s,v) \in M_\sigma$. If $\rho = \sigma$ assume $(t,w) \neq (s,v)$. If $\rho \neq \sigma$ and $t \neq 0$, assume $\gamma_{\sigma\rho}(t,w) \neq (s,v)$. If $\rho = \sigma$, take U and V open im M_ρ with $(t,w) \in U$ and $(s,v) \in V$ and $U \cap V = \emptyset$. Then $V \cap \gamma_{\sigma\rho}(U \cap A_{\sigma\rho}) = V \cap U = \emptyset$. If $\rho \neq \sigma$, let $\{U_n\}_{n\in\mathbb{N}}$ and $\{V_n\}_{n\in\mathbb{N}}$ be bases of open neighborhoods of (t,w) in M_ρ and $(s,v) \in M_\sigma$ respectively. Suppose that

$$V_n \cap \gamma_{\sigma\rho}(U_n \cap A_{\rho\sigma}) \neq \emptyset$$

for all $n \in \mathbb{N}$. Then

$$(t_n,w_n) \in U_n \cap A_{\sigma\rho} \qquad (s_n,v_n) \in V_n$$

with $(s_n,v_n) = \gamma_{\sigma\rho}(t_n,w_n) = (t_n,t_n^{\rho-\sigma}w_n)$ exists for each n. Then

$(s_n, v_n) \to (s,v)$ and $(t_n, w_n) \to (t,w)$ for $n \to \infty$. Moreover $s_n = t_n$. Hence $s = t \in D$. Moreover $w \neq 0 \neq v$ and $\rho \neq \sigma$. Hence

$$s = t = \lim t_n = (v_n/w_n)^{\frac{1}{\rho-\sigma}} \to (v/w)^{\frac{1}{\rho-\sigma}} \in \mathbb{C}^*.$$

Hence $t \neq 0$ and $t^{\rho-\sigma} w = v$. Therefore $\gamma_{\sigma\rho}(t,w) = (s,v)$, which is wrong. Hence assumption 4. of Theorem 17.1 is also satisfied.

By Theorem 17.2, a connected complex manifold X of dimension 2, an open covering $\{X_\lambda\}_{\lambda\in\mathbb{Z}}$ of X and a family $\{\alpha_\lambda\}_{\lambda\in\mathbb{Z}}$ of biholomorphic maps $\alpha_\lambda : M_\lambda \to X_\lambda$ exist such that

$$\alpha_\rho(A_{\lambda\rho}) = \alpha_\lambda(A_{\rho\lambda}) = X_\rho \cap X_\lambda \neq \emptyset$$

$$\alpha_\lambda^{-1} \circ \alpha_\rho = \gamma_{\lambda\rho} : A_{\lambda\rho} \to A_{\rho\lambda}$$

if $(\lambda, \rho) \in Z \times Z$. The manifold X has a countable base of open sets.

For each $\lambda \in \mathbb{Z}$, the projection $\pi_\lambda : M_\lambda \to D$ is a surjective, holomorphic, 1-fibering, regular, quasi-proper map of rank 1. Hence

$$\varphi_\lambda = \pi_\lambda \circ \alpha_\lambda^{-1} : X_\lambda \to D$$

is a surjective, holomorphic, 1-fibering, regular, quasi-proper map of rank 1.

Take $x \in X_\lambda \cap X_\rho$. Then $\alpha_\rho^{-1}(x) = (t,w)$ and $\alpha_\lambda^{-1}(x) = (s,v)$. Then

$$(s,v) = \alpha_\lambda^{-1}(\alpha_\rho(t,w)) = (t, t^{\rho-\lambda} w)$$

Hence t = s and

$$\varphi_\lambda(x) = \pi_\lambda(\alpha_\lambda^{-1}(x)) = \pi_\lambda(t,w) = t = s = \pi_\rho(s,v) =$$

$$= \pi_\rho \circ \alpha_\rho^{-1}(x) = \varphi_\rho(x).$$

Hence $\varphi_\lambda | X_\lambda \cap X_\rho = \varphi_\lambda | X_\lambda \cap X_\rho$ if $(\lambda, \rho) \in \mathbb{Z} \times \mathbb{Z}$. Therefore one and only one holomorphic map $\varphi: X \to D$ exists such that $\varphi | X_\lambda = \varphi_\lambda$. Then φ is surjective, holomorphic, 1-fibering, regular and has rank 1. Because $\varphi | X_\lambda = \varphi_\lambda$ is quasi-proper, φ is full.

The analytic set $S = \varphi^{-1}(0)$ is a pure 1-dimensional complex submanifold of X. Define $S_\lambda = S \cap X_\lambda$.

$$S_\lambda = \varphi_\lambda^{-1}(0) = \alpha_\lambda(\pi_\lambda^{-1}(0)) = \alpha_\lambda(\{0\} \times \mathbb{C}^*).$$

Therefore $S_\lambda \cap S_\rho = \emptyset$ if $\lambda \neq \rho$. Moreover $\{S_\lambda\}_{\lambda \in \mathbb{Z}}$ is the family of connectivity components of S.

Let $\mathfrak{H}(X)$ and $\mathfrak{H}(D)$ be the rings of holomorphic functions on X and D respectively. Then

$$\varphi^*(\mathfrak{H}(D)) = \{f \circ \varphi | f \in \mathfrak{H}(D)\} \subseteq \mathfrak{H}(X)$$

<u>Proposition 17.2.</u> $\mathfrak{H}(X) = \varphi^*(\mathfrak{H}(D))$.

<u>Proof:</u> Take $f \in \mathfrak{H}(X)$. Then $f \circ \alpha_\lambda: D \times \mathbb{C}^* \to C$ is holomorphic. Holomorphic function $a_{\lambda n} \in \mathfrak{H}(D)$ exist such that

$$f(\alpha_\lambda(t,w)) = \sum_{n=-\infty}^{+\infty} a_{\lambda n}(t) w^n$$

where this Laurent series converges uniformly and absolutely on every compact subset of $D \times \mathbb{C}^*$. Define $a_n = a_{0n}$.

If $\lambda \in \mathbb{Z}$, then

$$f \circ \alpha_\lambda = f \circ \alpha_0 \circ \alpha_0^{-1} \circ \alpha_\lambda = f \circ \alpha_0 \circ \gamma_{0\lambda}$$

on $D^* \times \mathbb{C}^*$. Hence

$$\sum_{n=-\infty}^{+\infty} a_{\lambda n}(t) w^n = \sum_{n=-\infty}^{+\infty} a_n(t) t^{\lambda n} w^n$$

if $(t,w) \in D^* \times \mathbb{C}^*$ because $\gamma_{0\lambda}(t,w) = (t, t^\lambda w)$. Hence $a_{\lambda n}(t) = t^{\lambda n} a_n(t)$ for all $t \in D^*$ and $\lambda \in \mathbb{Z}$. Hence, if σ_g is the zero multiplicity of g, then

$$\sigma_{a_n}(0) + \lambda n = \sigma_{a_{\lambda n}}(0) \geqq 0$$

for all $\lambda \in \mathbb{Z}$. Hence $\sigma_{a_n}(0) = \infty$ and $a_n \equiv 0$. Therefore $f \circ \alpha_\lambda(t,w) = a_0(t) = a_0(\pi_0(t,w))$ or $f \circ \alpha_0 = a_0 \circ \pi_0$ on X_0. Hence $f = a_0 \circ \pi_0 \circ \alpha_0^{-1} = a_0 \cdot \varphi_0 = a_0 \circ \varphi$ on the open subset X_0 of X. Hence $f = a_0 \circ \varphi$ on X, because X is connected. Therefore $f \in \varphi^*(\mathfrak{H}(D))$; q.e.d.

For $\lambda \in \mathbb{Z}$, define a meromorphic function ψ_λ on M_λ by

$$\psi_\lambda(t,w) = t^\lambda w.$$

Then ψ_λ is holomorphic on $D^* \times \mathbb{C}^*$ and holomorphic on $M_\lambda = D \times \mathbb{C}^\lambda$ if $\lambda \geqq 0$.

If $\lambda < 0$, then ψ_λ has a pole of order $|\lambda|$ on $\{0\} \times \mathbb{C}^*$. Define $\xi_\lambda = (\alpha_\lambda^{-1})^*(\psi_\lambda)$ as a meromorphic function on X_λ. Then ξ_λ is holomorphic on $X_\lambda - S$. If $\lambda \geqq 0$, ξ_λ is holomorphic on X_λ. If $\lambda < 0$, then ξ_λ has a pole of order $|\lambda|$ at every $x \in S \cap X_\lambda = S_\lambda$.

If $x \in X_\lambda \cap X_\rho$ with $\lambda \neq \rho$, then $a_\rho^{-1}(x) = (t,w) \in A_{\lambda\rho} = D^* \times \mathbb{C}^*$. Hence $x \in X_\rho - S$ and ξ_ρ is holomorphic at x with $\xi_\rho(x) = t^\rho w$. Also $\alpha_\lambda^{-1}(x) = A_{\rho\lambda} = D^* \times \mathbb{C}^*$ and ξ_λ is holomorphic at x with

$$\xi_\lambda(x) = \psi_\lambda \circ \alpha_\lambda^{-1} \circ \alpha_\rho(t,w) = \psi_\lambda(\gamma_{\lambda\rho}(t,w)) = \psi_\lambda(t, t^{\rho-\lambda} w)$$

$$= t^\rho w = \xi_\rho(x).$$

Therefore $\xi_\lambda | X_\lambda \cap X_\rho = \xi_\rho | X_\rho \cap X_\lambda$ if $\lambda \in \mathbb{Z}$ and $\rho \in \mathbb{Z}$ and $\lambda \neq \rho$. One and only one meromorphic function ξ on X exists such that $\xi | X_\lambda = \xi_\lambda$ for all $\lambda \in \mathbb{Z}$. The function ξ is holomorphic on $X - S$. If $\lambda \geqq 0$, then ξ is holomorphic at each point $x \in S_\lambda$ with $\sigma_\xi(x) = \lambda$. If $\lambda < 0$, then ξ has a pole $|\lambda|$ at each point $x \in S_\lambda$.

Let $\{f_n\}_{n\in\mathbb{Z}}$ be a sequence of meromorphic functions on Y. The series

$$\sum_{n\in\mathbb{Z}} f_n$$

is said to converge uniformly on compact subsets of Y, if and only if for every compact subset K of Y a number $n_0 \in \mathbb{N}$ exists such that f_n is holomorphic at every point of K if $n \in \mathbb{Z}$ and $|n| \geqq n_0$ and if the series

$$\sum_{n=n_0}^{\infty} f_{(-n)} \quad \text{and} \quad \sum_{n=n_0}^{\infty} f_n$$

converge uniformly on K. Then

$$f = \sum_{n\in\mathbb{Z}} f_n$$

is a well-defined meromorphic function on Y.

Recall that the holomorphic map $\varphi: X \to D$ is a bounded holomorphic function on X.

Proposition 17.3. The series

$$\eta = \sum_{n\in\mathbb{Z}} \varphi^{(n^2)} \xi^n$$

converges uniformly on compact subsets of X and defines a meromorphic function η on X which is holomorphic on X - S.

Proof. For $0 < r < 1$ and $R > 1$ define

$$C(r,R) = \{(t,w) \mid |t| < r, \tfrac{1}{R} < |w| < R\}.$$

If $\lambda \in \mathbb{Z}$, define $K_\lambda(r,R) = \alpha_\lambda(C(r,R))$. It suffices to prove that the series converges uniformly on compact subset of each X_λ. Hence it suffices to prove, that the series converges uniformly on $K_\lambda(r,R)$ for each (λ,r,R) with $\lambda \in \mathbb{Z}$, $0 < r < 1$ and $R > 1$. Hence take such a triple (λ,r,R). Take $n_0 \in \mathbb{N}$ such that

$$n_0 \geqq \mathrm{Max}(2|\lambda|, \frac{2 \log 2R}{\log {}^1/_r}).$$

If $n \in \mathbf{Z}$ with $|n| \geqq n_0$, then

$$n^2 + n\lambda \geqq n^2 - |n||\lambda| \geqq n^2 - \tfrac{1}{2}n^2 = \tfrac{1}{2}n^2$$

$$|n| \log \frac{1}{r} \geqq \log(2R)^2$$

$$(\frac{1}{r})^{|n|} \geqq (2R)^2$$

$$\frac{1}{2} \geqq R\, r^{\frac{|n|}{2}}$$

$$(\frac{1}{2})^{|n|} \geqq R^{|n|}\, r^{\frac{n^2}{2}} \geqq r^{n^2+n\lambda} R^{|n|}.$$

Take $x \in K_\lambda(r,R) - S_\lambda$. Then $\alpha_\lambda^{-1}(x) = (t,w) \in C(r,R)$ with $t \neq 0$. Then

$$|\varphi^{(n^2)}(x)\xi^n(x)| = |t^{n^2}(t^\lambda w)^n| = |t^{n^2+n\lambda}w^n|$$

$$\leqq r^{n^2+n\lambda}|R|^{|n|} \leqq (\tfrac{1}{2})^n$$

for all $x \in K_\lambda(r,R) - S_\lambda$. Hence $\varphi^{(n^2)}\xi^n$ is holomorphic at every point of $K_\lambda(r,R)$ and

$$|\varphi^{(n^2)}(x)\xi^{(n)}(x)| \leqq (\tfrac{1}{2})^n$$

for all $x \in K_\lambda(r,R)$ and all $n \in \mathbf{Z}$ with $|n| \geqq n_0$. Hence

$$\sum_{n=n_0}^{\infty} \varphi^{(-n)^2}\xi^{-n} \quad \text{and} \quad \sum_{n=n_0}^{\infty} \varphi^{(n^2)}\xi^{n}$$

converge uniformly on $K_\lambda(r,R)$; q.e.d.

Lemma 17.4. Take $\tau \in \mathbb{C}$ with $\operatorname{Im}\tau > 0$. Define

$$Q = \{x+\tau y \mid 0 \leqq x < 1 \quad \text{and} \quad 0 \leqq y < 1\}.$$

Let $f \not\equiv 0$ be an entire function. Suppose that a constant $c \neq 0$ exists such that

$$f(z+1) = f(z)$$

$$f(z+\tau) = c \cdot e^{-2\pi i z} f(z)$$

for all $z \in \mathbb{C}$. Define $M = \{z \in \mathbb{C} \mid f(z) = 0\} = f^{-1}(0)$. Then one and only one $z_0 \in Q \cap M$ exists and

$$M = \{z_0+m+n\tau \mid m \in \mathbb{Z} \text{ and } n \in \mathbb{Z}\}.$$

Moreover, $\sigma_f(z) = 1$ for each $z \in M$.

Proof. If G is a bounded subset of $\mathbb{C}$ then

$$n_f(G) = \sum_{z \in G} \sigma_f(z) < \infty.$$

If $1 > \varepsilon > 0$, define

$$Q_\varepsilon = \{x+\tau y \mid -\varepsilon \leqq x < 1-\varepsilon,\ -\varepsilon \leqq y < 1-\varepsilon\}.$$

The number ε can be taken such that

a) $f(z) \neq 0$ if $z \in \overline{Q}_\varepsilon - Q_\varepsilon$

b) $f(z) \neq 0$ if $z \in Q_\varepsilon - Q$

c) $f(z) \neq 0$ if $z \in Q - Q_\varepsilon$.

Then

$$n_f(\overline{Q}_\varepsilon) = n_f(Q_\varepsilon) = n_f(Q_\varepsilon - (Q_\varepsilon - Q)) = n_f(Q_\varepsilon \cap Q)$$

$$= n_f((Q_\varepsilon \cap Q) \cup (Q - Q_\varepsilon)) = n_f(Q).$$

Define the curves γ_λ on the unit interval by

$$\gamma_0(x) = -\varepsilon + x - \varepsilon\tau \qquad 0 \leqq x \leqq 1$$

$$\gamma_1(x) = (1-\varepsilon) + (y-\varepsilon)\tau \qquad 0 \leqq y \leqq 1$$

$$\gamma_2(x) = -\varepsilon + x + (1-\varepsilon)\tau \qquad 0 \leqq x \leqq 1$$

$$\gamma_3(x) = -\varepsilon + (y-\varepsilon)\tau \qquad 0 \leqq y \leqq 1.$$

Then $\gamma_2(x) = \gamma_0(x) + \tau \qquad \gamma_0'(x) = 1 = \gamma_2'(x)$

$$\gamma_1(x) = \gamma_3(x) + \tau \qquad \gamma_1'(x) = \gamma_3'(x).$$

Moreover, $f\circ\gamma_1 = f\circ\gamma_3$ and $(f'\circ\gamma_1)\gamma_1' = (f\circ\gamma_1)' = (f\circ\gamma_3)' = (f\circ\gamma_3)\gamma_3'$.

Hence

$$\int_{\gamma_1} \frac{f'}{f} dz = \int_0^1 \frac{f'\circ\gamma_1}{f\circ\gamma_1} \gamma_1' \, dy = \int_0^1 \frac{f'\circ\gamma_3}{f\circ\gamma_3} \gamma_3' \, dy$$

$$= \int_{\gamma_3} \frac{f'}{f} dz.$$

Now,

$$\frac{f'(z+\tau)}{f(z+\tau)} = -2\pi i + \frac{f'(z)}{f(z)}$$

Hence

$$\int_{\gamma_2} \frac{f'}{f} dz = \int_0^1 \frac{f'(\gamma_0(x)+\tau)}{f(\gamma_0(x)+\tau)} \gamma_0'(x) dx =$$

$$= \int_0^1 \frac{f'(\gamma_0(x))}{f(\gamma_0(x))} \gamma_0'(x) dx - 2\pi i$$

$$= \int_{\gamma_0} \frac{f'}{f} dx - 2\pi i.$$

The argument principle implies

$$n_f(Q) = n_f(Q_\varepsilon) = \frac{1}{2\pi i} \int_{\gamma_0} \frac{f'}{f} dz + \frac{1}{2\pi i} \int_{\gamma_1} \frac{f'}{f} dz - \frac{1}{2\pi i} \int_{\gamma_2} \frac{f'}{f} dz -$$

$$- \frac{1}{2\pi i} \int_{\gamma_3} \frac{f'}{f} dz$$

$$= 1.$$

Hence one and only one $z_0 \in Q \cap M$ exists and $\sigma_f(z_0) = 1$. By induction

$$f(z+m) = f(z)$$

$$f(z+m+n\tau) = c^n e^{-2\pi i n z} f(z)$$

for all $m \in \mathbb{Z}$, $n \in \mathbb{Z}$ and $z \in \mathbb{C}$. Hence

$$M = \{z_0+m+n\tau \mid (m,n) \in \mathbb{Z} \times \mathbb{Z}\}$$

and $\sigma_f(z) = 1$ for each $z \in M$; q.e.d.

Lemma 17.5. Let $\eta \in \mathfrak{K}(x)$ be the meromorphic function defined in Proposition 17.3. Then there exists a function

$$v\colon D^* \to \mathbb{C}^*$$

$\eta(x_{2\lambda}(t)) = 0$ if $x_{2\lambda}(t) = \alpha_{2\lambda}(t, v(t))$ for every $t \in D^*$ and $\lambda \in \mathbb{Z}$. (Observe that $x_{2\lambda}(t) \in \mathfrak{X}_{2\lambda} - S$ and that η is holomorphic at $x_{2\lambda}(t)$).

Proof. Take $t \in D^*$. Take $\tau \in \mathbb{C}$ with $t = e^{\pi i \tau}$. Then $\operatorname{Im}\tau > 0$. Define an entire function $f \not\equiv 0$ by

$$f(z) = \eta(\alpha_0(t, e^{2\pi i z}))$$

for $z \in \mathbb{C}$. Obviously, $f(z+1) = f(z)$ for all $z \in \mathbb{C}$. Then

$$\varphi(\alpha_0(t,e^{2\pi iz})) = t$$

$$\xi(\alpha_0(t,e^{2\pi iz})) = e^{2\pi iz}$$

$$f(z) = \eta(\alpha_0(t,e^{2\pi iz})) = \sum_{n\in\mathbb{Z}} t^{(n^2)} e^{2\pi izn}$$

$$f(z) = \sum_{n\in\mathbb{Z}} e^{\pi i\tau n^2+2\pi izn} .$$

Hence

$$f(z+\tau) = \sum_{n\in\mathbb{Z}} e^{\pi i\tau n^2+2\pi izn+2\pi in\tau}$$

$$= \sum_{n\in\mathbb{Z}} e^{\pi i\tau(n+1)^2+2\pi iz(n+1)-\pi i(\tau+2z)}$$

$$= \frac{1}{t}\, e^{-2\pi iz} f(z).$$

Define Q as in Lemma 8.4. Hence one and only one $z_0(t) \in Q$ with $f(z_0(t)) = 0$ exists. Define

$$v(t) = e^{2\pi iz_0(t)} \in \mathbb{C}^*.$$

Hence $v\colon D^* \to \mathbb{C}^*$ is defined. Take $\lambda \in \mathbb{Z}$. Define $x_{2\lambda}(t) = \alpha_{2\lambda}(t,v(t))$. Then $x_{2\lambda}(t) \in X_{2\lambda} - S$ and

$$\eta(x_{2\lambda}(t) = \eta(\alpha_{2\lambda}(t,v(t))$$

$$= \eta(\alpha_0(\alpha_0^{-1}(\alpha_{2\lambda}(t,v(t)))))$$

$$= \eta(\alpha_0(\gamma_{02\lambda}(t,v(t))))$$

$$= \eta(\alpha_0(t,t^{2\lambda}v(t)))$$

$$= \eta(\alpha_0(t,e^{2\pi i(z_0(t)+\lambda\tau)}))$$

$$= f(z_0(t) + \lambda\tau) = 0$$

because $\lambda \in \mathbb{Z}_0$; q.e.d. (Obviously, f is a theta function.)

Theorem 8.6 (Kas). The meromorphic functions ξ, η are φ-dependent but algebraically independent over $\varphi^*(\mathfrak{H}(D)) = \mathfrak{H}(X)$.

Proof. Because φ has fiber dimension 1, ξ, η are φ-dependent by Lemma 4.1. Suppose that ξ, η are algebraically dependent over $\varphi^*(\mathfrak{H}(D))$. Then holomorphic functions $b_{\mu\nu} \in \mathfrak{H}(D)$ exist such that

$$P(t,x,y) = \sum_{\mu=0}^{b} \sum_{\nu=0}^{q} b_{\mu\nu}(t)X^{\mu}y^{\nu}$$

and such that

$$0 \equiv P(\varphi,\xi,\eta) = \sum_{\mu=0}^{b} \sum_{\nu=0}^{q} (b_{\mu\nu}\circ\varphi)\xi^{\mu}\eta^{\nu}$$

on X where not all function $b_{\mu\nu}$ are the zero function on D.

Further it can be assumed that $P \in \mathfrak{H}(D)[x,y]$ is irreducible.

Suppose that $P(t,x,0) = 0$ for all $(t,x) \in D \times \mathbb{C}$. Then

$$\sum_{\mu=0}^{p} b_{\mu 0}(t)x^{\mu} = 0$$

for all $(t,x) \in D \times \mathbb{C}$. Hence $b_{\mu 0} \equiv 0$ for $\mu = 0,1,\ldots,p$. Hence

$$P(t,x,y) = y \sum_{\mu=0}^{p} \sum_{\nu=1}^{q} b_{\mu\nu}(t)x^{\mu}y^{\nu-1}$$

and P is reducible in $\mathfrak{H}(D)[x,y]$ which is wrong. Hence $t_0 \in D^*$ exists such that $b_{\mu 0}(t_0) \neq 0$ for at least one index μ. Define $\mathbf{x}_{2\lambda} = x_{2\lambda}(t_0)$. Then

$$\xi(x_{2\lambda}) = \xi_{2\lambda}(\alpha_{2\lambda}(t_0, v(t_0))) = \psi_{2\lambda}(t_0, v(t_0)) = t_0^{2\lambda}v(t_0)$$

$$0 = P(t_0, \xi(x_{2\lambda}), \eta(x_{2\lambda})) = P(t_0, t_0^{2\lambda}v(t_0), 0)$$

for each $\lambda \in \mathbb{Z}$. Hence

$$\sum_{\mu=0}^{p} b_{\mu 0}(t_0)(t_0^{2\lambda}v(t_0))^{\mu} = 0$$

for each $\lambda \in \mathbb{Z}$. Because $0 < |t_0| < 1$ and $v(t_0) \neq 0$, the set $\{t_0^{2\lambda}v(t_0) | \lambda \in \mathbb{Z}\}$ consists of infinitely many points. Hence $b_{\mu 0}(t_0) = 0$ for $\mu = 0,1,\ldots,p$, which is wrong. Therefore ξ and η are algebraically independent over $\varphi^*(\mathfrak{H}(D))$; q.e.d.

References

[1] Abhyankar, S.: Local analytic geometry. Academic Press, New York London 1964, pp. 484.

[2] Andreotti, A.: Theoremes de dépendance algébrique sur les espaces complexes pseudoconcaves. Bull. Soc. Math. France (91) (1963), 1-38.

[3] Andreotti, A.: - Grauert H.: Algebraische Körper von automorphen Funktionen, Göttinger, Nachrichten 1961, 39-48.

[4] Hörmander, L.: An introduction to complex analysis in several variables. D. von Nostrand Co. Princeton, N.J. 1966, pp. 208.

[5] Kodaira, K.: On the structure of compact complex analytic surfaces III. Amer. J. Math 90 (1968) 55-83.

[6] Kuhlmann, N.: Ueber holomorphie Abbildungen komplexer Räume Archiv d. Math 15 (1964), 81-90.

[7] Kuhlmann, N.: Algebraic function fields on complex analytic spaces Proc. Conf. on Compl. Analysis, Minneapolis, 1964, Springer Verlag (1965), 155-172.

[8] Kuhlmann, N.: Bemerkungen über holomorphe Abbildungen komplexer Räume. Festschr. Gedächtnisfeier K. Weierstr. 475-522. Westdeutscher Verlag Cologne 1966.

[9] Narasinhan, R.: Introduction to the theory of analytic spaces. Lecture Notes in Mathematics 25. Springer Berlin-Heidelberg-New York 1966, pp. 143.

[10] Remmert, R. and Stein, K.: Ueber die wesentlichen Singularitäten analytischer Mengen, Math. Annalen 126 (1953), 263-306.

[11] Remmert, R.: Meromorphe Funktionen in kompakten komplexen Räumen. Math Annalen 132 (1956), 277-288.

[12] Remmert, R.: Holomorphe und meromorphe Abbildungen komplexer Räume. Math. Annalen 133 (1957) 328-370.

[13] Remmert, R.: Analytic and algebraic dependence of meromorphic functions. Amer. J. of Math. 82 (1960) 891-899.

[14] Shiffman, B.: On the removal of singularities of analytic sets. Michigan Math J. 15 (1968) 111-120.

[15] Siegel, C. L.: Meromorphe Funktionen auf kompakten analytischen Mannigfaltigkeiten. Nachr. Akad. Wiss. Göttingen, 1955, 71-77.

[16] Stein, K.: Maximale holomorphe und meromorphe Abbildungen I. Amer. J. of Math 85 (1963), 298-313.

[17] Stein, K.: Maximale holomorphe und meromorphe Abbildungen II. Amer. J. of Math 86 (1964), 823-868.

[18] Siu, Y.-T.: A pseudoconcave generalization of Grauert's direct image theorem, 203 pp. ms.

[19] Stoll, W.: Ueber meromorphe Modifikationen IV. Math. Annalen 130 (1955) 147-182.

[20] Stoll, W.: Ueber meromorphe Abbildungen komplexer Räume I. Math. Annalen 136 (1958) 201-239.

[21] Stoll, W.: The multiplicity of a holomorphic map. Invent. Math. 2 (1966), 15-58.

[22] Thimm, W.: Ueber algebraische Relationen zwischen meromorphen Funktionen in abgeschlossenen Räumen. Thesis Königsberg 1939, pp. 44.

[23] Thimm, W.: Ueber meromorphe Abbildungen von komplexen Mannigfaltigkeiten. Math. Annalen 128 (1954), 1-48.

[24] Thimm, W.: Meromorphe Abbildungen von Riemann-schen Bereichen. Math. Zeitschr. 60 (1954), 435-457.

[25] Thimm, W.: Der Weierstrasssche Satz der algebraischen Abhangigkeit von Abelschen Funktionen and seine Verallgemeinerung. Festschr. Gedächtnisfeier K. Weierstrass 123-154. Westdeutscher Verlag Cologne 1966.

[26] Van der Waerden, B. L.: Moderne Algebra I. 1. ed. Die Grundl. d. Math. Wiss. in Einzeldarst. 33. Springer Berlin 1930, pp. 243.

[27] Zariski, O. and Samuel, P.: Commutative Algebra I. D. van Nostrand Co., Princeton, N.J. 1958, pp. 329.

Index

Footnotes

Page	No	
6	1)	In this paper, all complex spaces are reduced and have a countable base of open sets if not explicitly stated to the contrary.
6	2)	Since X is pure dimensional, the rank coincides with the pseudo-rank as defined in §1.
6	3)	A branch of an analytic subset A is an irreducible analytic subset B of A such that $B \subseteq C \subseteq A$ and C irreducible analytic imply $B = C$.
7	4)	As an open subset of Y, the set G is a complex space. As such, it is assumed to be irreducible.
19	5)	Compare Kuhlmann [8] page 501, 4β).
61	6)	Compare Hörmander [4], Theorem 7.4.6.
75	7)	This definition is due to Remmert [12].
136	8)	The number $\nu_f(x)$ is the multiplicity of f at x.
167	9)	This concept is defined on page 20.
176	10)	Let X be a complex space. Let E be a thin analytic subset of X. Let T be an analytic subset of $X - E = A$. Then $a \in E$ is said to be a regular point of T, if and only if an open subset X_a of X and an analytic subset T_a of X_a exist such that $A \cup \{a\} \subseteq X_a \subseteq X$ and such that $T_a \cap A = T$. The point $a \in E$ is said to be a singular point of T if and only if a is not a regular point of T.
365	11)	For the purpose of this theorem alone, a complex manifold does not need to have a countable base of open sets.
370	12)	See Kodaira [5].

Lecture Notes in Mathematics

Comprehensive leaflet on request

Vol. 38: R. Berger, R. Kiehl, E. Kunz und H.-J. Nastold, Differentialrechnung in der analytischen Geometrie IV, 134 Seiten. 1967 DM 12,–

Vol. 39: Séminaire de Probabilités I. II, 189 pages. 1967. DM 14,–

Vol. 40: J. Tits, Tabellen zu den einfachen Lie Gruppen und ihren Darstellungen. VI, 53 Seiten. 1967. DM 6.80

Vol. 41: A. Grothendieck, Local Cohomology. VI, 106 pages. 1967. DM 10,–

Vol. 42: J. F. Berglund and K. H. Hofmann, Compact Semitopological Semigroups and Weakly Almost Periodic Functions. VI, 160 pages. 1967. DM 12,–

Vol. 43: D. G. Quillen, Homotopical Algebra. VI, 157 pages. 1967. DM 14,–

Vol. 44: K. Urbanik, Lectures on Prediction Theory. IV, 50 pages. 1967. DM 5,80

Vol. 45: A. Wilansky, Topics in Functional Analysis. VI, 102 pages. 1967. DM 9,60

Vol. 46: P. E. Conner, Seminar on Periodic Maps. IV, 116 pages. 1967. DM 10,60

Vol. 47: Reports of the Midwest Category Seminar I. IV, 181 pages. 1967. DM 14,80

Vol. 48: G. de Rham, S. Maumary et M. A. Kervaire, Torsion et Type Simple d'Homotopie. IV, 101 pages. 1967. DM 9,60

Vol. 49: C. Faith, Lectures on Injective Modules and Quotient Rings. XVI, 140 pages. 1967. DM 12,80

Vol. 50: L. Zalcman, Analytic Capacity and Rational Approximation. VI, 155 pages. 1968. DM 13.20

Vol. 51: Séminaire de Probabilités II. IV, 199 pages. 1968. DM 14,–

Vol. 52: D. J. Simms, Lie Groups and Quantum Mechanics. IV, 90 pages. 1968. DM 8,–

Vol. 53: J. Cerf, Sur les difféomorphismes de la sphère de dimension trois ($\Gamma_4 = 0$). XII, 133 pages. 1968. DM 12,–

Vol. 54: G. Shimura, Automorphic Functions and Number Theory. VI, 69 pages. 1968. DM 8,–

Vol. 55: D. Gromoll, W. Klingenberg und W. Meyer, Riemannsche Geometrie im Großen. VI, 287 Seiten. 1968. DM 20,–

Vol. 56: K. Floret und J. Wloka, Einführung in die Theorie der lokalkonvexen Räume. VIII, 194 Seiten. 1968. DM 16,–

Vol. 57: F. Hirzebruch und K. H. Mayer, O (n)-Mannigfaltigkeiten, exotische Sphären und Singularitäten. IV, 132 Seiten. 1968. DM 10,80

Vol. 58: Kuramochi Boundaries of Riemann Surfaces. IV, 102 pages. 1968. DM 9,60

Vol. 59: K. Jänich, Differenzierbare G-Mannigfaltigkeiten. VI, 89 Seiten. 1968. DM 8,–

Vol. 60: Seminar on Differential Equations and Dynamical Systems. Edited by G. S. Jones. VI, 106 pages. 1968. DM 9,60

Vol. 61: Reports of the Midwest Category Seminar II. IV, 91 pages. 1968. DM 9,60

Vol. 62: Harish-Chandra, Automorphic Forms on Semisimple Lie Groups X, 138 pages. 1968. DM 14,–

Vol. 63: F. Albrecht, Topics in Control Theory. IV, 65 pages. 1968. DM 6,80

Vol. 64: H. Berens, Interpolationsmethoden zur Behandlung von Approximationsprozessen auf Banachräumen. VI, 90 Seiten. 1968. DM 8,–

Vol. 65: D. Kölzow, Differentiation von Maßen. XII, 102 Seiten. 1968. DM 8,–

Vol. 66: D. Ferus, Totale Absolutkrümmung in Differentialgeometrie und -topologie. VI, 85 Seiten. 1968. DM 8,–

Vol. 67: F. Kamber and P. Tondeur, Flat Manifolds. IV, 53 pages. 1968. DM 5,80

Vol. 68: N. Boboc et P. Mustată, Espaces harmoniques associés aux opérateurs différentiels linéaires du second ordre de type elliptique. VI, 95 pages. 1968. DM 8,60

Vol. 69: Seminar über Potentialtheorie. Herausgegeben von H. Bauer. VI, 180 Seiten. 1968. DM 14,80

Vol. 70: Proceedings of the Summer School in Logic. Edited by M. H. Löb. IV, 331 pages. 1968. DM 20,–

Vol. 71: Séminaire Pierre Lelong (Analyse), Année 1967 – 1968. VI, 190 pages. 1968. DM 14,

Vol. 72: The Syntax and Semantics of Infinitary Languages. Edited by J. Barwise. IV, 268 pages. 1968. DM 18,–

Vol. 73: P. E. Conner, Lectures on the Action of a Finite Group. IV, 123 pages. 1968. DM 10,–

Vol. 74: A. Fröhlich, Formal Groups. IV, 140 pages. 1968. DM 12,–

Vol. 75: G. Lumer, Algèbres de fonctions et espaces de Hardy. VI, 80 pages. 1968. DM 8,–

Vol. 76: R. G. Swan, Algebraic K-Theory. IV, 262 pages. 1968. DM 18,–

Vol. 77: P.-A. Meyer, Processus de Markov: la frontière de Martin. IV, 123 pages. 1968. DM 10,–

Vol. 78: H. Herrlich, Topologische Reflexionen und Coreflexionen. XVI, 166 Seiten. 1968. DM 12,–

Vol. 79: A. Grothendieck, Catégories Cofibrées Additives et Complexe Cotangent Relatif. IV, 167 pages. 1968. DM 12,–

Vol. 80: Seminar on Triples and Categorical Homology Theory. Edited by B. Eckmann. IV, 398 pages. 1969. DM 20,–

Vol. 81: J.-P. Eckmann et M. Guenin, Méthodes Algébriques en Mécanique Statistique. VI, 131 pages. 1969. DM 12,–

Vol. 82: J. Wloka, Grundräume und verallgemeinerte Funktionen. VIII, 131 Seiten. 1969. DM 12,–

Vol. 83: O. Zariski, An Introduction to the Theory of Algebraic Surfaces. IV, 100 pages. 1969. DM 8,–

Vol. 84: H. Lüneburg, Transitive Erweiterungen endlicher Permutationsgruppen. IV, 119 Seiten. 1969. DM 10.–

Vol. 85: P. Cartier et D. Foata, Problèmes combinatoires de commutation et réarrangements. IV, 88 pages. 1969. DM 8,–

Vol. 86: Category Theory, Homology Theory and their Applications I. Edited by P. Hilton. VI, 216 pages. 1969. DM 16,–

Vol. 87: M. Tierney, Categorical Constructions in Stable Homotopy Theory. IV, 65 pages. 1969. DM 6,–

Vol. 88: Séminaire de Probabilités III. IV, 229 pages. 1969. DM 18,–

Vol. 89: Probability and Information Theory. Edited by M. Behara, K. Krickeberg and J. Wolfowitz. IV, 256 pages. 1969. DM 18,–

Vol. 90: N. P. Bhatia and O. Hajek, Local Semi-Dynamical Systems. II, 157 pages. 1969. DM 14,–

Vol. 91: N. N. Janenko, Die Zwischenschrittmethode zur Lösung mehrdimensionaler Probleme der mathematischen Physik. VIII, 194 Seiten. 1969. DM 16,80

Vol. 92: Category Theory, Homology Theory and their Applications II. Edited by P. Hilton. V, 308 pages. 1969. DM 20,–

Vol. 93: K. R. Parthasarathy, Multipliers on Locally Compact Groups. III, 54 pages. 1969. DM 5,60

Vol. 94: M. Machover and J. Hirschfeld, Lectures on Non-Standard Analysis. VI, 79 pages. 1969. DM 6,–

Vol. 95: A. S. Troelstra, Principles of Intuitionism. II, 111 pages. 1969. DM 10,–

Vol. 96: H.-B. Brinkmann und D. Puppe, Abelsche und exakte Kategorien, Korrespondenzen. V, 141 Seiten. 1969. DM 10,–

Vol. 97: S. O. Chase and M. E. Sweedler, Hopf Algebras and Galois theory. II, 133 pages. 1969. DM 10,–

Vol. 98: M. Heins, Hardy Classes on Riemann Surfaces. III, 106 pages. 1969. DM 10,–

Vol. 99: Category Theory, Homology Theory and their Applications III. Edited by P. Hilton. IV, 489 pages. 1969. DM 24,–

Vol. 100: M. Artin and B. Mazur, Etale Homotopy. II, 196 Seiten. 1969. DM 12,–

Vol. 101: G. P. Szegö et G. Treccani, Semigruppi di Trasformazioni Multivoche. VI, 177 pages. 1969. DM 14,–

Vol. 102: F. Stummel, Rand- und Eigenwertaufgaben in Sobolewschen Räumen. VIII, 386 Seiten. 1969. DM 20,–

Vol. 103: Lectures in Modern Analysis and Applications I. Edited by C. T. Taam. VII, 162 pages. 1969. DM 12,–

Vol. 104: G. H. Pimbley, Jr., Eigenfunction Branches of Nonlinear Operators and their Bifurcations. II, 128 pages. 1969. DM 10,–

Vol. 105: R. Larsen, The Multiplier Problem. VII, 284 pages. 1969. DM 18,–

Vol. 106: Reports of the Midwest Category Seminar III. Edited by S. Mac Lane. III, 247 pages. 1969. DM 16,–

Vol. 107: A. Peyerimhoff, Lectures on Summability. III, 111 pages. 1969. DM 8,–

Vol. 108: Algebraic K-Theory and its Geometric Applications. Edited by R. M. F. Moss and C. B. Thomas. IV, 86 pages. 1969. DM 6,–

Vol. 109: Conference on the Numerical Solution of Differential Equations. Edited by J. Ll. Morris. VI, 275 pages. 1969. DM 18,–

Vol. 110: The Many Facets of Graph Theory. Edited by G. Chartrand and S. F. Kapoor. VIII, 290 pages. 1969. DM 18,–

Vol. 111: K. H. Mayer, Relationen zwischen charakteristischen Zahlen. III, 99 Seiten. 1969. DM 8,–

Vol. 112: Colloquium on Methods of Optimization. Edited by N. N. Moiseev. IV, 293 pages. 1970. DM 18,–

Vol. 113: R. Wille, Kongruenzklassengeometrien. III, 99 Seiten. 1970. DM 8,–

Vol. 114: H. Jacquet and R. P. Langlands, Automorphic Forms on GL (2). VII, 548 pages. 1970. DM 24,–

Vol. 115: K. H. Roggenkamp and V. Huber-Dyson, Lattices over Orders I. XIX, 290 pages. 1970. DM 18,–

Vol. 116: Sèminaire Pierre Lelong (Analyse) Année 1969. IV, 195 pages. 1970. DM 14,–

Vol. 117: Y. Meyer, Nombres de Pisot, Nombres de Salem et Analyse Harmonique. 63 pages. 1970. DM 6.–

Vol. 118: Proceedings of the 15th Scandinavian Congress, Oslo 1968. Edited by K. E. Aubert and W. Ljunggren. IV, 162 pages. 1970. DM 12,–

Vol. 119: M. Raynaud, Faisceaux amples sur les schémas en groupes et les espaces homogènes. III, 219 pages. 1970. DM 14,–

Vol. 120: D. Siefkes, Büchi's Monadic Second Order Successor Arithmetic. XII, 130 Seiten. 1970. DM 12,–

Vol. 121: H. S. Bear, Lectures on Gleason Parts. III, 47 pages. 1970. DM 6,–

Vol. 122: H. Zieschang, E. Vogt und H.-D. Coldewey, Flächen und ebene diskontinuierliche Gruppen. VIII, 203 Seiten. 1970. DM 16,–

Vol. 123: A. V. Jategaonkar, Left Principal Ideal Rings. VI, 145 pages. 1970. DM 12,–

Vol. 124: Séminare de Probabilités IV. Edited by P. A. Meyer. IV, 282 pages. 1970. DM 20,–

Vol. 125: Symposium on Automatic Demonstration. V, 310 pages. 1970. DM 20,–

Vol. 126: P. Schapira, Théorie des Hyperfonctions. XI, 157 pages. 1970. DM 14,–

Vol. 127: I. Stewart, Lie Algebras. IV, 97 pages. 1970. DM 10,–

Vol. 128: M. Takesaki, Tomita's Theory of Modular Hilbert Algebras and its Applications. II, 123 pages. 1970. DM 10,–

Vol. 129: K. H. Hofmann, The Duality of Compact Semigroups and C*- Bigebras. XII, 142 pages. 1970. DM 14,–

Vol. 130: F. Lorenz, Quadratische Formen über Körpern. II, 77 Seiten. 1970. DM 8,–

Vol. 131: A Borel et al., Seminar on Algebraic Groups and Related Finite Groups. VII, 321 pages. 1970. DM 22,–

Vol. 132: Symposium on Optimization. III, 348 pages. 1970. DM 22,–

Vol. 133: F. Topsøe, Topology and Measure. XIV, 79 pages. 1970. DM 8,–

Vol. 134: L. Smith, Lectures on the Eilenberg-Moore Spectral Sequence. VII, 142 pages. 1970. DM 14,–

Vol. 135: W. Stoll, Value Distribution of Holomorphic Maps into Compact Complex Manifolds. II, 267 pages. 1970. DM 18,–

Vol. 136: M. Karoubi et al., Séminaire Heidelberg-Saarbrücken-Strasbuorg sur la K-Théorie. IV, 264 pages. 1970. DM 18,–

Vol. 137: Reports of the Midwest Category Seminar IV. Edited by S. MacLane. III, 139 pages. 1970. DM 12,–

Vol. 138: D. Foata et M. Schützenberger, Théorie Géométrique des Polynômes Eulériens. V, 94 pages. 1970. DM 10,–

Vol. 139: A. Badrikian, Séminaire sur les Fonctions Aléatoires Linéaires et les Mesures Cylindriques. VII, 221 pages. 1970. DM 18,–

Vol. 140: Lectures in Modern Analysis and Applications II. Edited by C. T. Taam. VI, 119 pages. 1970. DM 10,–

Vol. 141: G. Jameson, Ordered Linear Spaces. XV, 194 pages. 1970. DM 16,–

Vol. 142: K. W. Roggenkamp, Lattices over Orders II. V, 388 pages. 1970. DM 22,–

Vol. 143: K. W. Gruenberg, Cohomological Topics in Group Theory. XIV, 275 pages. 1970. DM 20,–

Vol. 144: Seminar on Differential Equations and Dynamical Systems, II. Edited by J. A. Yorke. VIII, 268 pages. 1970. DM 20,–

Vol. 145: E. J. Dubuc, Kan Extensions in Enriched Category Theory. XVI, 173 pages. 1970. DM 16,–

Vol. 146: A. B. Altman and S. Kleiman, Introduction to Grothendieck Duality Theory. II, 192 pages. 1970. DM 18,–

Vol. 147: D. E. Dobbs, Cech Cohomological Dimensions for Commutative Rings. VI, 176 pages. 1970. DM 16,–

Vol. 148: R. Azencott, Espaces de Poisson des Groupes Localement Compacts. IX, 141 pages. 1970. DM 14,–

Vol. 149: R. G. Swan and E. G. Evans, K-Theory of Finite Groups and Orders. IV, 237 pages. 1970. DM 20,–

Vol. 150: Heyer, Dualität lokalkompakter Gruppen. XIII, 372 Seiten. 1970. DM 20,–

Vol. 151: M. Demazure et A. Grothendieck, Schémas en Groupes I. (SGA 3). XV, 562 pages. 1970. DM 24,–

Vol. 152: M. Demazure et A. Grothendieck, Schémas en Groupes II. (SGA 3). IX, 654 pages. 1970. DM 24,–

Vol. 153: M. Demazure et A. Grothendieck, Schémas en Groupes III. (SGA 3). VIII, 529 pages. 1970. DM 24,–

Vol. 154: A. Lascoux et M. Berger, Variétés Kähleriennes Compactes. VII, 83 pages. 1970. DM 8,–

Vol. 155: Several Complex Variables I, Maryland 1970. Edited by J. Horváth. IV, 214 pages. 1970. DM 18,–

Vol. 156: R. Hartshorne, Ample Subvarieties of Algebraic Varieties. XIV, 256 pages. 1970. DM 20,–

Vol. 157: T. tom Dieck, K. H. Kamps und D. Puppe, Homotopietheorie. VI, 265 Seiten. 1970. DM 20,–

Vol. 158: T. G. Ostrom, Finite Translation Planes. IV. 112 pages. 1970. DM 10,–

Vol. 159: R. Ansorge und R. Hass. Konvergenz von Differenzenverfahren für lineare und nichtlineare Anfangswertaufgaben. VIII, 145 Seiten. 1970. DM 14,–

Vol. 160: L. Sucheston, Constributions to Ergodic Theory and Probability. VII, 277 pages. 1970. DM 20,–

Vol. 161: J. Stasheff, H-Spaces from a Homotopy Point of View. VI, 95 pages. 1970. DM 10,–

Vol. 162: Harish-Chandra and van Dijk, Harmonic Analysis on Reductive p-adic Groups. IV, 125 pages. 1970. DM 12,–

Vol. 163: P. Deligne, Equations Différentielles à Points Singuliers Reguliers. III, 133 pages. 1970. DM 12,–

Vol. 164: J. P. Ferrier, Seminaire sur les Algebres Complètes. II, 69 pages. 1970. DM 8,–

Vol. 165: J. M. Cohen, Stable Homotopy. V, 194 pages. 1970. DM 16,–

Vol. 166: A. J. Silberger, PGL_2 over the p-adics: its Representations, Spherical Functions, and Fourier Analysis. VII, 202 pages. 1970. DM 18,–

Vol. 167: Lavrentiev, Romanov and Vasiliev, Multidimensional Inverse Problems for Differential Equations. V, 59 pages. 1970. DM 10,–

Vol. 168: F. P. Peterson, The Steenrod Algebra and its Applications: A conference to Celebrate N. E. Steenrod's Sixtieth Birthday. VII, 317 pages. 1970. DM 22,–

Vol. 169: M. Raynaud, Anneaux Locaux Henséliens. V, 129 pages. 1970. DM 12,–

Vol. 170: Lectures in Modern Analysis and Applications III. Edited by C. T. Taam. VI, 213 pages. 1970. DM 18,–

Vol. 171: Set-Valued Mappings, Selections and Topological Properties of 2^X. Edited by W. M. Fleischman. X, 110 pages. 1970. DM 12,–

Vol. 172: Y.-T. Siu and G. Trautmann, Gap-Sheaves and Extension of Coherent Analytic Subsheaves. V, 172 pages. 1971. DM 16,–

Vol. 173: J. N. Mordeson and B. Vinograde, Structure of Arbitrary Purely Inseparable Extension Fields. IV, 138 pages. 1970. DM 14,–

Vol. 174: B. Iversen, Linear Determinants with Applications to the Picard Scheme of a Family of Algebraic Curves. VI, 69 pages. 1970. DM 8,–

Vol. 175: M. Brelot, On Topologies and Boundaries in Potential Theory. VI, 176 pages. 1971. DM 18,–

Vol. 176: H. Popp, Fundamentalgruppen algebraischer Mannigfaltigkeiten. IV, 154 Seiten. 1970. DM 16,–

Vol. 177: J. Lambek, Torsion Theories, Additive Semantics and Rings of Quotients. VI, 94 pages. 1971. DM 12,–

Vol. 178: Th. Bröcker und T. tom Dieck, Kobordismentheorie. XVI, 191 Seiten. 1970. DM 18,–

Vol. 179: Seminaire Bourbaki – vol. 1968/69. Exposés 347-363. IV. 295 pages. 1971. DM 22,–

Vol. 180: Séminaire Bourbaki – vol. 1969/70. Exposés 364-381. IV, 310 pages. 1971. DM 22,–

Vol. 181: F. DeMeyer and E. Ingraham, Separable Algebras over Commutative Rings. V, 157 pages. 1971. DM 16.–

ISBN 3-540-05670-X
ISBN 0-387-05670-X